Tools for Mobile Multimedia Programming and Development

D. Tjondronegoro
Queensland University of Technology, Australia

A volume in the Advances in Wireless Technologies and Telecommunication (AWTT) Book Series

Managing Director: Lindsay Johnston
Editorial Director: Joel Gamon
Production Manager: Jennifer Yoder
Publishing Systems Analyst: Adrienne Freeland
Development Editor: Myla Merkel
Assistant Acquisitions Editor: Kayla Wolfe
Typesetter: Christina Henning
Cover Design: Jason Mull

Published in the United States of America by
Information Science Reference (an imprint of IGI Global)
701 E. Chocolate Avenue
Hershey PA 17033
Tel: 717-533-8845
Fax: 717-533-8661
E-mail: cust@igi-global.com
Web site: http://www.igi-global.com

Library of Congress Cataloging-in-Publication Data

Tools for mobile multimedia programming and development / Dian Tjondronegoro, editor.
pages cm
Includes bibliographical references and index.
Summary: "This book investigates the use of mobile platforms for research projects, focusing on the development, testing, and evaluation of prototypes rather than final products, which enables researchers to better understand the needs of users through image processing, object recognition, sensor integration, and user interactions"-- Provided by publisher.
ISBN 978-1-4666-4054-2 (hardcover) -- ISBN 978-1-4666-4055-9 (ebook) -- ISBN 978-1-4666-4056-6 (print & perpetual access) 1. Multimedia systems. 2. Mobile computing. 3. Multimedia communications. I. Tjondronegoro, Dian, 1979- editor of compilation.
QA76.575.T65 2013
006.7--dc23

2013003920

This book is published in the IGI Global book series Advances in Wireless Technologies and Telecommunication (AWTT) (ISSN: 2327-3305; eISSN: 2327-3313)

British Cataloguing in Publication Data
A Cataloguing in Publication record for this book is available from the British Library.

All work contributed to this book is new, previously-unpublished material. The views expressed in this book are those of the authors, but not necessarily of the publisher.

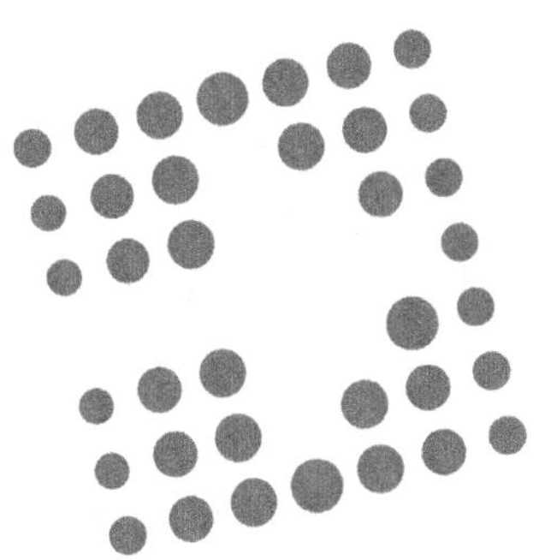

Advances in Wireless Technologies and Telecommunication (AWTT) Book Series

Xiaoge Xu
The University of Nottingham Ningbo China

ISSN: 2327-3305
EISSN: 2327-3313

Mission

The wireless computing industry is constantly evolving, redesigning the ways in which individuals share information. Wireless technology and telecommunication remain one of the most important technologies in business organizations. The utilization of these technologies has enhanced business efficiency by enabling dynamic resources in all aspects of society.

The **Advances in Wireless Technologies and Telecommunication Book Series** aims to provide researchers and academic communities with quality research on the concepts and developments in the wireless technology fields. Developers, engineers, students, research strategists, and IT managers will find this series useful to gain insight into next generation wireless technologies and telecommunication.

Coverage

- Cellular Networks
- Digital Communication
- Global Telecommunications
- Grid Communications
- Mobile Technology
- Mobile Web Services
- Network Management
- Virtual Network Operations
- Wireless Broadband
- Wireless Sensor Networks

IGI Global is currently accepting manuscripts for publication within this series. To submit a proposal for a volume in this series, please contact our Acquisition Editors at Acquisitions@igi-global.com or visit: http://www.igi-global.com/publish/.

The Advances in Wireless Technologies and Telecommunication (AWTT) Book Series (ISSN 2327-3305) is published by IGI Global, 701 E. Chocolate Avenue, Hershey, PA 17033-1240, USA, www.igi-global.com. This series is composed of titles available for purchase individually; each title is edited to be contextually exclusive from any other title within the series. For pricing and ordering information please visit http://www.igi-global.com/book-series/advances-wireless-technologies-telecommunication-awtt/73684. Postmaster: Send all address changes to above address.

Titles in this Series

For a list of additional titles in this series, please visit: www.igi-global.com

Cognitive Radio Technology Applications for Wireless and Mobile Ad Hoc Networks
Natarajan Meghanathan (Jackson State University, USA) and Yenumula B. Reddy (Grambling State University, USA)
Information Science Reference • copyright 2013 • 331pp • H/C (ISBN: 9781466642218) • US $190.00 (our price)

Evolution of Cognitive Networks and Self-Adaptive Communication Systems
Thomas D. Lagkas (University of Western Macedonia, Greece) Panagiotis Sarigiannidis (University of Western Macedonia, Greece) Malamati Louta (University of Western Macedonia, Greece) and Periklis Chatzimisios (Alexander TEI of Thessaloniki, Greece)
Information Science Reference • copyright 2013 • 348pp • H/C (ISBN: 9781466641891) • US $195.00 (our price)

Tools for Mobile Multimedia Programming and Development
D. Tjondronegoro (Queensland University of Technology, Australia)
Information Science Reference • copyright 2013 • 357pp • H/C (ISBN: 9781466640542) • US $190.00 (our price)

Cognitive Radio and Interference Management: Technology and Strategy
Meng-Lin Ku (National Central University, Taiwan, R.O.C.) and Jia-Chin Lin (National Central University, Taiwan, R.O.C.)
Information Science Reference • copyright 2013 • 354pp • H/C (ISBN: 9781466620056) • US $190.00 (our price)

Wireless Radio-Frequency Standards and System Design Advanced Techniques
Gianluca Cornetta (Universidad San Pablo-CEU, Spain) David J. Santos (Universidad San Pablo-CEU, Spain) and Jose Manuel Vazquez (Universidad San Pablo-CEU, Spain)
Engineering Science Reference • copyright 2012 • 422pp • H/C (ISBN: 9781466600836) • US $195.00 (our price)

Femtocell Communications and Technologies Business Opportunities and Deployment Challenges
Rashid A. Saeed (UIA, Malaysia) Bharat S. Chaudhari (International Institute of Information Technology, India) and Rania A. Mokhtar (Sudan University of Science and Technology, Sudan)
Information Science Reference • copyright 2012 • 295pp • H/C (ISBN: 9781466600928) • US $190.00 (our price)

Advanced Communication Protocol Technologies Solutions, Methods, and Applications
Katalin Tarnay (University of Pannonia, Hungary) Gusztáv Adamis (Budapest University of Technology and Economics, Hungary) and Tibor Dulai (University of Pannonia, Hungary)
Information Science Reference • copyright 2011 • 592pp • H/C (ISBN: 9781609607326) • US $195.00 (our price)

www.igi-global.com

701 E. Chocolate Ave., Hershey, PA 17033
Order online at www.igi-global.com or call 717-533-8845 x100
To place a standing order for titles released in this series, contact: cust@igi-global.com
Mon-Fri 8:00 am - 5:00 pm (est) or fax 24 hours a day 717-533-8661

Editorial Advisory Board

List of Reviewers

Table of Contents

Preface xiv

Acknowledgment xviii

Section 1
Methodology and Tools

Chapter 1
Going DEEP: Public, Iterative Release as a Mobile Research Strategy 1
Andrew Dekker, University of Queensland, Australia
Justin Marrington, University of Queensland, Australia
Stephen Viller, University of Queensland, Australia

Chapter 2
User-Centered Study on Quality of Mobile Video Services 18
Wei Song, Queensland University of Technology, Australia
Dian Tjondronegoro, Queensland University of Technology, Australia
Michael Docherty, Queensland University of Technology, Australia

Chapter 3
Modeling and Documenting Aspect-Oriented Mobile Product Lines 52
Camila Nunes, Pontifical Catholic University of Rio de Janeiro, Brazil
Uirá Kulesza, Federal University of Rio Grande do Norte, Brazil
Roberta Coelho, Federal University of Rio Grande do Norte, Brazil
Carlos Lucena, Pontifical Catholic University of Rio de Janeiro, Brazil
Flávia Delicato, Federal University of Rio de Janeiro, Brazil
Paulo Pires, Federal University of Rio de Janeiro, Brazil
Thais Batista, Federal University of Rio Grande do Norte, Brazil

Chapter 4
Developing a Multimodal Application for a Scientific Experiment on Smartphones: Case Study, Tools, and Results 68
José Rouillard, University of Lille, France

Section 2
Sensors-Based Interactivity

Chapter 5
Contextual and Personalized Mobile Recommendation Systems 82
Jitao Sang, Chinese Academy of Sciences, China
Tao Mei, Microsoft Research Asia, China
Changsheng Xu, Chinese Academy of Sciences, China
Shipeng Li, Microsoft Research Asia, China

Chapter 6
Gamifying Everyday Activities using Mobile Sensing 98
Zachary Fitz-Walter, Queensland University of Technology, Australia
Dian Tjondronegoro, Queensland University of Technology, Australia
Peta Wyeth, Queensland University of Technology, Australia

Chapter 7
A Context-Aware Smart TV System with Body-Gesture Control and Personalized Recommendation 115
Wei-Po Lee, National Sun Yat-sen University, Taiwan
Che KaoLi, National Sun Yat-sen University, Taiwan

Chapter 8
Intelligent Mobile Learning Systems for Learners with Style 131
Tracey J. Mehigan, University College Cork, Ireland
Ian Pitt, University College Cork, Ireland

Section 3
Accessible Technology

Chapter 9
Monitoring User's Emotions Using Brain Computer Interfaces 151
Katie Crowley, University College Cork, Ireland
Ian Pitt, University College Cork, Ireland

Chapter 10
WayFinder: A Navigational Interface for Visitors and Blind Students on Campus 168
Tracey J. Mehigan, University College Cork, Ireland
Ian Pitt, University College Cork, Ireland

Section 4
Health and Environment Monitoring

Chapter 11
Ubiquitous Multimedia Data Access in Electronic Health Care Systems 191
Muhammad H. Aboelfotoh, Queen's University of Kingston, Canada
Patrick Martin, Queen's University of Kingston, Canada
Hossam Hassanein, Queen's University of Kingston, Canada

Chapter 12
BioCondition Assessment Tool: A Mobile Application to Aid Vegetation Assessment...................... 228
Chin Loong Law, Queensland University of Technology, Australia
Paul Roe, Queensland University of Technology, Australia
Jinglan Zhang, Queensland University of Technology, Australia

Compilation of References 248

About the Contributors 269

Index 276

Detailed Table of Contents

Section 1
Methodology and Tools

This section provides an overview of the methodology and tools used for mobile multimedia research, including some case studies.

This chapter proposes a rapid and reflective model of real deployment of high-fidelity prototypes, borrowing the best habits of industry, where the researcher relinquishes tight control over their prototype in exchange for an opportunity to observe patterns of use that would be intractable to plan for in a controlled study. The approach moves the emphasis in prototyping away from evaluation and towards exploration and reflection. No other discipline but mobile design has by nature so little control over the variables in the context of its research.

This chapter firstly reviews the current user studies on quality of user experience of mobile videos in three themes: research scopes or focuses, user study methods, and data analysis. Each is discussed with examples. This helps researchers better understand the research area of mobile video and determine a clear research direction. The chapter then provides an example of conducting user study of mobile video research, along with the discussion on a series of relative issues, such as participants, materials and devices, study procedure, and results. This helps researchers conduct user studies properly and achieve good performance. Finally, this chapter explores challenges and opportunities in mobile video related research, and discusses potential improvement in future research.

Aspect-Oriented Software Development (AOSD) has evolved as a software development paradigm over the last decade. This chapter presents a strategy for modeling and documenting aspect-oriented variations by integrating two existing approaches: (1) use cases and (2) crosscutting. The synergy and benefits of the integration between these approaches are demonstrated by modeling and documenting MobileMedia, a software product line that provides support to manage different media (photo, music, and video) on mobile devices.

This chapter describes the development process of a scientific experiment, in which a mobile application is used to determine which modality (e.g. touch, voice, QRcode) is preferred for entering expiration dates of food products.

Section 2
Sensors-Based Interactivity

This section focuses on achieving an engaging and intuitive interaction by leveraging mobile devices' multimodal sensors.

Mobile devices are becoming ubiquitous. People are getting used to using their phones as a personal concierge to discover what is around and decide what to do. Mobile recommendation therefore becomes important to understand user intent and simplify task completion on the go. Since user intents essentially vary with users and sensor contexts (time and geo-location, for example), mobile recommendation needs to be both contextual and personalized. While rich user mobile data is available, such as mobile query, click-through, and check-in record, there exist two challenges in utilizing them to design a contextual and personalized mobile recommendation system: exploring characteristics from large-scale and heterogeneous mobile data and employing the uncovered characteristics for recommendation. In this chapter,

Section 3
Accessible Technology

This section provides some emerging mobile-multimedia applications for accessible computing.

This chapter discusses the use of Brain-Controller Interfaces to monitor the emotions and interactions of a subject as they use a system. The proliferation of mobile devices as an emerging platform offers scope for the development of the relationship between Brain-Controller Interfaces and mobile technology, towards ubiquitous, minimally invasive, mobile systems. Tracking how a user interacts with a system, and the emotion-based responses that are invoked as they interact with the system, yield very valuable datasets for the development of intelligent, adaptive systems.

Navigating a university campus can be difficult for visitors and incoming students/staff, and is a particular challenge for vision-impaired students and staff. A potential solution to these problems is to provide navigation data using wireless and mobile devices. Technologies such as Bluetooth, and the compass and accelerometer now included in many smart-phones, could be combined to accurately orient the user, providing feedback on their exact location on campus. This chapter describes the development of the system and ongoing user testing to assess the viability of the interface for use by visually impaired people.

Section 4
Health and Environment Monitoring

This section provides some emerging mobile-multimedia applications for health and environment monitoring.

This chapter provides an introduction to the issues and challenges in remote access to medical data in different settings, and then provides a background on the different software technologies used in the medical systems that have been proposed for use in these settings. It presents a comprehensive overview and qualitative analysis of the functional aspects of these different medical systems.

Chapter 12
BioCondition Assessment Tool: A Mobile Application to Aid Vegetation Assessment...................... 228
Chin Loong Law, Queensland University of Technology, Australia
Paul Roe, Queensland University of Technology, Australia
Jinglan Zhang, Queensland University of Technology, Australia

This chapter looks at BioCondition, a newly developed vegetation assessment framework by Queensland Department of Resource Management (DERM) and how mobile technology can assist beginners in conducting the survey. Even though BioCondition is designed to be simple, it is still fairly inaccessible to beginners due to its complex, time consuming, and repetitive nature. A Windows Phone mobile application—BioCondition Assessment Tool—was developed to provide on-site guidance to beginners and document the assessment process for future revision and comparison. The application was tested in an experiment at Samford Conservation Park with 12 students studying ecology in Queensland University of Technology.

Compilation of References 248

About the Contributors 269

Index 276

Preface

Rapid proliferations of mobile devices—as the preferred platform for people to work, socialize, and play—have created many new opportunities to develop innovative solutions. The rise of tablets and smart phones has enabled users across all age groups to easily create, share, and consume multimedia contents and services, anywhere and at anytime. Most of the current generation smart phones are equipped with high-quality image/video camera, high-speed Internet connectivity, and sensors for context awareness, such as time, location, lighting conditions, motion, and object proximity, thus promoting rich social and mobile multimedia interactions.

Mobile and multimedia researchers worldwide are starting to adopt and fully leverage the latest devices' capabilities to deliver innovative solutions for many real-world applications. For example, delivery of online education via mobile video streaming (Ulrich, et al., 2010), smart mobility via mobile transit information for public transports (Ferris, et al., 2010), and personalized health delivery via mobile interfaces (Liu, et al., 2011). Such applications can potentially solve grand challenges and improve current societies and environments, and will inspire new techniques for context-aware multimedia data analysis, mobile interaction design frameworks, and ultimately revolutionize the traditional mobile multimedia systems.

Mobile multimedia researchers and engineers need to design, develop, and evaluate mobile application prototype(s) to test new algorithms (e.g. face recognition for smartphone [Choi, et al., 2011]), and innovative paradigms for mobile interactions, such as gesture-based browsing in mobile devices (Wachs, et al., 2011) and substance abuse prevention using mobile app (Marsch, 2012). To gain the knowledge of tools and methodologies available for developing and programming on mobile devices, most of the available resources have focused on teaching general skills to equip aspiring programmers to develop applications that are ready to sell. Such resources may not be entirely relevant for researchers, as they need to primarily learn how to develop prototypes for testing and evaluation purposes, instead of a product. On the other hand, multimedia researchers require more in-depth knowledge on using specific components and tools for their research projects, such as image processing, objects recognition, sensors processing, and user-interactions logging.

While many mobile-multimedia innovations seem to be directed from technological perspectives, there is a need for a strong balance of attention to the user interaction issues as the emerging platforms create new ways for people to achieve tasks and activities. User-driven innovations lead to adaptations in techniques, methodologies, and approaches during research experiments, including the design, programming, development, and evaluation of mobile multimedia tools and applications.

This book aims to provide an intersection between the latest research findings in the area of mobile multimedia, and the latest tools and methodologies for developing prototypes for testing and evaluating theoretical frameworks. A particular focus will be given to understanding how to use mobile platforms for research development projects. The book compiles chapters written from the perspective of researchers who work in mobile platforms, and who share their case studies, tips, and tricks, in order to inspire and quick start researchers' skills in developing novel solutions for the emerging mobile platforms.

The book is organized into four sections: Methodologies and Tools; Sensor-Based Interactivity; Accessibility Technology; and Health and Environmental Monitoring.

Section 1: Methodology and Tools contains four chapters that provide an overview of the methodology and tools used for mobile multimedia research, including some case studies.

Chapter 1 (by Dekker et al.) proposes a rapid and reflective model of real deployment of high-fidelity prototypes, borrowing the best habits of industry, where the researcher relinquishes tight control over their prototype in exchange for an opportunity to observe patterns of use that would be intractable to plan for in a controlled study. The approach moves the emphasis in prototyping, away from evaluation, and towards exploration and reflection.

Chapter 2 (by Song et al.) will help researchers to understand the tools and methodologies for a user-driven study on the quality of experience in mobile videos, by providing examples on the scoping of issues, user study methods, and data analysis examples. Through the presented case study, readers will explore challenges and opportunities in mobile video research, and identify potential improvements for future research.

Chapter 3 (by Nunes et al.) presents a strategy for modeling aspect-oriented variations for product line software, which provides support to manage different media (photo, music, and video) on mobile devices. Their experiment demonstrates the synergy and benefits of integrating two existing approaches: use cases and crosscutting.

Chapter 4 (by Rouillard) describes the development process of a scientific experiment in which a mobile application will be used to determine which modality (e.g. touch, voice, QRcode) is preferred for entering expiration dates of food products. The experiment on "pervasive fridge" case study demonstrates the benefits and limitations of AppInventor framework as the visual development tool.

Section 2: Sensor-Based Interactivity contains four chapters that focus on achieving engaging, context-aware, and intuitive user interactions by leveraging mobile devices' multimodal sensors.

Chapter 5 (by Sang et al.) presents a framework and algorithms for contextual and personalized mobile recommendation systems. As people are increasingly using their phones as a personal concierge to discover what is around and deciding what to do, mobile recommendation becomes important to understand user intent and simplify task completion on the go. The presented algorithms exploit mobile query data for local business recommendation, and mobile check-in record to assist activity planning.

Chapter 6 (by Fitz-Walter et al.) explores opportunities and challenges in the use of mobile applications and sensing for *gamification*, which means turning day-to-day activities into fun and engaging tasks. The proposed framework and its case study will demonstrate the benefits of adding achievements to mobile applications for engaging university students to participate in events and accomplish tasks during an orientation period.

Chapter 7 (by Lee et al.) presents a context-aware smart TV system with body-gesture control and personalized recommendation. It focuses on the design of a body control system that recognizes and interprets human gestures as machine commands to control TV, new social tag-based method to recommend most suitable multimedia contents to users, and a context-aware platform that takes into account different environmental situations in order to make the best recommendations.

Chapter 8 (by Mehigan et al.) discusses the development of intelligent and personalized user models for mobile learning, along with the subsequent benefits for learners. The use of biometric technologies for the identification of visual or verbal learners in mobile learning environments is discussed, with a focus on the use of accelerometer sensors. A user interface model is presented, designed to intelligently identify the user's learning style and adapt learning content accordingly in mobile learning environments.

Section 3: Accessible Technology contains two chapters that discuss emerging mobile-multimedia applications for accessible computing.

Chapter 9 (by Crowley et al.) presents the opportunities in monitoring users' emotions using brain controller interfaces to build ubiquitous, minimally invasive, mobile systems. Tracking how a user interacts with a system, and the emotion-based responses that are invoked as they interact with the system, yield very valuable datasets for the development of intelligent, adaptive systems.

Chapter 10 (by Mehigan et al.) describes the development of a Navigational Interface for Visitors and Blind Students on Campus, and the ongoing user testing to assess the viability of the interface for use by vision-impaired people. Technologies such as Bluetooth, and the compass and accelerometer are combined to accurately orient the user, providing feedback on their exact location on campus.

Section 4: Health and Environment Monitoring contains two chapters that explore emerging mobile-multimedia applications for health and environment monitoring.

Chapter 11 (by Aboelfotoh et al.) focuses on ubiquitous multimedia data access in electronic health care systems. It introduces issues and challenges in remote access to medical data in different settings, and provides a background on the different software technologies used in the medical systems that have been proposed for use in these settings. It presents a comprehensive overview and qualitative analysis of the functional aspects of these different medical systems.

Chapter 12 (by Law et al.) presents the framework, real-world experiment, and evaluation for a mobile application that aids vegetation assessment and assists beginners with on-site guidance for conducting the survey. A Windows Phone mobile application, BioCondition Assessment Tool, was developed to provide on-site guidance to beginners and document the assessment process for future revision and comparison. The application was tested at Samford Conservation Park, Australia.

The target audience of this book includes professionals and researchers working in multiple disciplines, especially in mobile multimedia fields, as the case studies demonstrate how to rapidly develop prototypes for image, video, audio, and other sensor-related research projects in emerging mobile devices.

Dian Tjondronegoro
Queensland University of Technology, Australia
March 2013

REFERENCES

Choi, K., Toh, K.-A., & Byun, H. (2011). Realtime training on mobile devices for face recognition applications. *Pattern Recognition*, *44*(2), 386–400. doi:10.1016/j.patcog.2010.08.009.

Ferris, B., Watkins, K., & Borning, A. (2010). OneBusAway: Results from providing real-time arrival information for public transit. In *Proceedings of the SIGCHI Conference on Human Factors in Computing Systems* (pp. 1807–1816). New York, NY: ACM. doi:10.1145/1753326.1753597

Liu, C., Zhu, Q., Holroyd, K. A., & Seng, E. K. (2011). Status and trends of mobile-health applications for iOS devices: A developer's perspective. *Journal of Systems and Software*, *84*(11), 2022–2033. doi:10.1016/j.jss.2011.06.049.

Marsch, L. A. (2012). Leveraging technology to enhance addiction treatment and recovery. *Journal of Addictive Diseases*, *31*(3), 313–318. doi:10.1080/10550887.2012.694606 PMID:22873192.

Ullrich, C., Shen, R., Tong, R., & Tan, X. (2010). A mobile live video learning system for large-scale learning—System design and evaluation. *IEEE Transactions on Learning Technologies*, *3*(1), 6–17. doi:10.1109/TLT.2009.54.

Wachs, J. P., Kölsch, M., Stern, H., & Edan, Y. (2011). Vision-based hand-gesture applications. *Communications of the ACM*, *54*(2), 60–71. doi:10.1145/1897816.1897838 PMID:21984822.

Acknowledgment

I thank the editorial advisory board and the additional reviewers who have helped in the double blind reviewing process.

I also would like to thank all the authors for contributing their chapters, and IGI Global for giving us the opportunity of publishing this book and for all the assistance and support.

Dian Tjondronegoro
Queensland University of Technology, Australia
March 2013

Section 1
Methodology and Tools

This section provides an overview of the methodology and tools used for mobile multimedia research, including some case studies.

Chapter 1
Going DEEP:
Public, Iterative Release as a Mobile Research Strategy

Andrew Dekker
University of Queensland, Australia

Justin Marrington
University of Queensland, Australia

Stephen Viller
University of Queensland, Australia

ABSTRACT

Unlike traditional forms of Human-Computer Interaction (such as conducting desktop or Web-based design), mobile design has by its nature little control over the contextual variables of its research. Short-term evaluations of novel mobile interaction techniques are abundant, but these controlled studies only address limited contexts through artificial deployments, which cannot hope to reveal the patterns of use that arise as people appropriate a tool and take it with them into the varying social and physical contexts of their lives. The authors propose a rapid and reflective model of in-situ deployment of high-fidelity prototypes, borrowing the tested habits of industry, where researchers relinquish tight control over their prototypes in exchange for an opportunity to observe patterns of use that would be intractable to plan for in controlled studies. The approach moves the emphasis in prototyping away from evaluation and towards exploration and reflection, promoting an iterative prototyping methodology that captures the complexities of the real world.

DOI: 10.4018/978-1-4666-4054-2.ch001

INTRODUCTION

The field of Human-Computer Interaction (HCI) is seen to be in a perpetual state of identity crisis. To a point, this is something that can be expected. Contributors to the shared body of research come from a startlingly diverse set of backgrounds: software engineering, operating system development, ethnography, sociology, cognitive science, the arts, design, journalism, and media theory, each authoring important works in our pantheon. Each contribution has its own perspective on the field.

In 2009, long-time SIGCHI contributor James Landay published a frustrated missive on what he saw as the fundamentally incorrect approaches of reviewers for CHI (ACM SIGCHI Conference on Human Factors in Computing Systems) and UIST (ACM Symposium on User Interface Software and Technology) conferences (Landay, 2009). Landay asserts that "systems work"—that is, design research involving the design, build, and release of actual software—is significantly harder to publish in comparison to short, artificial deployments aimed directly at evaluations. The unproductive focus on evaluation as the yardstick for research success, he said, was a strong disincentive to do real work, since more significant work requires more exhaustive evaluation. One commenter summed up the problem perfectly: "I encounter more innovation scanning Techmeme these days than I do at the average conference."

Landay's post echoed a familiar cry: in recent years many well-established academics in the field have expressed their concerns that a fatal combination of the pressure to publish and the obsession with evaluation leads too many graduate students towards trivial deployments and away from actual innovation. When examining user evaluation methods, Buxton and Greenberg invoked the classical "considered harmful" maxim against applying usability evaluation thoughtlessly, and blast the conferences for explicitly including "evaluation validity" as a guideline for publication (Buxton & Greenberg, 2009). Lieberman rails against the "tyranny" of evaluation: "There is no ISO standard human," he says, and trying to design studies around that assumption has given user interface research a "bad case of physics envy" (Lieberman, 2003). At UIST, Olsen spoke plainly: without the added context of a real system, a usability evaluation is a trap that reduces all interactions into standardized problems reduced to the minimum of complexity and scale (Olsen, 2007).

This methodological miasma is nowhere more apparent than when trying to design software for mobile applications. It is difficult enough to bridge gaps between simulated and real interaction within the well-defined boundaries of the desktop or workplace contexts. As with other areas within the ubiquitous computing discipline, mobile applications must contend with the additional complexity of a perpetually shifting context coupled with the difficult-to-predict relationship between their interaction patterns and the shifting reliability of the actors that drive them. It would be difficult to describe a design problem less suited to artificial deployment, but when we examine mobile-focused publications, we find traditional evaluation methods in use: emphasis on getting to the evaluation, often misdirected evaluation, as quickly as possible.

As mobile designers and developers, our discipline is uniquely situated to demonstrate the plausibility of real systems work as an effective research approach. Mobile applications are best when they are limited in scope (due to the modal nature of mobile operating systems demanding that each application do one thing well), and their deployments are therefore easier to manage than complex systems. Furthermore, as researchers we can make use of the same high-level toolsets and distribution platforms that have allowed self-taught mobile developers to push their ideas to millions of users.

The focus of this chapter is threefold: first, to demonstrate how artificial evaluations are more of an issue in Mobile HCI than they are in other more definable contexts; second, to extract what must go into a piece of mobile design research to give it a chance of being both innovative and rigorous; and finally, to propose a model for mobile HCI research that uses early deployment as a facilitator for the more traditional design methods, enabling them to go further and explore the subtle problems that can only be revealed by actual, uncontrolled use.

BACKGROUND

Traditional Design and Evaluation Methods in Mobile Application Design

It is crucial to understand how the dynamic context of mobile development differs from more traditional forms of application design such as desktop and screen based computing. In traditional application design, designers will for the most part follow either a waterfall or iterative user-centered design process. These processes are similar, in that a large part of the process is dedicated to understanding the context in which the software is to be used, and how it is to be used. This need finding and requirements gathering is a critical part to any application design, and usually involves collaboration with a number of stakeholders, such as clients, potential users and other people within the environment. To help this explorative need-finding process, designers and researchers employ a number of methods to help us better understand the context in which the application is to be used (Guindon, 1990; Hix & Hartson, 1993; Buur & Soendergaard, 2000; Tacchi, Foth, & Hearn, 2007). These methods are not used in isolation, but rather are used in conjunction with each other to gather a holistic understanding of the context, using various methods of triangulation of both quantitative and qualitative data (Dick, 1979). An important aspect of this process is to not only hear from the user and stakeholder perspectives, but to also in parallel conduct ethno-focused methods, to reveal the connection between the social and the technical, and understand the process from an observer and participant perspective (Martin & Sommerville, 2004).

This process of requirements gathering and need-finding is for the most part constrained in traditional application contexts (such as work, home and living room environments), but becomes a lot more complicated and wicked (Buchanan, 1992) when considering design for devices which change environments and contexts with the user. Context aware computing is a design and implementation method where the changing of context cannot only be designed for in this conceptual stage, but also allows the application to adapt itself through the ability to sense the change in context. This idea of context aware computing has become more and more integral to the design of software as mobile/portable computing has become increasingly ubiquitous (Abowd et al, 1999).

Despite this increase in focus on understanding the nature of the context we are designing for, the methods that are deployed to understand this context are done either in limited trials, or built into the application for runtime contextual understanding. This becomes a problem when we have devices that have a limited understanding of their context, but have also been created with the understanding that they can be used in an unknown context (in that they could be used in any place or situation at any time. Instead of these devices being used in a variety of constrained environments, they instead become used in an infinite number of possible environments: while driving in a car, while sitting on a toilet, while giving a presentation. While it is valid to predict that the conceptual design of an application may limit the contexts in which it is useful, many applications are not necessary limited to the context in which they are anticipated to be used. In these situations, it is important to know where they are used and how they are used *in-situ*.

Fixed-location computers are clearly used for a variety of tasks and are set within a rich social and organizational context. However, this is at best realized within individual applications and the nature of the device as a whole is fixed and a-contextual. In contrast, the very nature of mobile devices sets them within a multi-faceted contextual matrix, bound into the physical nature of the application domain and closely meshed with existing work settings (Rodden et al, 1998).

The real world use of an application cannot be fully understood until it is in the wild. There are many benefits for gaining an authentic an understanding as possible of how the technology that has been designed functions *in-situ*, in specific contexts. Context aware computing not only shows us how we can design applications which have the ability to adapt to a continually shifting context, but it also has the potential to leverage these methods early during an iterative design process (Abowd et al, 1997).

Rodden et al. (1998) discuss the importance of the understanding and detection of context for mobile devices.

Making use of the context of a device is important for two reasons. Firstly, it may allow us to produce new applications based on the special nature of the context, for example interactive guide maps. Secondly, and equally important, it can help us tailor standard applications for mobile devices, for example when a sales rep visits a company, the spread sheet can have a default files menu which includes the recent ordering history for the company. Such tailoring is not just an added extra, limited screen displays mean that highly adaptive, contextual interfaces become necessary for acceptable interaction.

Rodden et al detail various types of context that can be considered when designing and developing mobile devices (and applications):

- **Infrastructure Context:** Understanding which environments the system will be used, and what limits on infrastructure must be supported (such as lack of Internet or GPS information).
- **Application Context:** Understanding the situations in which the application will be used.
- **System Context:** Especially when designing mobile applications that may be either distributed or collaborative, it is important to understand the change of context and infrastructure supporting this collaboration.
- **Location Context:** Contextual cues that may arise from specific locations (such as at a train station) may change how the user interacts with the application.
- **Physical Context:** How does the mobile device interact with the physical environment (such as the volume of the audio based on the physical location or the brightness of the screen).

The use of using mobile devices for prototyping is not new. Previous work has focused on custom hardware, which can emulate mobile device hardware for prototyping mobile applications. Raento et al (2005) discussed the ContextPhone, which allows prototypes to be designed with context aware information and functionality, such as location and marker detection. The evolution of development environments for commercial mobile platforms has enabled this prototype-based platform to be accessible not only to developers, but also their users. Additionally, the phones themselves provide some sophisticated levels of context that can take away the need for consideration on the device. For instance, most mobile devices utilize light level sensors, to detect the amount of light within the environment, and adjust the lighting of the screen accordingly.

ALPHA BETA GAGA

If we are to acknowledge that short-term lab-based prototype studies are an incomplete way of exploring the mobile design context, other feasible alternatives must be investigated. The obvious approach—system design-driven studies—can certainly provide great value, but simply extending the average study length does little but reduce the researcher's potential output. After all, this is why short-term studies are so prevalent in major HCI conferences. Recently, industry has developed the method of the public beta - where services are deployed into the wild prior to the product and design completion. This method has proven successful not only for single developers, but also business supported start-ups, as well as major industry players.

Minecraft

Minecraft (http://minecraft.net/), the popular sandbox game, is well known as an independent game development success story. Sole developer Markus "Notch" Persson released the game as an early alpha from his Website in 2009. Two years later, when Minecraft 1.0 was released, the game had already sold over 6 million copies, all during the alpha and beta stages of development. More interesting than the business model itself, is the approach taken to assist in the evolution of the concept - based on a continual guiding feedback loop with the users.

Due to its rapid evolution via customer feedback, we can consider Minecraft as one of industry's most successful longitudinal prototyping studies, and a near-perfect application of viral/snowball participant recruitment. When Notch released Minecraft Alpha 1 as a humble Java applet, it was a very basic sandbox: a procedural tile generating engine, a few mining and crafting mechanics, and very little else. Notch opened a thread at TIGSource, an independent game development forum, and asked his fellow "indies" for feedback (Various authors, 2009). 125 pages and several thousand posts later, this original Minecraft alpha release thread is still going strong, and many of the snippets of feedback or ideas presented made it into beta releases of the game.

The experimental and playful nature of the game made it ideal for observatory research. By virtue of the artifact being enticing and full of possibility on its own, Notch was able to use the Internet as a feedback mechanism: as well as TIGSource and his own bursting-at-the-seams forum, the windowed Java applet nature of the Minecraft executable made it trivial for screen-recording play. YouTube began to overflow with videos of Minecraft players showing off things they built in the game, or different ways they've managed to accomplish things, break things, or glitch the game out. Of course, many YouTube videos began with "Notch, can you add <insert item here>?". This method of feedback may be considered different to traditional research practices, in that rather than having well defined channels of feedback; users were encouraged to share their findings not only back to the developers, but to the community at large independently of a specified medium.

As the beta evolved in line with his vision and his users' enthusiastic suggestions, Notch released a set of mod tools to truly allow his fanbase to appropriate Minecraft into whatever they could imagine. Players were no longer limited to simply begging for their ideas in forum posts: with the mod tools and a few pieces of free open source software, they could put them into the game itself (and then, typically, release them as a "Please add this, Notch" YouTube video). It's important to note that Minecraft modding (performing modifications or customization to the software) wasn't open only to developers or game designers: it was players of all sizes (from 14 year old boys to cynical grandmothers) who began to mod: the vast amount of community support smoothed out the learning curve. This is in-the-wild participatory design method: giving users the tools to actually build their ideas, and carefully observe what falls out.

Certainly, we're not suggesting that every prototype research project can be as successful as Minecraft. Instead we offer this approach as what should be the vision of every research project that hopes to propose new designs or ways of working: make something fantastic (and unfinished) first, and then worry about how you will evaluate it as it grows. Plant the seed before you try to pick the apples.

Nnub

Neighbourhood Nub (Nnub), is an exploratory prototype of a digitized local community notice board system. Nnub's primary artifact was an interactive touchscreen notice board, installed into community centres (Redhead et al., 2010). Paired with a public social Web application that allowed any community member to post to the public display, Nnub was an experiment in designing community based software in-situ and in public. Community members were involved from the very early participatory design sessions, through the first public prototype release, and then in ongoing studies and probes that Nnub's researchers conducted periodically through the extended period that it ran in public.

There are a number of aspects of the project that make Nnub interesting from a design perspective. The first is that Nnub is a HCI/Ubicomp research project that made the leap from controlled evaluations to "in the wild" longitudinal study, situated within the public's eye. A research goal of this system was to reveal emergent patterns of use that could only be possible to discover through allowing a community to appropriate the system (Redhead & Brereton, 2008). Nnub was by nature "continually available": its Web-technology implementation meant that it wasn't necessary to retract prototypes or take the system down in order to change it. This property meant that Nnub's users were able to "trust and come to rely on it" (and therefore find more and more varied uses for it). The physical notice board also provided an ongoing cue of use, as the usage of the system was visible without members of the community having to visit the Website. Nnub was intended from the beginning as an organically evolving system: a perpetual beta. While the researchers maintained a central vision for the goals of the research, the system itself was redesigned in an evolutionary fashion in response to the aforementioned emergent and unintended patterns of use: for example "PDF notices in Nnub was not implemented until a government agency began using the system and had exacting requirements on how notices were to be presented." (Heyer & Brereton, 2010).

Google Wave

Not all prototypes deployed in this method are popular success stories. Google Wave was a product developed to explore the idea of "What would email look like if we set out to invent it today?" (O'Reilly, 2009). The process followed by the Wave team was to quickly develop an unfinished prototype, which was released to developers in 2009. In this release, many of the product's features were not available, and for the most part the prototype was a collection of technologies rather than a unified user experience. The focus of the initial release was on developers being given access to the system so that they could build new tools and interactions on top of the Wave platform—defining the user experience for themselves. However the deployment of the system, presented as the future of email, was actively marketed to the general public. Although clearly stated as a Beta product, the deployment approaches of the system led many to believe it to be a final product (similar to how Google conducted its Gmail deployment). Wave as a product was cancelled within two years of its launch (primarily due to the negative feedback from the large user base), however the technologies and interactions that Wave had developed have since been incorporated into a number of different concepts, such as Google Docs and Google Plus. It can be seen that while Wave as a product was a commercial failure, Wave framed as an exploratory prototype was successful - in helping to understand the problems associated with new methods of communication, as well as in-situ evaluation of technologies.

A critical factor of prototyping in public can be seen from the Wave process, that it is important to target the user experience when deploying to an unknown user base (such as on an AppStore) to be both transparent and clear on the purpose of the prototype, and engage users as testers rather than portray the prototype as a final product, similar to the approach taken when deploying technology probes (Hutchinson et al., 2003).

REFLECTIVE, AGILE, ITERATIVE DESIGN

RAID (Heyer & Brereton, 2008) is a design-researcher's approach that focuses on the idea of release early, release often. The thesis behind RAID is that short-term exploratory studies fail to evaluate the emergent qualities of technology, which can only be revealed through long periods of use in the wild. In order to meet that challenge, RAID proposes a familiar model wherein the researchers and developers release the software under study at the very beginning with a base level of functionality, and then collect information and feedback while it's in use.

With a live, publicly accessible prototype as their artifact, researchers are able to employ a range of methods both quantitative (usage statistics, logs, etc.) and qualitative (participant interviews, journal studies, feedback forms, observations, and case studies) to collect information on how participants are "using, misusing or under-using" the prototype. Using a live system allows these methods to reveal unexpected patterns of use that would slip through the gaps of a controlled study, and the iterative and longitudinal nature of the study allows redesigning the system around such emergent uses.

This approach enables researchers to work very flexibly: as with agile lifecycles in the software industry, the researcher can easily change the direction of their project without abandoning the work they've already completed. It also allows the researcher to conduct "probes" with very little effort: introduce a change to the running system, and then observe how users react.

Developing "in public" makes RAID an excellent approach for long-term studies of social software systems, since it provides much greater opportunity for snowball recruitment to meet the challenges of network effects and the prisoner's dilemma (Grudin, 1994). It is exactly these properties that make such an approach ideal for exploring mobile systems: mobile design studies must not only contend with network effects but must also contend with the emergent properties revealed by imperfect sensors and use in many contexts. A long-term iterative design approach allows us as mobile design researchers to respond both to our users' appropriation of the application, but also to the in-the-moment and refined workarounds they find for the limitations of their hardware. How does a user compensate for inaccuracies in the compass or GPS system? How do they continue when the device loses connectivity? Iterative in-situ deployments allow us to reveal these ad-hoc interactions and to observe how our reflective changes improve or detriment our participants' abilities to meet their goals.

It's important to note that RAID differs from the typical agile development methodologies on which it is based by being adaptive to the needs of researchers: its focus is not on iterating organically towards polished products, but rather on the reflective process and "emergent use and appropriation". In other words, RAID prototypes don't seek the "correct" solution to a software problem, but rather to reveal more interesting problems.

DEPLOY EARLY, EVALUATE PROGRESSIVELY

Deconstructing RAID

Applying a RAID-like approach to mobile systems research projects allows for both a streamlined and in-the-wild evaluation process. However, it is important to disassemble RAID to better understand the kind of design problems that RAID is best suited for in the design of mobile software, and why RAID best suits the kind of project that must deal with network effects or deal with uncertainty, such as:

- An innovation project, where the researcher aims to create or explore a new idea and it's difficult to predict how users will react.
- Social or community-driven exploratory research, where the goal is to reveal unexpected use by allowing a group to appropriate a system over time.
- Interventions where the users' and researchers' understanding of the problem space and therefore their requirements will evolve as their investment in it grows (i.e. a longitudinal PD study).

When considering mobile design research, we can add the following considerations:

- Real-time Sensor-dependent applications where the value of the system to a user will shift unpredictably depending on signal qualities.
- Applications designed to engage or add value to many different contexts: such as locative discovery applications, locative social networks, and locative games.

The key challenge that scenarios such as these present lies in the difficulty of data collection: to conduct qualitative observation on the use of a mobile application necessitates the embedded researcher approach: and long-term embedded observations are intractably difficult (and expensive). Methods associated with a RAID-like approach tend towards just-in-time feedback mechanisms, making it straightforward for a user to provide information to the researcher as it occurs to them in the context of use, rather than during a set workshop or interview session. RAID projects have used a combination of usage logging, social groupware tools, and embedded feedback systems to provide this avenue for feedback. The maturity of mobile software platforms also offers a number of advantages to researchers, such as distribution and data-capture (discussed later in the chapter).

A summary of the RAID approach to system research is:

- **Prototype:** The interactive system.
- **Release:** This system into the wild (and to participants).
- **Collect Data:** Using the methods and tools available to you.
- **Reflect:** And analyze on the data.
- **Iterate:** Continuously on this project over an extended period.

Our take on RAID puts even further emphasis onto high-fidelity prototypes, and typically moves from paper prototyping quickly to implementation prototyping before it's released to users.

Going DEEP

We propose a rapid-iteration version of RAID called DEEP: Deploy Early, Evaluate Progressively. The approach we propose differs from RAID primarily in its emphasis on high-fidelity prototyping from the very onset, and in its rapidity. While both RAID and DEEP approaches work well with longitudinal studies, the individual iterations of a DEEP prototype should be very short: ideally working with nightly or weekly beta releases. This tight prototyping release cycle forces the prototype into a state of flux: where the prototype authors are not simply implementing a preset set of requirements, but are participating in a conversation with their testing subjects. Mobile development is strongly suited to this rapid release cycle because of its mature deployment platforms. While deployment is not as effortless as with server-based systems such as Web applications, deployment is also not as nearly as transparent—users are notified when the application has been updated.

As with RAID, each tight release cycle should be matched against a focused period of use, in which researchers collect and analyze feedback on the release. At this stage we've used informal interview, written feedback forms, and observation effectively as rapid feedback mechanisms.

It is important to note that DEEP is not strictly a development methodology, but rather a research method for exploration. Development methodologies such as agile development process can be well suited to this research method, encouraging rapid deployment. However, the focus on these processes is to work towards a final product, while DEEP is focused on getting user feedback to better understand a problem space, occasionally regressing the design or development cycle to answer newly formed questions.

Deploy Early

When we say deploy, we mean deploy in the industry sense. The tools provided with modern mobile platforms—app store/market release, over the air ad-hoc release, and Web-application deployment—mean that it is trivial to release early versions beyond a select group of testers, and pushing updates is similarly straightforward. Whichever method is used should be public enough to allow for snowballing of testers beyond those that you explicitly recruit. Henze et al. (2010) suggest that the ability to deploy prototypes through these means has the potential to attract a much larger user base for testing with unknown audiences. Through our experiences, we recommend beginning with ad-hoc releases to a series of known users or stakeholders, which can be done without AppStore approval (which can delay the deployment for weeks at a time). As the application evolves, a deployment can be conducted to a larger user base through app stores. This allows a dual-deployment approach where newer features or concepts can be quickly tested by a smaller number of known users (to test for critical issues to stability, usability or functionality), while more iterative and wide-scale feedback can be gathered concurrently.

Evaluate Progressively

With a live prototype in the hands of explicit and self-recruited participants, and an aggressive release cycle, the prototype is ready for an extended period of rapid evaluation and experimentation. The focus of this release cycle is not only evolve the prototype from a technical level, but to directly engage with the testing audience, to continue and reinforce the conversation around the prototype. It is important to not ignore the benefits of throw away prototypes (Gordon et al., 1995), where a design can be forfeited based on user feedback. Mobile development is heavily based on the Model-View-Controller architecture (Krasner et al., 1988), where the View (user interface) can be redesigned without affecting the core programming investment of the prototype. Additionally, it is crucial to be transparent about the purpose of the prototype to its user base. Engaging with users on the explorative nature of the prototype will not only focus user feedback, but also develop stronger communication ties with the test user group.

The following case study describes an application of DEEP to an early stage mobile locative application prototype, over several months from January-May 2012.

RIVER CITY RADAR: A CASE STUDY

We tested the DEEP approach in the context of a location based mobile application, focused on direction finding. The focus for this application was to explore the situations when a person is looking for a specific type of location, but the need is not great enough to open, configure and use traditional mapping/turn by turn apps like Google Maps.

Our initial investigation found that within sandbox environment computer games (where the world is open for non-linear exploration), players see paths/directions to various locations through the game user interface (usually through a heads up display). Grand Theft Auto 4 (Rockstar Games) is an example where the nature of the open world leads to continual navigation to find locations. In this genre of game, the player has two views that can be used to support this interaction: the map view (which is a full-screen view similar to Google Maps); and a mini-map, which is abstracted to only show relevant information for the current goal. There are two main differences between these visualizations (See Table 1).

The initial design was based on the idea of a mini-map in your pocket, to assist the user in navigating the environment to provide on-the-go cues about locations of interest. User scenarios were developed to understand the context of use of such an application, to help understand when and where the application would be used.

Barry is out with friends, and suddenly realises he needs to use the restroom. Unfortunately it is late at night and he does not know the area well. He opens up the application, and is presented with a menu of location categories. He selects restrooms, and a mini-map is presented with his location in the middle. As he moves around, the mini-map changes, both in orientation and position. He notices that there is a toilet icon at the corner of the mini-map, with "200 meters" written underneath it. As he moves in the direction of the toilet, the distance counts down. While following the compass interface towards the toilet, he finds himself walking towards a closed building. Using the satellite view, he looks and see's that there is a path around the building to the left. After some walking, he finds the restroom and puts his phone back in his pocket.

Table 1. Comparison of map and minimap utility

	Map	Mini-map
Purpose	Directions and exploration	Directions
Use	Primary focus	Ambient

From this example scenario, paper prototypes were developed to assist in demonstrating the concept. While this worked well for helping figure out the screen interaction and understanding the nature of the app, a lot of the discussion around the paper prototype was based on hypothetical situations. These paper prototypes were not used as a formal user testing method to refine the interface, but rather as a method of exploration within the internal team to discuss ideas (See Figure 1).

To further explore the concept, we developed a high fidelity prototype that we could walk around with to test. The high fidelity prototype used a mash-up of existing data from Foursquare (http://foursquare.com/) and Yelp (http://www.yelp.

Figure 1. Paper prototypes/sketches

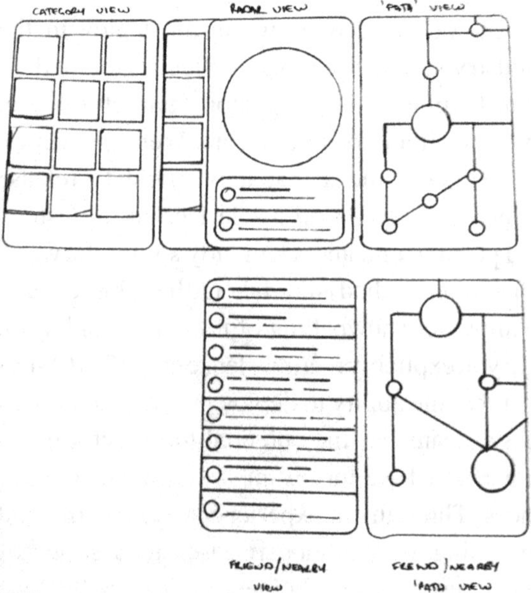

com), and was focused on testing the mini-map concept. The implementation was around emulating the existing functionality of mini-maps in games, where users could choose a category for a menu, and the closest would show up. All categories from these systems could be selected (Foursquare and Yelp offer not only food-based categories, but also categories such as clinics, toilets and parks), to help explore which services were sought for in use (See Figure 2).

While the implementation took longer than the paper prototype (approximately one week), it was possible to test the core functionality of the application in-situ. We enlisted 12 users to assist us in evaluating the prototype. We found a number of issues immediately, both in the concept and the user interface design. We also discovered a major technical issue in the nature of the concept, in the unreliability of GPS - a technical issue that directly affected the feasibility of the concept. We found that although in the lab environment the GPS was reliable, as our users walked around in some locations, the user's position would jump around, sometimes leading to completely unreliable information (being positioned on the wrong side of a river). Our paper prototyping had not taken this into consideration. Additionally, while only the closest of each category was being shown, there were times where the closest was not appropriate (for instance the closest coffee shop was often closed during testing, or was situated in an area that was not accessible on foot). In this respect our prototype was hugely successful in that we uncovered issues that we may have not otherwise thought about. Additionally, we also were able to better understand when users were likely to use the application. For instance, a lot of the time the application was not only used for direction finding, but also in exploring a location, "what's near me right now." This led to ad-hoc decisions being made such as "I could go a coffee right now." A valuable insight that we discovered was that by allowing users to have the prototype on them at all times, their expertise and knowledge of their local surroundings could be used to validate the data being provided by the prototype.

Figure 2. First high-fidelity iOS prototype

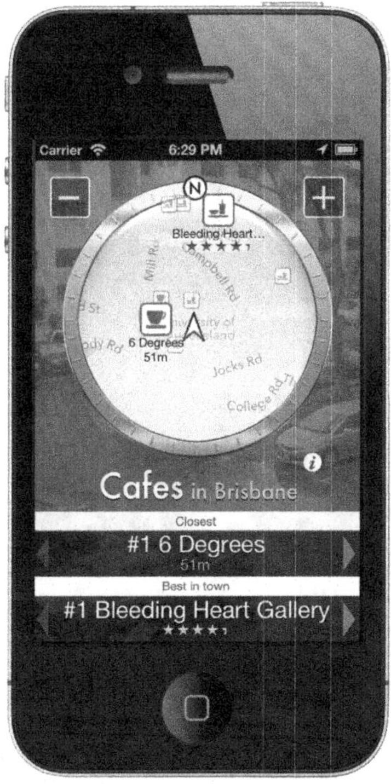

From this first iteration, we incorporated the feedback from our users and began to proceed through further iterations of the design. In this we focused on streamlining the interface to include multiple closest options that the user could swipe through. We also included the idea of exploration, and added the best of categories around the city. This allowed users to further explore their surroundings, informing their decision "hey, the best coffee shop in town is close by!" (See Figure 3).

From this prototype, we gathered more fine-grained user interface feedback. While the satellite view proved useful in some situations, a lot of the time users would change their direction to follow roads and paths to help them navigate. We decided to try a different visual style for the map to focus on this. We also changed some styling, and took away certain features (like zooming), which made the experience cumbersome, opting

Figure 3. First feedback-driven iteration

instead for an adaptive zooming mechanism where the map would zoom based on the distance to the location. Through feedback, we found that this better meshed with the users' local knowledge of an area - and allowed users to reflect on the information, rather than blindly following the arrow.

As we implemented different prototypes, we began to streamline the process of getting feedback and pushing out new updates. We found that by iterating quickly and based on user feedback, the feedback we received was more honest and more exploratory (rather than minor feedback on colors or layout). Testers began to provide ideas, rather than just functional feedback or problems. A lot of feedback suggested that we remove the map entirely, and focus more on an abstract form (See Figures 4 and 5).

Our final prototype (thus far), tested whether the concept still worked successfully without the map interface. Instead, we focus on a more information-centered design, where a compass points to the location. We even tested the idea of a 3D

Figure 4. Rough ideas from testers become further interface experiments

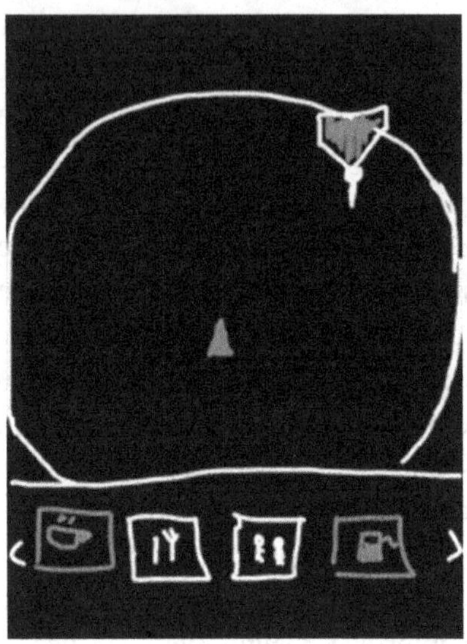

Figure 5. Rough ideas from testers become further interface experiments

compass, based on the vertical orientation of the phone, where the compass would swing and provide perspective. We found quickly though that users took this affordance literally, thinking that the indicator was providing information about whether the location was up hill. Feedback also frequently discussed the problems with only having a mini-map to view the location, as some users wanted the option to see the whole map along with routing information. We incorporated this feedback into our design, and created a card metaphor where the front of the card was information based, and the back of the card would have a postcard style map showing the location (with optional link out to the native Maps application for directions). We found that this was more flexible, as it meant that you could switch between how much information is presented based on the changing situation (such as walking down a road versus being stuck at a dead end).

While we used a number of methods (brainstorming, informal interviews, observation, paper prototyping) throughout this project, the emphasis was on rapid prototyping (DEEP) as an artifact for feedback. We found that this method was an excellent way to gather data from users that is focused on evaluating and iterating on the design concept. While initially there were concerns about the practicalities of using high fidelity prototypes as an exploratory method, the tools that are now available for deployment and tracking made it a lot easier than originally anticipated. In this study we used Test Flight (http://testflight.com/), which allowed us to wirelessly push out updates directly to users, and also recruit anonymous user testers from a global pool. As users interacted with the application, we were able to gather information about what they were looking for, and whether they were jumping between categories (exploring) or keeping the application active while moving (path finding). We also had the ability to better understand where and when our users would open the application, rather than making assumptions. This led us to better understand the importance of some potential features, in particular to be able to see whether locations were open at the time (a lot of the applications use was performed out of business hours).

Overall, we found that while the DEEP method took longer than prior investigations we had performed with similar designs, the level of insight that we gained on the concept was far deeper and further-reaching than following a more traditional iterative design process of observing, sketching, and prototyping. It should be reiterated that this case study uses the DEEP method as a form of design exploration, rather than a hypothesis-based method for gathering quantitative research data. Deployment frameworks such as Test Flight provided a number of capabilities in this regard, allowing us to automatically track usage of the application, as well as provide a built in feedback mechanism for users (See Figure 6). We were also able to assign "checkpoints" within the application, where a notification was triggered when a user performed a specific action, without requiring the user to perform extra action. Through this, designers have the ability to better understand the flow of their application in use, and understand which features are being used in specific situations.

PRACTICAL CONCERNS

Systems work is simply unavoidable when attempting to evaluate mobile software design, but as we've discussed there are a multitude of tools and frameworks to make the work more easy and better suited to the user interface researcher. We suggest that the ideal team for a DEEP-style research project be multidisciplinary: collaboration between mobile developers and user-interface and mobile design researchers. We also stongly suggest that all team members should have direct involvement in the design process, as this is crucial to understanding what information can be collected from the deployment, and be able to choose appropriate methods for engaging with the test users.

Another concern with these methods is to acknowledge and understand the potential difficulties with evaluating mobile software in the wild. In particular, the collection and validation

Figure 6. Test flight data collection

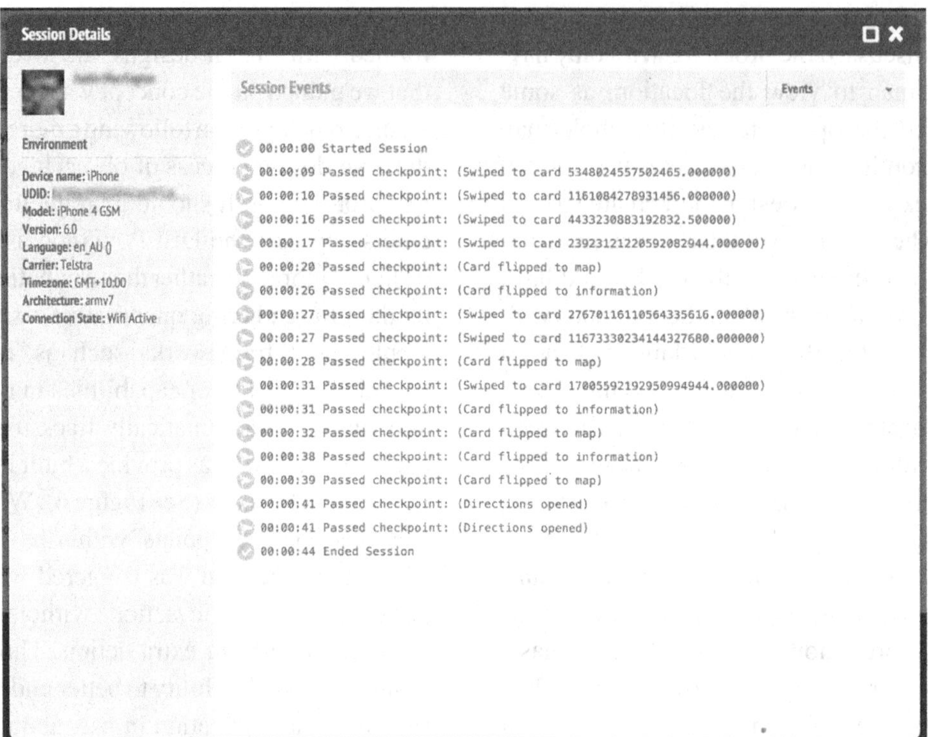

of the data can be difficult to gather with limited control of users. Miluzzo, (2010) highlights many of these issues, including the potential problems associated with identifying the demographics of the user group. "How do we collect and validate our research data when we have limited control over users and lack real ground truth? How do we make sure we have a good cross section of users to validate our study?" With this perspective, we do not recommend DEEP to replace existing research evaluation methods, but rather to augment them, to provide more insight from the outside of the lab environment – especially in the early stages of design iterations (rather than at the later evaluation stages). Through the dual deployment approach (Deploy Early), design researchers can also target their research, collecting highly verifiable data from a known group of users while still deploying to a large user base for a wider variety of feedback.

An alternative that has not been discussed is the use of creating a Website designed for mobile devices (called a Web-app). Web-apps provide some benefits of creating a native application, in particular that they can a) be run on a multiple of platforms (such as Android and iPhone), and b) be deployed to a large number of users without requiring them to sign up. In previous studies, we have found that while Web-apps appear to be a lightweight method of deploying these prototype applications, there are a number of issues with the technology that need to be considered. While the level of knowledge required to create Web applications appears lower at the outset, we have found that it requires a lot more time to create a Web-app that has sufficient levels of functionality to be tested in unknown contexts - and the disconnection from a specified platform means that iterative development slows due to testing of the software across platforms. In particular, due to the sandboxed nature of Web-apps (in that they are limited in which phone sensors can be accessed), Web-apps often involve working with a limited feature set on the phone. Tools such as Phonegap (http://phonegap.com) can alleviate this by providing Web apps with device sensor information, however this can be intermittent and also requires users to deal with the issues that occur with native applications.

CONCLUSION

The value of high-fidelity prototypes to mobile projects is self-evident: having access to the real application's features and real hardware interactions will always provide richer feedback than an artificial simulation. The question is whether it is tractable to focus on systems work if simulated deployments would be "good enough", and whether the problems inherent with real-world testing are manageable. There are a number of future directions we can see in this work—in particular how these studies are structured in a way which can provide rigor to the data gathering process in the wild—becoming a method not only exploration, but also for evaluation. Our argument is, for all the reasons we have discussed, that for all but the most trivial mobile applications simulated deployments cannot reveal the whole story of how your users will appropriate, use, and experience your application. The DEEP method is not proposed as a replacement for existing methods, but rather a way to augment existing investigative methods.

It is fortunate for mobile researchers that there is a booming industry of application development kit manufacturers—Apple and Google at the very least—who have made it their priority to provide easy to learn, accessible, and powerful toolkits for developing and deploying novel systems on their platforms. Mobile researchers simply need to stop avoiding the implied issue of developing on these platforms and use them to their advantage.

REFERENCES

Abowd, G. D., Anind, D. K., Brown, P. J., Davies, N., Smith, M., & Steggles, P. (1999). Towards a better understanding of context and context-awareness. In *Proceedings of the 1st International Symposium on Handheld and Ubiquitous Computing* (pp. 304-307). London: ACM.

Abowd, G. D., Atkeson, C. D., Hong, J., Long, S., Kooper, R., & Pinkerton, M. (1997). Cyberguide: A mobile context-aware tour guide. *Wireless Networks*, *3*(5), 421–433. doi:10.1023/A:1019194325861.

Buchanan, R. (1992). Wicked problems in design thinking. *Design Issues*, *8*(2), 5–21. doi:10.2307/1511637.

Buur, J., & Soendergaard, A. (2000). Video card game: an augmented environment for user centred design discussions. In *Proceedings of DARE 2000 on Designing Augmented Reality* (pp. 63-69). New York: ACM.

Buxton, B., & Greenberg, S. (2009). Usability evaluation considered harmful (some of the time). In *Proceedings of the Twenty-Sixth Annual SIGCHI Conference on Human Factors in Computing Systems –CHI 2009* (pp. 111-120). New York: ACM.

Dick, T. J. (1979). Mixing qualitative and quantitative methods: Triangulation in action. *Qualitative Methodology*, 602-611.

Gordon, V. S., & Bieman, J. M. (1995). Rapid prototyping: lessons learned. *IEEE Software*, *12*(1), 85–95. doi:10.1109/52.363162.

Grudin, J. (1994b). Groupware and social dynamics: Eight challenges for developers. *Communications of the ACM*, *37*(1), 92–105. doi:10.1145/175222.175230.

Guindon, R. (1990). Designing the design process: exploiting opportunistic thoughts. *Human-Computer Interaction*, *5*(2), 305–344. doi:10.1207/s15327051hci0502&3_6.

Henze, N., & Boll, S. (2010). Push the study to the app. store: Evaluating on-screen visualizations for maps in the android market. In *Proceedings of MobileHCI '10: Human-Computer Interaction with Mobile Devices and Services* (pp. 373–374). Lisbon, Portugal: ACM. doi:10.1145/1851600.1851671.

Heyer, C., & Brereton, M. (2008). Reflective agile iterative design. In *Proceedings of the Social Interaction with Mundane Technologies Conference*. Brisbane, Australia: QUT.

Heyer, C., & Brereton, M. (2010). Design from the everyday: continuously evolving, embedded exploratory prototypes. In *Proceedings of 8th ACM Conference on Designing Interactive Systems* (pp. 1-10). Aarhus, Germany: ACM.

Hix, D., & Hartson, R. H. (1993). *Developing user interfaces: Ensuring usability through product & process*. New York: John Wiley & Sons.

Hutchinson, H., Mackay, W., Westerlund, B., Bederson, B. B., Druin, A., & Plaisant, C. Sundblad, Y. (2003). Technology probes: Inspiring design for and with families. In *Proceedings of the ACM CHI 2003 Human Factors in Computing Systems Conference* (pp. 17-24). New York: ACM.

Krasner, G. E., & Pope, S. T. (1988). A description of the model-view-controller user interface paradigm in the smalltalk-80 system. *Journal of Object Oriented Programming*, *1*(3), 26–49.

Landay, J. (2009). *I give up on CHI/UIST*. Retrieved 06 2012, from http://dubfuture.blogspot.com.au/2009/11/i-give-up-on-chiuist.html

Lieberman, H. (2003). *The tyranny of evaluation*. Retrieved 05 2012, from http://Web.media.mit.edu/~lieber/Misc/Tyranny-Evaluation.html

Martin, D., & Sommerville, I. (2004). Ethnomethodology, patterns of cooperative interaction and design. *ACM Transactions on Computer-Human Interaction*, *11*(1), 59–89. doi:10.1145/972648.972651.

Miluzzo, E., Lane, N. D., Lu, H., & Campbell, A. T. (2010). *Research in the app. store era: Experiences from the cenceme app. deployment on the iPhone*. Paper presented at the First Workshop on Research in the Large at UbiComp 2010. Copenhagen, Denmark.

O'Reilly. (2009). *Google wave: What might email look like if it were invented today*. Retrieved 08 2012, from http://radar.oreilly.com/2009/05/google-wave-what-might-email-l.html

Olsen, R. D., Jr. (2007). Evaluating user interface systems research. In *Proceedings of the 20th Annual ACM Symposium on User Interface Software and Technology* (pp. 251-258). Providence, RI: ACM.

Raento, M., Oulasvirta, A., Petit, R., & Toivonen, H. (2005). ContextPhone: A prototyping platform for context-aware mobile applications. *IEEE Pervasive Computing Special Issue on Smartphone*, *4*(2), 51–59. doi:10.1109/MPRV.2005.29.

Redhead, F., & Brereton, M. (2008). Nnub: A display for local communications. In *Proceedings of the Workshop on Public and Situated Displays to Support Communities at the 22nd Conference of the Computer-Human Interaction Special Interest Group of Australia on Computer-Human Interaction 2008*. Cairns, Australia: ACM.

Redhead, F., Dekker, A., & Brereton, M. (2010). NNUB: The neighbourhood nub digital noticeboard system. In *Proceedings of the 22nd Conference of the Computer-Human Interaction Special Interest Group of Australia on Computer-Human Interaction 2010* (pp. 418-419). Brisbane, Australia: ACM.

Rodden, T., Chervest, K., Davies, N., & Dix, A. (1998). *Exploiting context in HCI design for mobile systems*. Paper presented at the Workshop on Human Computer Interaction with Mobile Devices. Glasgow, UK.

Tacchi, J., Foth, M., & Hearn, G. (2007). *Ethnographic action research*. Brisbane, Australia: QUT.

TIGSource. (2009). *Minecraft (alpha)*. Retrieved 06 27, 2012, from http://forums.tigsource.com/index.php?PHPSESSID=8a7200c9319b16be007dabdc9d31706b&topic=6273.0

Chapter 2
User-Centered Study on Quality of Mobile Video Services

Wei Song
Queensland University of Technology, Australia

Dian Tjondronegoro
Queensland University of Technology, Australia

Michael Docherty
Queensland University of Technology, Australia

ABSTRACT

Mobile video, as an emerging market and a promising research field, has attracted much attention from both industry and researchers. Considering the quality of user-experience as the crux of mobile video services, this chapter aims to provide a guide to user-centered studies of mobile video quality. This will benefit future research in better understanding user needs and experiences, designing effective research, and providing solid solutions to improve the quality of mobile video. This chapter is organized in three main parts: (1) a review of recent user studies from the perspectives of research focuses, user study methods, and data analysis methods; (2) an example of conducting a user study of mobile video research, together with the discussion on a series of relative issues, such as participants, materials and devices, study procedure, and analysis results; and (3) a conclusion with an open discussion about challenges and opportunities in mobile video related research, and associated potential future improvements.

INTRODUCTION

Thanks to the rapid advance of multimedia and mobile technologies, an increasing number of people are watching videos through the Internet using mobile devices, such as smart phones, tablets, and laptops. The increasing trend of consuming videos (and TV) on mobile devices, that is, mobile video, has been seen in many parts of the world. According to a survey by QuickPlay Media in 2012 (Burger, 2012), 35% of American respondents have tried mobile video services primarily viewed by means of smartphones and tablets. In Australia, video content consumption

DOI: 10.4018/978-1-4666-4054-2.ch002

on extended screens of computers (e.g., tablets and smartphones) is also showing a 11% increase of mobile video streaming over the end of 2011 for teens, according to the first quarter report of 2012 by the Nielsen Co. (2012). The global growth of mobile video consumption brings pressure to the network data traffic. Allot's surveys in 2011 (Lawson, 2011) showed that video streaming dominated mobile traffic, accounting for 39% of all mobile data traffic worldwide and increasing 93% of use over the year. To reduce the network load and smooth video streaming, a high compression ratio of the videos is needed. This, however, may result in a low video quality, especially when compressing the video to a low bitrate for wireless network transmission. Research has found that the consumers' Willingness To Pay (WTP) is closely associated with the delivered quality of the mobile video stream (Ries, Nemethova, & Rupp, 2008). It has also been recognized that video content and video quality viewed on a small screen importantly affect user's willingness to watch mobile video (Orgad, 2006; Song & Tjondronegoro, 2010). In fact, although mobile video is demanding, the actual uptake of mobile video is still low. There is evidence that the monthly time spent on viewing videos on a mobile phone per American user only increased from about 3:37 minutes in 2010 to 4:20 minutes in 2011 (The Nielsen Company, 2011a). To sum up, the previous investigations indicate a promising future of mobile video, as well as a huge pressure for the mobile video vendors in content and quality delivery.

To achieve the success of mobile video service, user experience (i.e., the way of users feel about or perceive the mobile video) has to be considered. The challenge of improving user satisfaction mainly comes from two aspects. On the one hand, due to the complexity of user experience (UX) and the recent emergence of the mobile video, the understanding of UX of mobile video is insufficient, which leads to an incapability of designing for good UX when providing video content to mobile users. On the other hand, the resource constraints, such as the limited display capability of mobile devices, the limited bandwidth of mobile networks, and the big size of video, bring forth difficulties to deliver satisfactory quality to mobile video users.

To understand and improve user experience of mobile video, including user perception, acceptance, satisfaction and needs, a great deal of user-centered research has been made. The focus of the research varies: some studies investigated users' attitudes toward and their experiences with using the mobile video; some aimed to find which technical factors influence user perceived video quality or overall acceptability; and others attempted to understand how users evaluate the quality of the service. In order to help future research on mobile video (including mobile TV and other mobile video formats) or relevance area such as mobile multimedia, and mobile HCI and applications, this chapter will firstly review the current user studies on quality of user experience of mobile videos in three themes: research scopes or focus, user study methods, and data analysis. Each will be discussed with examples. This will help researchers better understand the research area of mobile video and determine a clear research direction. The chapter then provides an example of conducting user study of mobile video research, along with the discussion on a series of relative issues, such as participants, materials and devices, study procedure, and results. This will be helpful for researchers to conduct user study properly and achieve good performance. Finally, this chapter explores challenges and opportunities in mobile video related research, and discusses potential improvement in future research.

FOCUS OF USER STUDIES IN MOBILE VIDEO QUALITY

Concentrating on the quality of the mobile video service, this section summarizes three main research focus of user studies: (1) understanding of mobile video users; (2) impacts of user-perceived mobile video quality; and (3) understanding of user requirements for mobile video quality. Other aspects of mobile video, such as user interface design and video content creation, are not encompassed in this chapter.

Understanding Mobile Video Users

As users are the subject of a mobile video service, their attitudes, needs and expectations are important to service providers. Therefore, a large number of studies have been made to understand existing users and potential users. The scope of the studies covers different aspects, such as distinguishing user types, investigating their motivations and preferences, and exploring contextual, cultural, and social impacts on user's consumption and usage of mobile video.

Generally, researchers tend to distinguish users by age, gender, work, education background or relationship with modern technology. According to Orgad (2006), adults aged 18 to 34 will be the primary mobile TV users who are familiar with technologies built into mobile devices and who are interested in new technologies; teenagers and children are likely to form another important user group who may think it is "fashionable to watch television on small mobile screens instead of big static screens" (2006, p. 11); the third group expected is business people who spend more time on the go and for whom consuming information is an essential part of their lives. Södergård's Mobile TV trial study in Finland (2003), however, showed a different user group pattern: children under 12 and adults from 41 to 50 were more enthusiastic than young adults from 21 to 30, because families could use the service the most at home for entertainment, while workers and students' usage were limited by the environment and time. The different findings may be influenced by the time and location when conducting the studies. Nevertheless, according to most common findings from trials, studies and investigations (Carlsson & Walden, 2007; Frank N. Magid Associates Inc., 2009; Jumisko-Pyykkö, Weitzel, & Strohmeier, 2008; The Nielsen Company, 2011b), the young adults at the age range of 18–40 are the major or typical mobile video users; meanwhile, there is a trend that older people are becoming more interested in mobile video. Studies also indicate a gender effect: males watch mobile videos more than females (Frank N. Magid Associates Inc., 2009; Jumisko-Pyykkö, Weitzel, et al., 2008; Södergård, 2003; The Nielsen Company, 2011b).

Motivation for or purpose of using mobile video (TV), as a part of user's internal state, affects the user experience (Hassenzahl & Tractinsky, 2006). Bunchger et al. (2009) have summarized a set of motivations for watching mobile TV. Simplifying those, the five major motivations of viewing mobile videos are: consuming time, being entertained, staying up to date (e.g., with news or popular events), sharing with others, or isolating oneself from the surroundings. The motivations are, in fact, highly related to the content and context of viewing mobile video. For example, short news videos are suitable for watching when waiting for a bus, whereas movies are more likely to be watched at home for entertainment.

Content has been defined as the "King" of the mobile video service (Chipchase, Yanqing, & Jung, 2006). Previous studies indicate that users prefer concise (short), up-to-date and focused video content in mobile devices, as a result, the most popular content on Mobile TV include news, general entertainment, drama, sports, cartoons, music, film and documentary (Carlsson & Walden, 2007; Chipchase, et al., 2006; Mäki, 2005; Petrovic, Fallenböck, Kittl, & Langl, 2006; Södergård, 2003). However, recent studies show that mobile video is not necessary to be short and an increasing number of people are watching a full TV show or a full movie (Song & Tjondronegoro, 2010; The Nielsen Company, 2011b).

The mobility and portability of mobile devices determine that the physical location of and the time spent on mobile video could be anywhere, such as on transportation when commuting, at home before sleeping, at a public location when queuing or waiting, and in a cafe for a work break. However, the popularity ratings of physical locations of watching mobile video are culture dependent (Buchinger, Kriglstein, Brandt, & Hlavacs, 2011; Buchinger, et al., 2009). A study in Belgium (Vangenck, Jacobs, Lievens, Vanhengel, & Pierson, 2008) found that people tended to use mobile TV at home rather than on the move; however, a study in Japan (Miyauchi, Sugahara, & Oda, 2008) stated that main consumption of mobile TV was on the go. Studies in Finland (Södergård, 2003) and South Korea (Chipchase, et al., 2006) noted that home, commute, and short waiting time were the main contexts for using Mobile TV. Despite this popularity discrepancy, it is still clear that at home, moving, and short waiting are the main physical contexts.

In addition, people use media not just for the purpose of using it (functionalism or instrumentalism), but rather in a social context composed of the surrounding world and other people (Falk, 1992). When solitary viewing or video sharing happens, people are using mobile video to manage relationships with others in shared or public settings. They are trying either to cut off the outside setting or to enjoy others' attendance (O'Hara, Mitchell, & Vorbau, 2007). User's viewing experience under the two situations, individual watch and shared watch, is completely different. Generally speaking, user-generated contents are more inclined to be shared. As a sharing watch generally involves more complicated public surroundings and other people's impact, more comprehensive research is needed to discuss the issues.

Regarding user-preferred consumption modes, three were mostly mentioned: a fixed monthly fee, time-based payment, and event-based payment (Buchinger, et al., 2009; Carlsson, Carlsson, Puhakainen, & Walden, 2006; Carlsson & Walden, 2007). Additionally, a free model financed by ads may attract users (The Nielsen Company, 2011b; Winder, 2001); it is also possible that advertisements are consumed as other content for entertainment (Orgad, 2006). For the success of mobile TV, the right pricing approach should be to give users a choice of various payment options (Trefzger, 2005).

Understanding how people use mobile video is necessary for designing better services to improve user experience. For example, providing more suitable content for different user groups, designing the service for intuitive interaction and convenient access, and developing strategies to promote its use and quality.

Impacts of Mobile Video Quality Experienced by Users

Perceived (or experienced) quality describes the quality from a user's point of view (Jumisko-Pyykkö, Strohmeier, Utriainen, & Kunze, 2010) and this has a critical impact on the usage of mobile video. Therefore, many studies have been made to find how the experienced quality is affected and how it can be improved. In different studies, the user-experienced quality has been evaluated from different angles, such as perceptual quality, perception, satisfaction, acceptability, and enjoyment. Here, we collectively refer to these terms as the "user-experienced quality" for convenience of description. The best way of knowing the user-experienced quality is to directly ask for the user's opinions. This kind of method is called subjective quality assessment. Technically, the user-experienced quality will be affected by the whole process of delivering a video from the server to the user's mobile device. Therefore, subjective quality assessments are often made to identify influencing factors in three parts of the process: video coding, video transmission, and video display.

In the aspect of video coding, encoding parameters—video resolution, Frame Rate (FR), Quantization Parameter (QP), and encoding bitrate—are found to play an important role in influencing the user-experienced quality (Agboma & Liotta, 2012; Buchinger, et al., 2009; Knoche & Sasse, 2009). However, the findings from previous research show that the impact of encoding parameters is complicated. For example, one study found that user's acceptability declined with decreasing image resolution at a higher bitrate; while at a very low bitrate (<64kbps) there was no significant impact of image resolution reduction on acceptability (Knoche, McCarthy, & Sasse, 2005). This study also found an interactive impact of between image resolution and video content type on acceptability, that is, the change of acceptability with resolutions varied with the content types, such as news, sport, and music. As for the effect of FR, it was found that the relation between the FR and perceived quality is not linear, and also associated with the content. McCarthy et al. (2004) found that when viewing a sports video on a handheld device, sports fans could tolerate as low as 6fps if only the content was personally favourite and the image quality was enough high. Moreover, the minimum bitrate that users can accept for mobile devices is also different with image resolution, frame rate and video content. According to Knoche et al.'s study (2005) and Agboma and Liotta's study (2006), at 320×240 pixels resolution and 12.5fps, a bitrate around 200kbps is acceptable by 80% of participants, and for news and cartoon videos the minimum bitrate can even be much smaller, but football over 300 kbps is needed. Complicatedly, the acceptable bitrate seems increasing due to the development of mobile technology and the growth of a user's viewing experience. It was found that the minimum acceptable bitrate was about 1.5 times higher in a study in 2010 than those in 2005 (Song, Tjondronegoro, & Docherty, 2011b).

The performance of video transmission has been considered as crucial for the quality of mobile video services, because the effects of jitter, delay, and packet loss during wireless network transmission may cause a serious negative impact on the user-experienced quality. Many studies have investigated how the user-experienced quality is affected by the network performance and how to eliminate the negative effect. A study revealed that the overall quality of experience was linearly related to audio and video packet loss rate, audio packet jitter, and received signal strength indicator (RSSI) (Ketyko, De Moor, Joseph, Martens, & De Marez, 2010). Another study showed that the QoE through Audio-video IP transmission could be a non-linear model on basis of error concealment ratio and media unit loss ratio (Tasaka, Yoshimi, & Hirashima, 2008). In addition, another study adopted a less than 10-second buffering before play to reduce jitter delay, to smooth throughput fluctuation, and to help with error recovery for Mobile TV channel switching (Robitza, Buchinger, Hummelbrunner, & Hlavacs, 2010).

Regarding the video display, a vital impact comes from the mobile device. Researchers have studied the effect of screen features (e.g., size and resolution) and processing features (e.g., CPU, memory, and battery) of a mobile device. It was found that a bigger screen is preferred but people do not want a big mobile device (Jumisko-Pyykkö, Weitzel, et al., 2008; Knoche & McCarthy, 2004; Knoche & Sasse, 2008a); high resolution content has to be presented at a sufficient size; otherwise it leads to a poorer acceptability than lower resolution content at the same size (Knoche & Sasse, 2008b). It was also found that the limited battery life has become a bottleneck of mobile video consumption (Chipchase, et al., 2006; Knoche & McCarthy, 2004; Knoche & Sasse, 2008a; O'Hara, et al., 2007). Researchers also studied the effects of viewing distance and viewing angle on the lowest acceptance, which is related to how the object of a video is displayed on the mobile screen (Knoche & Sasse, 2008b).

Identifying the impacts of user-experienced quality helps determine the relationship between the user value and the main influencing factors. Some research has an eventual goal to establish quality assessment models that can predict the user values (user-experienced quality) based on a set of measurable variables. These models can then be utilized to optimize the video quality delivery (Agboma & Liotta, 2012; Ketyko, et al., 2010).

Understanding User Requirements for Mobile Video Quality

Studies on which factors significantly influence the user-experienced quality and how they impose that influence (e.g., a positive or negative effect) have difficulty in explaining why the user experience is affected by these factors in those ways. Therefore, some researchers have attempted to understand how users think a video quality is good or bad in a descriptive manner.

Jumisko-Pyykkö et al. (2010) conducted five studies to collect descriptive data after the psycho-perceptual quality evaluation experiment on Mobile 3D TV. Through analyzing qualitative interviews and written attribute description tasks from over 90 naïve participants, they found that people derive two different kinds of quality descriptions: "Low- level factors" are directly derived from the characteristics of the presented stimuli, whereas "high-level factors" represent attributes that take into account users' goals of using the system or their knowledge about the system. Another study has tried to understand how mobile users would like to accept a mode of delivering proper video quality (Song, Tjondronegoro, & Docherty, 2011a). From the user's perspective, the study showed that how user requirements for video quality are related to personal preference, technology background and video viewing experience, and how users preferred different quality-delivery modes and interactive modes.

A thorough understanding of the complex user requirements is supportive to the user-centred design for meeting the requirements, such as providing more flexible and personalized quality delivery and interaction to the mobile video users.

USER STUDY METHODS

Along with diverse purposes and focuses, research can be generally categorized into qualitative research and quantitative research. Qualitative research is inductive, aiming to explore attitudes, behaviours, needs and experiences and to get an in-depth understanding of participants. Collected qualitative data are usually in the form of words, pictures, audios, videos, or objects, which are rich but less able to be generalized; the data analysis relies on researchers' interpretation, which may involve in personal bias. Quantitative research is deductive, aiming to classify features, confirm acknowledged concepts, and generate statistical models in an attempt to explain what is observed. Quantitative data, which is numerically or statistically measurable, can be used to test specific hypotheses objectively and may allow a researcher to draw reliable inferences from findings; however, it may miss contextual details (Bazeley, 2004; Creswell, 2003; Denzin & Lincoln, 2000).

There are a variety of methods for qualitative and quantitative research that have been employed in user studies in mobile video field (Buchinger, et al., 2009; Obrist, Meschtscherjakov, & Tscheligi, 2010). Qualitative methods, such as interviews (Jumisko-Pyykkö, et al., 2010), focus groups (Eronen, 2001; Knoche & McCarthy, 2004), observation (Chipchase, et al., 2006; Södergård, 2003), diary and probe (Eronen, 2006; Miyauchi, et al., 2008; O'Hara, Black, & Lipson, 2006), have been applied for exploring user needs and attitudes and understanding user experience of the service and user criteria on video quality. And quantitative methods, such as survey (Jumisko-Pyykkö, Weitzel, et al., 2008; Song & Tjondronegoro, 2010), questionnaire (Schatz & Egger, 2008), and subjective quality assessment (Knoche, McCarthy, & Sasse, 2008; Song, Tjondronegoro, & Docherty, 2010), have been used to classify user groups, to examine factors influencing user-perceived quality of mobile video, and to evaluate usability of a product. The following content will introduce some commonly used qualitative and quantities methods, and discuss how to determine a proper study method.

Qualitative and Qualitative Research Methods

The frequently used qualitative and quantitative methods in mobile video user studies are listed in Table 1. Each method has its specific features, and thus is suitable for different study situations and purposes.

Interviews

Interview(s) is one of the most widely used qualitative methods. By face-to-face communication, the interviewer aims to obtain the interviewee's opinions and experiences about a certain thing, such as a product/service/system. In structured interviews, detailed questions are predetermined in order to gain more focused answers. In contrast, in unstructured interviews, questions are more open, with more opportunity for variations in answers. In between, semi-structured interviews use a fairly open framework and allow for focused, conversational intercommunication, which allows participants the time and scope to talk about their opinions on a particular subject (Lindlof & Taylor, 2002). The semi-structured interviews have been used in Jumisko-Pyykkö Häkkinen and Nyman's study (2007), where the main question was designed ahead of time, but was presented variedly and supported by other clarifying questions during the interview, allowing both the interviewer and the interviewee flexibly to probe for details.

A good interview requires the interviewer skilled in listening, understanding, clarifying, and probing when necessary. It also requires appropriate time and place where both the interviewer and the respondent can concentrate or where the investigated thing mostly occurs. For example, in many studies, interviews for understanding how people select a video quality were conducted in a quiet meeting room or a lab (Knoche & Sasse, 2009; Song, et al., 2011b), whereas in a study on personalised television by Chipchase et al. (2006), unstructured interviews were conducted in where participants used Mobile TV and in-depth interviews were at home where the use cases happened primarily.

Focus Groups

Focus groups are group interviews, guided by a moderator, discussing topics raised by the interviewers within a small group (Morgan, 1998). Usually, focus groups need to run over multiple groups (3-5); each group requires 6–10 participants and often lasts 60–90 minutes. Its goal is to gather perceptions, information, experiences, and understandings through discussion on a specified topic. The strength of focus groups lies in the synergy of group discussions, which create sharing and comparing process and self-motivated communication among the participants. In this way, valuable information can be collected quickly and cheaply. However, the group discussion may overwhelm individual opinions (Edmunds, 1999).

Focus groups are a suitable practice for initial exploration and discovery at the early stage of a study, even when the researchers have minimal knowledge about the topic. An example of using focus groups is the study of exploring users' interactive requirements for a mobile TV application (Knoche & McCarthy, 2005), conducted in 2004 when the Mobile TV just launched in Europe. Based on the results from focus group, a set of principles of designing the functionalities of the Mobile TV application was made to address these user requirements. Another example of using focus groups is for user interface design of mobile video application (Huber, Steimle, & Mühlhäuser, 2010). Focus groups provide an opportunity to test an existing interface (including content, features and design elements) and/or explore alternate design concepts being considered.

Table 1. Features of qualitative and quantitative methods in user studies of mobile video

Method	Description	Applied Study Examples
Interviews	The interviewer asks questions of the interviewee(s). Interviews may be structured, semi-structured, or unstructured. Data is captured through audio or video recording or in hand written notes. It is time consuming.	Interviews were conducted in possible use places of mobile TV and at home to understand: motivations, barriers and contexts of use, and design implications for future products and services. (Chipchase, et al., 2006)
Focus groups	Group interviews, guided by a moderator, discuss topics raised by the interviewers within a small group. Data collection is quicker than individual interviews, and group discussions encourage interactivity within participants. However, individuals are not easy to identify.	Usage scenario-based focus groups were used to investigate users' thoughts on product concepts or ideas of device and service design for mobile 3D TV (Jumisko-Pyykkö, Weitzel, et al., 2008).
Direct observation	People are studied (observed) by the researcher(s) in their natural environment. It collects rich qualitative data of human behaviours, avoiding contrived research arrangement; but is time consuming and subject to observer bias.	Observations were made with the consideration that user experience of mobile video might be unique to the Korean context. (Chipchase, et al., 2006) A study was conducted to discover social context on basis of observation in public spaces (Södergård, 2003)
Probe study (diary, self-documenting, self-report, and Experience Sampling method (ESM))	A probe study is one in which participants keep records of their activities, feelings and thoughts, at regular intervals or immediately (e.g., ESM), over a period of time. It offers insight into imperceptible details for understanding users, but it is hard to be complied by participants.	A study adopted dairy techniques to better understand how people were consuming videos on mobile devices in their everyday lives and private environment (O'Hara, et al., 2007) At the beginning of design process for Mobile 3D TV, probe studies have been used to discover users' needs, values and feelings (Jumisko-Pyykkö, Weitzel, et al., 2008)
Survey (& Questionnaire)	A survey study collects data by asking people to fulfil a questionnaire, which generally includes a combination of closed questions (predetermined responses, either yes/no or multi choice), Likert scales to indicate strength of agreement with a statement, and open questions (free text, which must be coded for analysis). It is easy to collect data from a large number of samples but weak in generating new ideas.	An online questionnaire was used to collect user requirement for mobile 3D television and video, including background information, user's motivations, requirements for content, and system and service functionalities, and context of use (Jumisko-Pyykkö & Häkkinen, 2005). An online survey study was run in Australia for investigating users' attitudes, needs, and concerns to mobile video and mobile TV (Song & Tjondronegoro, 2010).
Laboratory Experiment (for subjective quality assessment)	A quantitative study asks participants to evaluate video quality using scales or psychological judgments in a lab or controlled environment. It is efficient and convenient for data collection and analysis, but ignores the effect of its natural settings.	Lab experiments compared whether two subjective assessment approaches, simplified continuous assessment (during viewing) and retrospective assessment of overall quality (after viewing), were different in evaluating acceptance and satisfaction (Jumisko-Pyykkö, Vadakital, & Hannuksela, 2008)
Experiment in Field/Context (for subjective quality assessment)	Subjective evaluation on video quality is conducted in a real or mimic use context. It aims to capture more realistic user experience, but its data may be impacted by unknown factors.	A study to evaluate the quality of mobile TV in parallel to three different tasks: waiting for a friend in a railway station, traveling by local bus, and spending time in a café (Jumisko-Pyykkö & Hannuksela, 2008). A quality acceptance evaluation on mobile video on a campus and on a bus (Song, et al., 2010).

Direct Observation

When conducting direct observations, the researcher is observing certain sampled situations or people in the entire natural context, in order to find people's spontaneous behaviours and activities. For mobile video studies, the observation environments that have been selected include transportation, cafes, bars, restaurants, shopping malls, streets, and mobile phone shop and service centers, etc. (Chipchase, et al., 2006; Södergård, 2003). To collect data, tools such as pen and notebook or camera are needed. As the meanings and interpretation of observations highly depend on the observers, an observation study is often compensated by other methods such as interviews to minimize human bias. For example, at the beginning of a study on Mobile TV, researchers observed people's everyday media use in 24 public spaces in three cities. Then, the researchers conducted interviews to ask people how they use media in the everyday lives. The findings of observations benefited the interview design, and interviews confirmed or questioned the observations made earlier (Södergård, 2003). In a south Korea Mobile TV field study, however, observations were conducted after interviews to compliment the interview study (Chipchase, et al., 2006).

Probe Study

In user-centred design, probe or diary studies have been used to discover users' needs, values and feelings, and cultural or technology impacts. Typically, participants are given documentation tools (e.g., a diary/notebook and pens, a voice recorder, a camera, or a mobile phone) and asked to freely record their own thoughts about new products and services. The probe studies provoke implicit responses from users without observing or asking them directly; what's more, they provide access to participants' everyday lives and private environment, which are not easily accessible through conventional studies (Jumisko-Pyykkö, Weitzel, et al., 2008).

In O'Hara et al.'s study (2007), to better understand how people are using commercially available mobile video technologies in their everyday lives, the researchers conducted a diary study to ask participants to write down all mobile video related behaviors over three weeks. In another probe study (Jumisko-Pyykkö, Weitzel, et al., 2008), participants were given self-documentary tools (diary, disposable camera) and a projective task (collage); and then they were asked to take pictures of any situations in which they could imagine watching mobile 3D TV and wrote a short note about each image in the diary. This helped the researchers to discover potential usage context of mobile 3D TV.

Survey and Questionnaire

A questionnaire is a list of questions. Although some researchers regard it as one element of a well-done survey that is a complete methodological approach (Dellman, 2000), questionnaires and surveys are interchangeably used in mobile video research. Questionnaire-like surveys have been used in the exploratory phase to identify user groups, current practices, and user needs and attitudes to the mobile video services. They are easy to collect data from a large number of people at a relatively low cost, but they are weak in generating new ideas (Lazar, Feng, & Hochheiser, 2010). For example, Jumisko-Pyykkö and Weitzel et al. (2008) designed a questionnaire about mobile 3D television and video, which collected 342 responses through online data collection in Finnish and German. The survey results revealed user backgrounds, users' motivations for viewing mobile 3D television and video, user requirements for content and functionalities of system and service, and typical context of use. Another typical use of questionnaire is for usability and satisfaction evaluation. A good example can be found in the report of the Mobile-TV project (Södergård, 2003).

Survey is not necessarily a quantitative method. Survey questions can ask for people's comments and opinions, which purpose is not for measurement. In this case, we should treat these comments as qualitative data and analyze them in a qualitative manner.

Laboratory Experiment

When examining user's perception or experience of video quality, most studies are designed as laboratory experiments. The frequently used approaches are called subjective quality assessments.

This kind of assessments allows testers to rigidly control the conditions of viewing settings (e.g., illumination, distance, and monitor), test clips (e.g., specific contents, codecs, and encoding parameters), and evaluation procedure (e.g., fixed steps and ad-hoc test tool/software). Therefore, it is efficient and convenient for data collection and easy for data analysis, but potentially less realistic.

The well-known subjective quality assessment methods are those proposed by International Telecommunication Union–Telecommunication Standardization Sector (ITU-T) and Radiocommunication Sector (ITU-R), including the Absolute Category Rating (ACR), the Degraded Category Rating (DCR) (also called DSIS), the Single Stimulus Continuous Quality Evaluation (SSCQE) and the Double-Stimulus Continuous Quality Scale (DSCQS) (ITU-R, 2004; ITU-T, 1999). Using a 5/11 scale and a 100 scale, the ACR and SSCQE methods request the testers to rate the test video's quality individually, while the DCR or DSIS methods ask the testers to judge the quality impairment of the test video compared to a reference video, and the DSCQS method requires the testers to rate the qualities of a test video and a reference video each time. The results of these subjective assessments are Mean Opinion Scores (MOS) or Differential Mean Opinion Scores (DMOS). Comparing these methods used for mobile video applications, Tominaga et al. (2010) clarified the ACR and DSIS (or DCR) with 5 scales performed better than the other methods.

Notwithstanding that the scaled assessments are widely used, they are prone to overburdening participants, who struggle to determine a proper score for the quality of a video (Sasse & Knoche, 2006). Furthermore, the scaled assessments cannot sufficiently answer the question of which quality level is acceptable to the end users (Schatz, Egger, & Platzer, 2011). Binary measure is therefore suggested to use in assessing acceptability of mobile TV (videos) by McCarthy and Knoche et al. (Knoche, et al., 2005; McCarthy, et al., 2004). The idea of acceptability is to identify the lowest acceptable quality level or threshold. A psychological method used to determine threshold is known as the Method of Limits created by Gustav Theodor Fechner. It is often done through asking participants to simply decide whether or not they accept the quality of a displaying video in successive, discrete steps in either an ascending or a descending series (Agboma & Liotta, 2006).

Another psychometric way to qualify intensity of video stimuli is called the two-alternative forced choice (2AFC) methods. The 2AFC methods present the person with a choice of two stimuli and ask him to discriminate their intensity. These types of tests are considered to have less bias and variability than the conventional subjective rating methods (e.g., ACR and DCR) because "the procedure is more natural and direct to a person; no internal mapping is necessary" (Menkovski, Exarchakos, & Liotta, 2011). However, the 2AFC testing such as Maximum Likelihood Difference Scaling (MLDS) often requires a large number of subjective tests for all combinations of four samples for a set of video materials. An adaptive MLDS approach was proposed by Menkovski and Liotta (2012). It enables an adaptive test selection procedure for MLDS to improve the learning efficiency, whereby the subjective tests can stop early when they cannot bring significant improvement in the accuracy of the psychometric curve.

Experiment in Field (Context)

Subjective quality assessment can also be done in field (or in natural surroundings), which is able to capture more realistic user experience, but the data collection is difficult and the data may be impacted by unknown factors. Only a few researchers have studied the effect of usage context on mobile video quality evaluation. In Jumisko-Pyykkö and Hannuksela's study (2008), they asked participants to evaluate the quality of mobile TV in parallel to three different tasks: waiting for a friend in a railway station, traveling by local bus, and spending time in a café. The results showed that the usage contexts did not cause a difference in acceptance or satisfaction of mobile TV quality, but affected users' enjoyment and information recognition. The comparison between the context study and their previous laboratory study revealed that people accepted higher transmission error rates in real context. Another study found that people were more likely to accept a low quality video under a relaxed context (e.g., on a campus) than a tense usage context (e.g., on a bus), and the acceptable level for high quality videos would be much higher under the relaxed context (Song, et al., 2010).

Determination of User Study Methods

For designing a user study, it is important to determine proper methods considering feasibility, validity and performance. Considering the feasibility, time and resources are essential in any research. For instance, a researcher has to deliberate how much time he/she can offer for observations or individual interviews, or how much money he/she can offer for employing participants in a probe study. Apart from this, several principles are suggested in order to determine the user study methods for good validity and good performance.

Firstly, the selection of user study method depends on the research purpose.

As exampled for each method in Section 3.1, interviews, focus groups, observations and probe studies are useful for understanding user behaviours, exploring user needs and experiences, and discovering implicit impacts. Surveys and subjective quality assessments are suitable for identifying user groups, user preferences, and major issues or problems and for measuring users' satisfaction, perception and acceptance to the quality of mobile video.

Secondly, it is beneficial to combine quantitative research and qualitative research.

In terms of the different characteristics of the quantitative and qualitative research, it is clear that a combination of both methods in one research can develop their advantages and compensate for their disadvantages. The benefits include: (1) multiple objectives such as discovery and explanation can be targeted; (2) multiple focuses (e.g., learning widely/deeply and testing specific hypothesis) can be included; and (3) both subjective and objective realities can be observed. Mixed research methods have been used in many studies in video fields. In Eronen's study of understanding viewers' experience on digital television, the quantitative study (questionnaire) was firstly adopted to formulate user groups; and then the qualitative study (focus group) was used to reveal different user groups' expectations and preferences (Bazeley, 2004; Eronen, 2001). In Södergård's study on mobile TV (2003), qualitative methods (observation and interview) were initially used for defining users, observing their behaviours, and exploring their feelings and requirements. Then, a quantitative method (questionnaire) was used to evaluate a product/service's performance of satisfying the users.

Moreover, the mixed use of quantitative and qualitative research often claims greater validity of the results. When different approaches are used for the same phenomenon and they provide the same result, the result gains superior evidence. Also, researchers often use one set of results to complement another, because some phenomena discovered by a quantitative method need further explanation via a qualitative method (Creswell, 2003). To maximize the outcome of one user study, quantitative research and qualitative research can sometimes be simultaneously conducted in the same study. In a subjective audio-visual quality evaluation study, before and/or after the quantitative video quality assessments, researchers usually request interviews with participants. The qualitative results can help understand how participants made their evaluations, thus explaining or confirming some quantitative results (Jumisko-Pyykkö, et al., 2007; Knoche & Sasse, 2006).

Thirdly, using multiple methods (triangulation) is beneficial for achieving inclusive outcomes.

As each method provides different and limited visibility into human behaviour and user experience, it is needed to use several methods together in order to acquire a comprehensive view of the user experience of mobile video (TV) services. For example, Chipchase et al. (2006) have used contextual interviews and observations to investigate the usage of mobile TV. O'Hara et al. (2006) investigated people's everyday use of mobile video telephony using diary techniques and ethnographic interviews. More effective combination is to mix multiple qualitative and quantitative methods. For example, Jumisko-Pyykköet, Weitzel et al. (2008) applied triangulation methodology on the basis of a survey, focus groups and a probe study to obtain explicit and implicit user requirements. In a long-term project of studying mobile TV usage compared to traditional TV in Belgium during 2006-2008, Vangenck et al. (2008) also used multiple methods—combining structured questionnaires, probing cards, diaries, auto collecting pictures, logging data and in-depth interviews—to generate a comprehensive user-oriented picture of the potential experience of the usage of (mobile) television.

It should be noted that this kind of multiple methods combination is well appropriate for a long-period research.

Fourthly, the selection of proper user study methods depends on research stages.

When studying user experience of mobile video, it is suggested to use different methods for different user experience levels (Obrist, et al., 2010). At the pre-experience level, inquiry-oriented methods like interviews and questionnaires might be able to explore contextual information (e.g., the social and culture effects) and investigate people's pre-experiences with similar product/services. At the visceral level, where people have the first impression (i.e., perception) of a product through its appearance and their feelings, e.g., like or dislike, observation will be a good way to inspect how the users spontaneously utilize the mobile video service for the first time. At the behavioural level, where the users are engaged in the use of a product/service/system, the study methods should be able to provide comprehensive understanding of how well the user needs are fulfilled, how easily the product/service can be used, and how satisfied the users are. Typically, usability test, interviews and probe studies can be adopted. An experiment combined with an interview can also be applied to examine how the users' satisfaction is affected by certain factors.

Lastly, the reflective level, which is relative to the product's interactivity and aesthetic quality, is best evaluated using questionnaires and interviews.

It is essential choosing apposite methods to collect useful and sufficient user data. However, this is only the first step towards a success research. The next tough step is to analyze the data properly and to get the most information out of the collected data.

DATA ANALYSIS METHODS

The difficulty of data analysis is how to select appropriate methods and provide valuable and reliable interpretation. This section will review some data analysis methods that have been used for analysing collected user data in mobile video study.

Qualitative Data Analysis

Qualitative data exists in various forms (e.g., text, voice, image, and video) and contains rich information. Since analyzing this sort of data involves human coding, therefore it is more susceptible to interpretation bias. Therefore, the analysis process should be highly deliberate and systematic. In this chapter, we introduce two methods that are well-established and widely-used for qualitative data analysis: Grounded Theory (Strauss & Corbin, 1990) and Content Analysis (Holst, 1969).

The qualitative analysis based on Grounded Theory (Strauss & Corbin, 1990) is an inductive approach, which is developed from the data and moves from the specific to the more general. This analysis method is essentially composed of three elements: concepts, categories and propositions (or "theory"). The derived typical procedures from this theory are open coding, axial coding, and selective coding. Open coding conceptualizes the describing phenomena found in the data; axial coding categorizes the related codes in an advanced stage of development; and selective coding identifies the core category and relates other categories to the core category (Strauss & Corbin, 1990).

Jumisko-Pyykkö and Weitzel et al. (2008) have applied the principles (or procedures) of Grounded Theory analysis into the study of understanding the nature of people's experience of Mobile 3D TV. The data collected in this study included text transcriptions from interviews and focus groups and multimedia materials such as images from a probe study. The first step of analysis was open coding to identify the concepts and their properties. Then, the concepts were organized into categories and grouped under the three main user experience factors: user, system and service, and context. Also, frequencies in each category were determined by counting the number of the participants who mentioned it. Finally, the theory was developed in the form of building blocks of user experience. Later, Jumisko-Pyykkö et al. (2010) used the Grounded Theory principles into a more systematical analysis across five studies to identify components of quality of experience of mobile 3D video.

Another method, *content analysis*, is an historical and extensively used analytical tool for textual data. Nowadays, it is also applicable for multimedia materials. Content analysis can describe systematically and semantically the questions about who, what, where and how with formulated rules (Holst, 1969). It examines meanings, themes and patterns manifested in particular information to allow researchers to understand social reality in a subjective but scientific manner (Zhang & Wildemuth, 2009). The analysis processes include: (1) to prepare data and represent them (i.e., define analysis unit) so that they can better comprehend or relate to the ideas the researchers wish to make; (2) to develop coding scheme and perform iterate coding process to generate categories and their properties with good consistency and reliability; (3) to discover patterns and relationships with the categories to present your reconstructions of meanings derived from the data; and (4) to relate analysis conclusions to data obtained from other methods or other situations to either validate the involved methods or provide missing information (Krippendorff, 2004; Zhang & Wildemuth, 2009).

The qualitative data confronts many of the reliability challenges in data analysis. For instance, the same word may have different meanings in different contexts; different expressions may mean the same thing; the data from body language, facial expressions, or artwork bring more ambiguous meanings. In order to reduce the interpretation bias, it is necessary to check reliability, including intra-coder reliability and inter-coder reliability. The intra-coder reliability examines whether the same coder conducts the coding process in a stable way; while the inter-coder reliability examines whether different coders code the same data in a consistent way. The commonly used reliability measurements are Cohen's Kappa (1960) and Krippendorff's alpha (2004, pp. 211-256). A well-accepted interpretation of Cohen's Kappa is that a value above 0.60 indicates satisfactory reliability. And a well-accepted interpretation of Krippendorff's alpha is that a value greater than 0.7, the index of agreement among coders, indicates that the coding process is reliable and acceptable.

Quantitative Data Analysis

Due to the diversity of data collection methods (e.g., questionnaire and subjective quality assessment) and required information, the collected quantitative data will be shown in different scales, including nominal (including dichotomous nominal), ordinal, interval, and ratio. The nominal scale represents categorical information, such as video content types and user groups. If there are only two categories, a nominal scale is described to be dichotomous or binary. Such a variable is often coded using "0" and "1" (e.g., gender: male-0, female-1; and acceptance: unacceptable-0, acceptable-1). The ordinal scale deals with the categories where one is bigger than another (e.g., the duration of working in a department: 1–less than 1 year, 2–1 to 2 years, 3–more than 2 years). The interval scale is continuous numerical quantity. Likert responses and subjective quality scales belong to the interval, where the increments of the interval scales are the same throughout the scale (e.g., video quality: 1-bad, 2-poor, 3-fair, 4-good, 5-excellent). The ratio scale is similar to the interval scale, but without assuming an equal increment. Variables such as age, income and revenue are in ratio scale.

To acquire correct analysis results, it is very important to choose an appropriate statistical test for a specific question with available data. A decision about which statistical test should be used for which analysis depends strictly on the corresponding relationship between the test hypotheses (or questions), the data types, and the test requirements. Table 2 lists frequently used statistical analysis tests in mobile video studies, with detailed information about required variable types, examples of applied studies, and applicable research question or hypothesis style (Aron, Aron, & Coups, 2009; Manning & Munro, 2007).

According to the relevant research questions, these statistical tests are used for the following purposes: correlation analysis, cause-and-effect analysis, and main effects determination.

Correlation Analysis

When examining whether there is a significant relationship between two variables, either Pearson Correlation or Spearman's Rank-order Correlation, depending on the variable type, can be performed to generate a correlation coefficient, Pearson *r* or Spearman's *rho*. The *r* or *rho* ranges from 0 to 1(-1), indicating a relationship from no to positive perfect (negative perfect).

The McNemar test assesses the significant difference between the two types of dependent samples in cases where the variable of the interest is dichotomous. It is applied to a 2x2 cross table with a dichotomous trait that matches pairs

Table 2. Description of statistical analysis tests and their applied examples

Analysis methods	Variables	Example Studies	Applicable Question (Q) & Hypothesis (H)
Pearson Correlation	Two interval or radio variables with satisfied normality assumption	To clarify the correlation between rankings with different quality assessment methods (Tominaga, et al., 2010)	Q: Is one variable associate with another?
Spearman Rank-order Correlation	Two ordinal, interval or radio variables		
McNemar test (chi-square test for within-subjects)	Two nominal variables	To test whether two categories (for motivations or contents) are equally likely to occur (Jumisko-Pyykkö, Weitzel, et al., 2008; Song & Tjondronegoro, 2010)	Q: Is it different between paired proportions?

continued on following page

Table 2. Continued

Analysis methods	Variables	Example Studies	Applicable Question (Q) & Hypothesis (H)
Analysis of Variance (ANOVA)	IV: nominal, or ordinal. DV(s): interval or ratio with satisfied normality assumption	To identify main effects of test variables and differences among datasets (ITU-T, 2004); To examine the effect of satisfaction and entertainment by contexts, error rates or content types (Jumisko-Pyykkö & Hannuksela, 2008)	Q: Does the change of independent variable (IV) result in a change in the dependent variable (DV)? H: It is predicted that change in IV cause a change in DV.
Independent-samples *t*-test	IV: dichotomous nominal DV: interval or ratio with satisfied normality assumption	To test the significance of the difference between two sample means	
Mann-Whitney U test	IV: dichotomous nominal DV: ordinal, interval or ratio	To compare the differences in quality satisfaction ratings given by two independent sample groups: one using simplified continuous assessment parallel to retrospective ratings and the other using only retrospective ratings (Jumisko-Pyykkö, Vadakital, et al., 2008)	
Repeated Measures ANOVA	IVs: interval or ratio with satisfied normality assumption; DV(s): the same variables is measured a second time (or more) with satisfied normality assumption	To examine the main effects and interactions for satisfaction and entertainment for the combinations of different contexts, error rates and content types (Jumisko-Pyykkö & Hannuksela, 2008) To compare the navigation speeds and errors with several interaction modes for browsing videos on a mobile phone (Huber, et al., 2010)	Q: Have things changed since last time they were measured? Q: Are the values of the DV changed in another measurement situation? H: It is predicted that the values of the DV(s) will have changed over time or in another situation.
Paired-samples t-test	IV: interval or ratio with satisfied normality assumption; DV: the same variables is measured a second time with satisfied normality assumption	To compare the navigation speeds and errors with two interaction modes between inter-related videos (Huber, et al., 2010)	
Wilcoxon Signed-Rank Test	IV: ordinal, interval or ratio; DV: the same variable is measured a second time	To examine whether the acceptable degree of mobile video quality are differently evaluated by the same samples under different spatial resolution and frame rate conditions (Song, et al., 2010)	
Logistic Regression	IVs: nominal, ordinal, interval or ratio; DV: nominal or ordinal	To test for main effects and interactions on whether a quality of mobile video is acceptable (Knoche & Sasse, 2009; Song, et al., 2011b)	Q: Are two or more variables causally related to another variable? H: It is predicted that two or more variables (IVs) are casually related to the DV.
Discriminant Analysis	IVs: interval or ratio with satisfied normality; assumption, or dichotomous nominal DV: nominal or ordinal	To determine whether a chosen set of QoS parameters lead to a reliable prediction of user ratings of service quality, and the relative importance of the QoS parameters in predicting the group membership (Agboma & Liotta, 2012) To classify the video quality into "good" and "not good" on the basis of different encoding parameter combinations (Song, et al., 2010)	Q: Are two or more variables causally related to another variable? H: It is predicted that two or more variables (IVs) are casually related to the DV. Q: Can a subset of predictors work equally well to as a larger set of predictors?
Multiple Linear Regression	IVs: interval or ratio with satisfied normality assumption, or binary; DV: interval or ratio with satisfied normality assumption	To model spatial quality, temporal quality and general quality of experience (QoE) based on significant influencing parameters (Ketyko, et al., 2010)	

Note: the table is modified based on the information from (Manning & Munro, 2007) of subjects, and determines the correlation based on marginal homogeneity of the cross table. The McNemar test does not provide evidence of how the two variables are related. All the significance tests for two dependent samples are assumed to be nonparametric tests, and therefore do not assume normal distribution or any other distribution of the population.

Cause-and-Effect Analysis

In order to test the effect of one factor on another, two types tests are used: parametric and non-parametric (Manning & Munro, 2007). Parametric methods such as ANOVA and *t*-test can be used if the variables meet the normality assumption. (One-way) ANOVA provides a statistical test of whether or not the means of several groups are all equal. When the sample is exposed to several conditions in turn and the measurement of the dependent variable is repeated, a repeated measures ANOVA is more appropriate to be used to compare the difference between different conditions. In addition, if one or more dependent variables and one or more independent variables are involved, multivariate analysis of variance (MANOVA) may be used. The independent and repeated *t*-test only work for testing dichotomous groups.

Without assumption of normality, non-parametric methods can be used for the same purpose (Aron, et al., 2009). The Mann-Whitney *U* test can be used to assess the difference between two independent samples, while the Wilcoxon test can be used to measure the difference between two related data sets.

Main Effects Determination

Logistic regression is able to find multiple variables' effects on one categorical variable. Specifically, when the tested dependent variable is dichotomous (e.g., acceptable or unacceptable), binary logistic regression was applied (Knoche & Sasse, 2009; Song, et al., 2011b). There are several advantages of using logistic regression analysis. Firstly, the independent variables do not have to be normally distributed, or interval, or have equal variance in each group. Secondly, a linear relationship between the independent variables and dependent variable is not assumed; therefore, it may handle nonlinear effects. Thirdly, there is no homogeneity of variance assumption. However, the flexibility of logistic regression comes at a cost: it requires more data to achieve stable and meaningful results – at least 50 data points per predictor. The effect size of logistic regression is indicated by Pseudo R-squared, such as Cox and Snell pseudo R-square and Nagelkerke pseudo R-square. They vary between 0 and 1 and are interpreted as weak, moderate, or strong effect size. Odds ratios are also measured in logistic regression for dichotomies, categorical variables, or continuous variables, with values above 1.0 reflecting positive effects and those below 1.0 reflecting negative effect.

Multiple Linear Regression (MLR) and Discriminant Analysis (DA) allow us to relate more than one independent variables (or predictors) to a single dependent variable (categorical variable for DA) with a linear function. In this function, the predictors (effects) and their coefficients reflect their relationships with the dependent variable. An alternative analysis of MLR or DA is stepwise MLR or stepwise DA. The stepwise methods are best justified in exploratory studies where a relatively large number of independent variables exist, and its goal is to identify the best subset of them, which may be considered as the most important influences. In this analysis, the first step is to identify the independent variable with the highest bivariate correlation with the dependent variable, and calculate the correlation coefficient of determination (R^2) of the regression equation with the single predictor. The second step is to add another independent variable whose entry would cause the greatest increase in R^2. At the third step, another independent variable may be added if it increases R^2 and also one of variables added previously may be removed if without affecting R^2. The process of adding and deleting continues until there are either no independent variables left or the inclusion of another independent variable will not significantly increase R^2 (Manning & Munro, 2007).

CASE STUDY: CONDUCTING A USER STUDY ON ACCEPTANCE OF MOBILE VIDEO QUALITY

In the previous sections, we have undertaken a comprehensive review of research methodology on user-centred study in the mobile video field. In this section, we will provide a concrete case study to demonstrate how to take the research methodology into consideration and how to conduct a user study properly.

The purpose of this case study is to identify which factors affect user acceptance of mobile video quality significantly and understand how their effects from user's perspective. To undertake this study, we need to determine study methodology, design the user study, and conduct the user study logically and systematically.

Study Methodology

The determination of study methodology included to specify study focus and to select appropriate methods.

First of all, this user study was at the middle stage of an entire research. Prior to it, an online survey study (Song & Tjondronegoro, 2010) and a pilot user study in field (Song, et al., 2010) had been completed to investigate the use of mobile video locally and explore relative issues and potential influences of user experience. Based on the preceding work,, this study was decided to focus on two aspects of user acceptance: (1) acceptability that refers to the probability of a video quality being just accepted; and (2) pleasantness that refers to the probability of a video quality being accepted for pleasant (i.e., long-term and comfortable) viewing. Strictly, the pleasantness is the acceptability serving a higher level of viewing purpose of pleasing users.

Furthermore, based on the preliminary studies and literature reviews, a set of hypothesis could be established. It was that the acceptability and pleasantness might be impacted by video encoding parameters (e.g., image resolution, frame rate, and quantization quality), encoding bitrate, video content, display mobile device, and user profile (e.g., prior relevant experience and user preference for video consumption). At this stage of distinguishing the important influencing factors, we used a controllable method – lab-based subjective quality assessment, in parallel to semi-structured interviews to have a deep understanding of their impacts. Using these methods, quantitative data and qualitative data were collected. Congruently, statistic data analysis and content analysis were used for data analysis.

Once the study focus and study methods were defined, the next step would be tangible and concrete study design.

User Study Design

With respect to the user study design on subjective assessments of mobile video quality, several aspects should be carefully considered, they are: participants, testing materials, equipment and application, test settings, test procedure.

Participants

It is always a critical and difficult question: who should be involved in a user study? For a survey study, it is usually desired to get as many responses as possible so as to achieve much diversity in demographic traits. However, in this focused study, our principle of selecting the participants was to choose them from the group that represents the majority of mobile video users. Based on the literature reviews and the prior survey, young adults were targeted as the main users. Gender balance and study/career background were also considered. In the end, we recruited a total of 80 people, who were aged between 17 and 40 (with an average of 26.24); 42 females and 38 males; with different technology backgrounds, including business, nursing, accounting, art and design, and information technology.

In addition, as we defined that people's preferences for video content types and their prior experience in watching mobile video might be potential factors influencing user experience, we collected this information from the participants, shown in Table 3.

Equipment

In mobile video-related studies, the necessary equipment is a real or simulated mobile device that can display videos. However, the available equipment is constraint to time, budget, and resources that can be handled. Obviously, it is impossible to use an unaffordable mobile device to conduct a study. Sometimes, researchers use a desktop to simulate the size of a mobile screen in specific experiments. But putting the budget limit aside, we suggest using a timely mobile device if possible. The reason is that the rapid technology revolution leads to a dramatic improvement of the mobile device, especially in displaying capability; thus, using an out-dated mobile device may significantly influence the study results due to its low display resolution. Similarly, the user-perceived quality through a simulated screen may be considerately different from the quality perceived through a real mobile device.

Table 3. Participants' profiles

Items	N (% out of 80)
Experience in watching mobile video • Frequency: at least once a week • Duration: longer than six months	 42 (52.5%) 38 (47.5%)
Preference for content types • Animation • Movie • Music • News • Sports	 43 (53.75%) 66 (82.50%) 46 (57.50%) 23 (28.75%) 25 (31.25%)

Device selection also depends on how easy for the researchers to handle the device. In many cases, to aid the user evaluation, a customized test software or application needs to be developed for the test mobile device. This requires knowledge on the development for the specific type of mobile devices, such as iOS and Android. Furthermore, we suggest choosing the popular or mainstream mobile devices that own the major user groups and have most general features.

In this study, we used two state-of-the-art mobile devices, iPhone 3GS and iPhone 4, which were the most popular mobile phones in Australia during the period of the first part and the second part of the user study, respectively. The main features relevant to this study are:

- **iPhone 3GS:** 3.5-inch multi-touch display, 480x320-pixel resolution at 163ppi; supporting H.264/AVC Baseline profile, up to 480p and 30fps; supporting AAC-LC audio format (up to 160 Kbps, 48kHz).
- **iPhone 4:** 3.5-inch multi-touch display, 960×640-pixel resolution at 326ppi; supporting H.264/AVC Main profile level 3.1, up to 720p and 30fps; supporting AAC-LC audio format (up to 160 Kbps, 48kHz).

Materials

Preparing video materials is another important task in a video-related study. With regard to what kind of video resources should be used, the main principle is to select representative videos; meanwhile, the decision-making should cooperate with the study purpose.

- If desiring to study the impact of video content characteristics on perceptual (or user-perceived) video quality, the video materials should cover various content features, such as high and low spatial complexity, fast and low motion, global and local movement, and colourfulness.

- If aiming to understand user attitude or social impact on mobile video usage, practical content genres should be used to represent the real usage of mobile video. Don't limit the number of videos and use as many genres as possible.
- If planning a lab experiment that involves real-world practice, such as user acceptability of viewed videos on mobile phones, it had better include real-world videos as testing sources and cover mostly common genres so as to bring a close realism into the study.

Our user study used two types of high definition (>= 1280×720 pixels) video sources, namely standard videos and real videos, respectively. The seven standard videos were from public databases (Sveriges Television AB (SVT), n.d.; Technical University of Munich, n.d.), which were commonly used for video (TV) quality testing and comparison. They were uncompressed, 10-20 seconds, covering nature scenes and crowd. These standard videos had different levels of motion and texture complexity defined with ITU-T P.910 Recommendation (1999) and thus, were classified into two groups, "stdH" (high complexity) and "stdL" (low complexity). The eight real videos were video footages with a length of 1–4 minutes (trimmed), which were from recorded news and soccer game, movie trailers and segments, music MTVs, and a open movie project (Foundation, 2008), covering five typical content genres of mobile video: news, sports, movie, music, and animation (Carlsson & Walden, 2007; O'Hara, et al., 2007; Song & Tjondronegoro, 2010). The reason of using both types of video sources was due to the consideration of both benchmark and realism. One the one hand, the standard video sources that are freely accessed for research purposes can make benchmarking the results easy. On the other hand, the real video materials are needed to obtain a more truly user experience than the standard videos (Sasse & Knoche, 2006).

Depending on the study purpose or the hypothesis, further processing may be required on these video resources. For example, to examine how the transmission error affects user-experienced quality, the videos need to be processed with simulated transmission errors. Since our study would like to examine the effects of encoding parameters, the video resources were encoded into test video clips with a combination of different encoding parameters, including 3 spatial resolutions (SR), 2 frame rates (FR), 10 Quantization Parameters (QP). In addition, considering the media format supported by the display devices iPhones, the video resources were encoded into the format of H.264/AVC for video and AAC for audio.

Study Procedure and Application

As mentioned early, this study would include both subjective assessments on quality acceptance and semi-structured interviews. The data collection was started by gathering the participant's background information such as age, gender, preference, and prior viewing experience of mobile video; followed by the subjective quality assessments; and ended with the interviews to ask how the participants had made the assessment. Using the test iPhones, the entire process was completed through a customized iPhone application (See Figure 1) for the first two steps and the build-in "voice memo" application for recording the interviews.

Differing from other acceptance evaluation studies where the subjects were asked to assess whether or not they accept the quality of each test clip (Agboma & Liotta, 2007; Jumisko-Pyykkö, Vadakital, et al., 2008; Knoche, et al., 2005), this study set up a scenario-based evaluation task for the participants, which was described as:

Figure 1. The test iPhone application: (a) user information collection; (b) a list of video content with a demo video has been demonstrated to the participant; (b) an example of recorded data; (c) an instruction of how to use the application; (d) a screenshot of the video quality determination

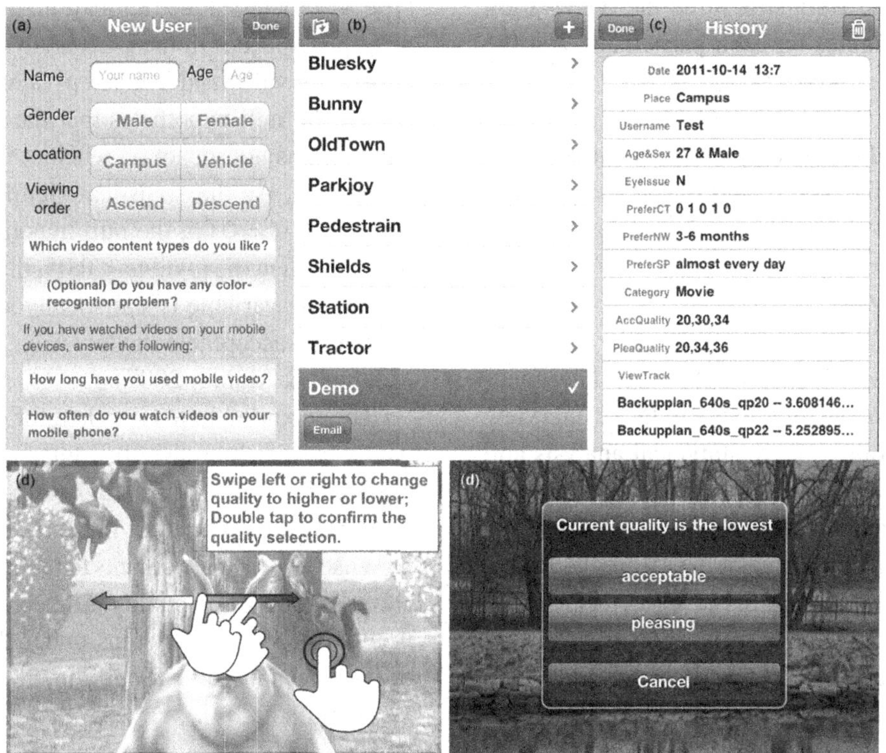

You are able to adjust the video quality that you will watch. Your task is to select the lowest acceptable quality and the lowest pleasing quality from each video quality group. The lowest acceptable quality means that below this quality, you are not willing to watch; while the lowest pleasing quality refers to the quality that you feel just good enough for long-term/regular and comfortable watch in order to reduce unnecessary cost in money or data flow.

Guided by the customized test application, the evaluation was easy and initiative. At the beginning, the participants entered personal profile information into the test application (Figure 1a); then they randomly chose one of video contents (Figure 1b) to watch, where the video played starting from the lowest quality or the highest quality within one test video group (including 10 quality levels in terms of QP at the same SR and FR). By swiping left or right on the screen, they could adjust the displayed video quality higher or lower gradually within the same group (Figure 1d). Once they determined that the current displaying quality was the lowest acceptable/pleasing quality, they could confirm their choices by double tapping on the screen and clicking the relative confirmation button from a pop-up message window (Figure 1e). During the process, a test video was playing in a loop mode. When switching the quality, the next quality of the same video content would start to play from the break point of the content (allowing up to 1-second overlap). The iPhone application automatically recorded the participants' decisions and stored into the iPhone device (Figure 1c).

The entire study involved two parts: part 1 using iPhone 3GS and part 2 using iPhone 4. As different numbers of test video clips were evaluated in each part, the assessment session lasted about 20 minutes in the part 1 and 45 minutes in the part 2. The semi-structured interviews took around 15 minutes for each participant, and the main question asked was:

- How did the participants decide the lowest acceptable quality and the lowest pleasing quality?

Data Analysis

The user study collected two classes of data: quantitative assessment data and qualitative interview data. What follows were statistical analysis and content analysis applied for the quantitative and qualitative data, respectively.

The original assessment data were the lowest acceptable/pleasing quality levels determined by the subjects. Thus, they were firstly transformed into binary data to denote whether a certain video quality is acceptable/pleasing or not. For each participant's records, video clips with lower quality (in terms of bitrate) than the selected lowest acceptable quality in the same SR and FR group were regarded as "unacceptable" and represented with "0", and the others with equal and greater quality of the lowest acceptable video were regarded as "acceptable" and represented with "1". The same transformation was made for the pleasing quality evaluation. After shifting missing data and outliers, a total of 76870 binary assessments (0/1) were obtained for the 870 test clips from the 80 participants. With the binary data (acceptable or unacceptable) as the dependent variable, *Binary Logistic Regression* (BLR) analysis was conducted to examine the effects of the user acceptance. Based on our hypotheses, the independent variables (effects) were composed of three categories of factors: (1) encoding parameters: SR (video resolution), FR (frame rate), QP (quantization parameter) and BR (bit rate); (2) CT (video content type); (3) DT (display device type); and (4) user profiles: gender, FreqGroup (whether a frequent user of mobile video), DurationGroup (whether a user over 6 months), and LikeX (whether like the content type X, which includes animation, movie, music, news and sports). In the BLR analysis, the Wald chi-square value and 2-tailed p-value are used in testing the null hypothesis that the independent variable has no impact on whether the quality is acceptable. As the statistical significance level was set as $\alpha = 0.05$, the null hypothesis will be rejected when the $p\text{-value} < 0.05$, that is, the corresponding variable is statistically significant to the overall acceptable model (Aron, et al., 2009; Manning & Munro, 2007).

The recorded interview data for each participant were firstly transcribed into text. Then, we used content analysis to analyze the transcription, for the purpose of identifying and describing criteria used by the participants to make judgements about mobile video quality. The content analysis was defined into four steps under the guidance of the relevant works (Schilling, 2006; Zhang & Wildemuth, 2009).

Step 1: Defining the unit of analysis. To make sure the data provide meaningful or valuable information, the analysis should focus on the research purpose or the key questions that the analysis is trying to answer (Dey, 1993). In this qualitative research, the focus was on how all individuals respond to each main question or topic that was asked in the semi-structured interviews. Therefore, the basic unit of the analysis is the individual themes defined by the main topics, and the analysis look across all respondents and their answers in order to identify consistencies and differences for each theme.

Step 2: Coding data into categories. Categorising here was identifying concepts and ideas and organising them into coherent categories. The process involved: (1) reading and re-reading the text; (2) highlighting meaningful words/sentences/paragraphs and summarising them into keywords or phrases as themes; (3) placing the themes into relevant categories or new categories; and (4) merging categories or breaking into subcategories if necessary, based on the consistency and discrimination.

Step 3: Identifying patterns and connections within and between categories. The next step of analysis was discovering the relationships between the categories and assessing the relative importance of different themes. Within a category, the similarities and differences in people's responses (to one theme) were summarised and distinguished. Between two or more categories, the relationship between themes might be identified as having either relative importance or consistent connection.

Step 4: Interpreting and concluding the data. In this step, the important findings discovered as the result of categorising and sorting data were concluded. These conclusions provided a description of some phenomena relevant to the mobile video users' behaviours and explained or gave an evidence for some quantitative results. Typical quotations were used to justify the conclusions.

In step 2, data coding for all respondents' answers was conducted in the same process. To validate the coding scheme early in the process, the clarity and consistency of category definition had been tested through an inter-coder reliability check (Zhang & Wildemuth, 2009). A sample of data was double-coded by two qualified researchers, and then the inter-coder reliability was measured in terms of Krippendorff's alpha (2004, pp. 211-256). The inter-coding agreement of 0.875 (> 0.7) indicated that the coding process of this content analysis was reliable and acceptable.

Results and Interpretation

Since the purpose of this chapter is not to show a study's outcomes, we will only demonstrate a few of the results from both qualitative and quantitative data analysis, and illustrate how to interpret their meanings.

Statistic Analysis Results

The result of the binary logistic regression analysis showed that the full prediction model containing all predictors (independent variables) was statistically significant ($p<.001$), indicating that the model was able to distinguish between the acceptable and unacceptable. For each of the independent variable, it provided significant contribution to the acceptance prediction according to the Wald criterion ($p<.05$). For example, a significant impact on user acceptance was given by CT ($\chi^2(7)=1885.6$, $p<.001$), Freqgroup ($\chi^2(1)=856.4$, $p<.001$), and Durationgroup ($\chi^2(1)=18.8$, $p<.001$) (Table 4).

To get an insight into how different these content types impose the impact, a series of McNemar tests was conducted, which compared the user assessments (0 or 1) for each pair of these content types under the control of the same encoding parameters (QP, SR and FR). The results showed that the acceptability was not significantly different among the types of "Animation" & "Cartoon", "News" and "StdL" ($p>.05$), but the acceptability for "Movie" and "Sports" and "StdH" was found to differ significantly from the others ($p<.001$).

Figure 2a and 2b illustrate how the characteristics of user's prior experience with mobile video influence user's pleasant level (similar graphs can be drawn for the acceptable level). It can be observed that the pleasantness of the low-frequent viewers was higher than that of the high-frequent viewers (Figure 2a), and that of the short-term viewers was higher than that of the long-term viewers (Figure 2b).

It may not be surprise that participant's preference for video content types had a significant impact on the overall quality acceptance ($p<.001$). However, what is interesting is that how the quality of video content is accepted differently by the people who like it and the people who dislike it. According to the coefficient of the independent

Table 4. A part of variables in BLR regression equation

	B	S.E.	Wald	df	Sig.
CT	.056	.007	1885.601	7	.000
likeMovie	-1.892	.066	812.166	1	.000
likeMusic	.487	.046	111.092	1	.000
likeSport	-.257	.052	24.161	1	.000
FreqGroup	.783	.027	856.388	1	.000
DurationGroup	-.112	.026	18.786	1	.000

Figure 2. Impacts of prior experience of using mobile video: (a) frequency, (b) duration

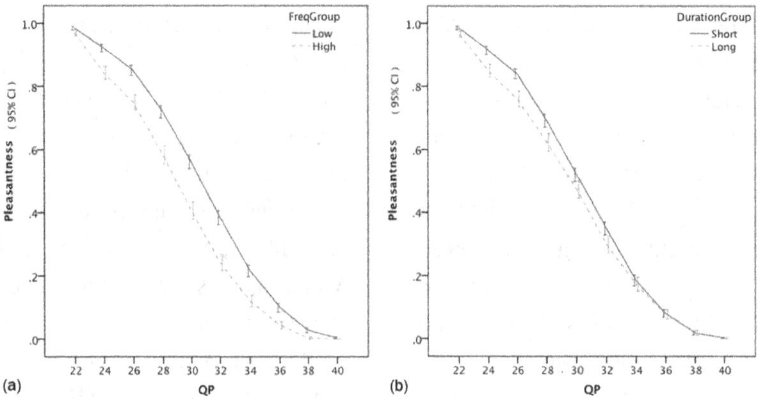

variable LikeX in the logistic regression equation (B value in Table 4), the negative value for likeMovie and likeSport means that being in the class of liking the content decreases the log odds of quality acceptance; while the positive value for likeMusic means that being in the class of liking music increases the log odds of quality acceptance. In other words, "Movie" and "Sports" lovers were less likely to be satisfied with the video quality than their counterparts were; in contrast, the party of "Music" lovers was more positive to the video quality than the opposite party. Further analysis using the Mann-Whitney U test for each of five video content types confirmed that the

Figure 3. Impacts of user preference for content type on pleasantness: (a) "sports," (b) "music"

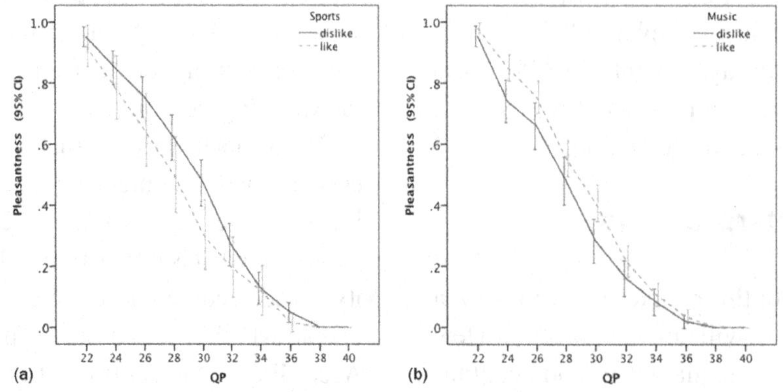

difference between "like" a content and "dislike" a content was significant for "Movie", "Music" and "Sports" ($p<.05$). Figure 3a and 3b illustrate the effects of people's preference for "Sports", and "Music", respectively. With regard to the opposite effects of "liking" for such "Sports" and "Music" content, further discussion will be offered through interview data analysis.

Content Analysis Results

The content analysis on the interview transcription identified the categories of user's criteria on assessing the acceptable video quality to meet their requirements. It revealed that the user-desired quality was highly dependent on video content and users' individualities as well.

A part of criteria is listed in Table 5, where N indicates the number of participants who gave meaningful opinions on a particular category; %

Table 5. Categories and criteria of quality acceptance

Categories of Criteria	Description	N %
Movie	**Specific judgement for movies**	**N=38**
Higher quality requirement compared to other contents	Need clearer and higher quality than other contents	42.1
Faces and expressions	Focusing on faces and expression details	36.8
Expectation	Good quality expectation due to relative viewing experience	26.3
Enjoyment	Enjoyable watching; interesting content	13.2
Music	**Specific judgement for music videos**	**N=36**
Faces	Good clarity of singers' faces	33.3
Unimportance of image quality	Picture quality is not very important	30.6
Sound focused	Focusing on listening to music	27.8
Clearness of characters	Clear people's profiles and movement	22.2
Enjoyment	Enjoyable watching; interesting content	5.6
News	**Specific judgement for news videos**	**N=30**
No demand for video quality	Need not high quality; low quality is acceptable; do not care about quality	66.7
Sound and AV sync	Focusing on listening and sync of audio and video	30.0
Faces	Clear close-up faces	13.3
Information	Concentration on news information	10.0
Sports	**Specific judgement for sports, esp. team and ball videos (e.g., soccer/football)**	**N=41**
High quality demand due to small objects	Hoping see small players and ball better	36.6
High quality demand due to fast motion	Fast moving makes the video quality low	36.6
Basic information receiving	General quality is enough to see what's happening	22.0
No demand due to being uninterested	No interest no demand	12.2
High quality demand due to importance	Important events and moments need high quality	7.3
Standard	**Specific judgement for standard videos**	**N=19**
Direct visual perception for boring content	Purely depending on clarity, blockiness and colour due to feelings of boring and uninteresting content	42.1
No care for scenery quality	Do not really care about the quality of scenery videos	31.6
Want of seeing people	Want to see people more clearly or bigger in videos involving human beings	26.3
Unapparent quality difference	It is difficult to see the differences between quality levels	21.1

represents the percentage of each item mentioned by the N participants within the corresponding category; the description for each attribute is summarised based on the relevant narration.

The information shown in Table 5 has confirmed that people have distinctive assessment criteria on video contents (Jumisko-Pyykkö & Häkkinen, 2006; Knoche, et al., 2005). Moreover, it shows why users make different judgements and how their criteria can be used to explain the results from the quantitative analysis.

For "News" video, audio quality and the synchronization of audio and video were more important than video quality. Relevance to the news' utility, receiving information was also regarded as a factor of quality selection. Several citations are given as follows:

... Look at face and mouth ... I don't like it if the speaker's lips movement doesn't match the speech.

...It (news) needn't be as clear as movie.

For news, voice is more important than image.

With respect to the judgements of "Movie" video, one participant's words were representative and conclusive. He said,

I have a higher quality requirement for movie than other video types. I need the actors or actresses' faces clear enough to recognize their expressions. Since movie is commonly watched in a HD quality on a cinema screen or a TV/PC screen, it should also be vivid on a mobile device screen; and HD is good if possible.

It seems that, with the experience of watching high definition (HD) films, users attained an expectation for a high quality movie even on a small-screen device. The high quality requirement might also be affected by people's viewing purpose for entertainment. In addition, the difference of quality selection criteria between "Movie" and "News" could explicate why "Movie" is the most difficult content type to satisfy users, while "News" is the easiest one.

Establishing user criteria is also helpful in discovering some implicit issues. Regarding the standard videos, the chief comment was "very boring". The videos hardly provoked the participants' interests; as a result, the quality acceptance was mainly related to the visual perception (42.1%). For the videos with global movement and/or complex images (e.g., "Parkjoy" and "Shields"), the participants felt that it was hard to tell the quality differences (21.1%); hence they could accept the relative low quality. On the other hand, many did not really care about the quality of scenery videos (31.6%). For example, one said: "In the video with some trees, I did not really care about the quality. It doesn't matter whether the quality is low or high, because anyway I won't watch it, or I will ignore this part in a video." With respect to the disability of the standard video in attracting people's attentions, it is concerned that using this kind of standard videos as test clips in user-centred study, the collected data may not be able to reflect the users' real feelings.

Crossing the user criteria with their profiles, it would be understandable that why the impact of user preference turns out to be in different directions, depending on the content types.

For the videos such as "Music" with mainly human figures and audio as an important part, people who liked music more allowed a lower video quality compared to those who disliked music (among people who had lower quality requirements, 62.5% came from those who liked music and 37.5% from those who did not, respectively). Because they paid more attention to what they could hear, they thought that "the image is not necessary to be very good" and "the singers' faces are big". This helps explain the phenomenon that those who liked music had higher acceptability or pleasantness than those who did not.

For "Sports" featured by fast movement, ball and team, such as soccer, people had a difficulty of viewing it on a small screen. To delight the people who were fond of sports, higher quality was needed, compared to the people who did not like sports (100% of 10 likes and 40% of 30 dislikes requested high quality). The soccer fans longed for very high quality, with big image resolution and smooth motion: "I want to have the best quality" (M, 18, likes sports); "is it possible to get higher (than the given highest test quality)?" (M, 21, likes sports). They also thought that "this is an important event, I want to see good quality". There were two different attitudes from the people who did not like sports. Some (14 out of 30) did not care about the quality because they thought that the players were too small to be recognised on the small screen even if at a high quality; thus a high quality was not necessary for sports video. They accepted the quality in which they could see enough of the game to understand what was going on and could see movement without jerks. But the others (12 out of 30) required a high quality because they did not understand the soccer match and only tried to see the small players and the ball clearly.

...I just keep selecting (quality) higher and higher because I hope I can see the small players clearer in the next higher quality (F, 20, dislike sports).

The interview data analysis could also explain the influence of people's prior viewing experience on their desired video quality. In general, participants who had rich experience in viewing mobile videos (frequent and long-term viewing) on a small screen were most likely to request a high quality. This is probably because the experienced users are more sensitive to the quality change. For instance, a one-year user said,

I feel the (mobile) video quality is becoming better and better. Nowadays I can't stand the previous (quality) anymore.

Moreover, people who often watched high quality videos such as HD 720p and 1080p on a big screen expected a similar visual perception on a mobile device. Since the commonly watched high quality videos were movies from DVD, theatre, and so on, the main impact was embodied in the content type "Movie".

I can't accept a movie with a low clarity because that's not correct, I mean, that's not what a movie should be. (F, 23)

CONCLUSION AND DISCUSSION

Mobile video as a promising service and an emerging research area has attracted great attention from both industry and academia. In less than a decade, a large number of lab and commercial trial studies have been done in this field. In particular, user-centred studies have been thriving due to the importance of user values, such as the quality of user experience, satisfaction, and perceived quality, in the interaction design and service supply.

This chapter has presented a comprehensive review of user-centred studies in the quality of mobile video. It focused on defining the main research foci, exhibiting various research methods, and illustrating the relative data analysis methods that were commonly used in the studies. To promote a deeper understanding of how to conduct an entire user study, this chapter then demonstrated a case study on user acceptance assessment of mobile video quality. In this case study, apart from describing the process of determining the research methodology from an overall perspective, each part of the user study was elaborated in detail, including participants, equipment, materials, study procedure, data analysis, and result interpretation.

Although great achievements have been made in understanding user needs and improving the quality of mobile video, there are still many challenges. These challenges include filling research gaps, developing effective user study methods, conducting efficient studies, as well as effectively utilizing the study results. Regarding these challenges as opportunities, further research can contribute to the following open issues.

1. How can we develop a more comprehensive understanding of mobile video users?

Current studies in understanding users are deficient in two main aspects. On the one hand, there is a lack of cross-cultural or cross-societal study, due to the fact that the user-centred studies were generally limited to particular geographical regions (e.g., conducting in certain cities and countries) and particular user groups (e.g., involving mostly young people). If possible, collaborative research among regions or countries will be beneficial for comparing differences between cultures and societies and identifying the impact. Also, as older people are going to join in the troop of mobile video users, it is becoming important to understand this new user group. On the other hand, with the rapid development of mobile and video technologies, the users have been experiencing a revolution of consuming video and audio service. This leads to a continuous change in user needs, user perceptions and user behaviours. However, lack of a longitudinal and systematic study to address this aspect makes the time-related effect unclear.

Apart from these two aspects, there are also insufficient studies on the effect of use context on user experience. In spite of a little exploration, the context-involved research is still in the beginning stage and needs to delve further into the complex correlations with user needs.

2. How can user studies on mobile video quality assessment be more effective?

It is always desirable to spend as little time and as a low budget as possible to collect sufficient and useful data. Researchers have compared various typical subjective assessment methods (e.g., ARC, DCR, etc.) and made a lot of efforts to establish new and easy methods (e.g., acceptable assessment). Still, the user studies on video quality assessment is time-consuming and faces the difficulty of gathering the data close to the real user feelings and perceptions.

For experimental user studies, the test videos are one critical element. However, the available video databases publically for research are very limited and the contained video contents are relatively dull, as a result, using them as the test videos will be ineffective to reflect the hedonic aspect of user experience. Some researchers have thus used self-prepared videos, but they might not be able to publish these video databases due to copyright issues. It would be ideal if various research centers, organizations, or video providers produced and provided a variety of video resources for public use.

Lab-experiment based study is essential when a controlled environment is required, but probably a more effective way of understanding the real users' needs and their opinions is to interact directly with them in a real use environment. An attempt has been made by De Moor et al. (2010), who proposed a framework to evaluate quality of experience in a mobile, testbed-oriented Living Lab Setting by monitoring real use context and gathering user feedback instantly.

Another process to improve study effectiveness is to enhance research collaboration. By collaborative work in both study method development and data collection and sharing, each research team will benefit from saving research time, acquiring a large sample of data, and obtaining more valuable results through comparison and contrast. Researchers in this field look forward to better solutions for this issue.

3. How can we better utilize the strengths of mobile devices to facilitate more efficient user study?

Mobile devices are becoming more powerful and more personalized, equipped with assorted sensors and being closely engaged in people's daily lives. However, their capabilities have not been utilized adequately for research. The roles of a mobile device in a user study can be multiple: a test instrument, a training tool, an A/V recorder, a diary, etc. It can be more useful when enabling its computing ability. For example, instead of manually self-documenting in a probe study, a mobile device can help to automatically log immediate usage information (e.g., time and location of watching videos) without requiring user intervention (Froehlich, Chen, Consolvo, Harrison, & Landay, 2007). It can also detect technical parameters (e.g., terminal screen size, battery level, network conditions, content quality, etc.) and contextual entities (e.g., location, mobility, sensors, and other running applications), and gather feedback with questionnaires and other forms (De Moor, et al., 2010). With a customized application on it, the mobile device is able to guide participants to complete a user study quickly and smoothly. What will make the mobile device more powerful will be implementing algorithms on it to perform required data analysis.

REFERENCES

Agboma, F., & Liotta, A. (2006). *User centric assessment of mobile contents delivery*. Paper presented at 4th International Conferences on Advances in Mobile Computing and Multimedia. Yogyakarta, Indonesia.

Agboma, F., & Liotta, A. (2007). Addressing user expectations in mobile content delivery. *Mobile Information Systems*, *3*(3-4), 153–164.

Agboma, F., & Liotta, A. (2012). Quality of experience management in mobile content delivery systems. *Telecommunication Systems*, *49*(1), 85–98. doi:10.1007/s11235-010-9355-6.

Aron, A., Aron, E., & Coups, E. J. (Eds.). (2009). *Statistics for psychology* (6th ed.). Harlow, UK: Pearson Education.

Bazeley, P. (2004). Issues in mixing qualitative and quantitative approaches to research. In Buber, R., Gadner, J., & Richards, L. (Eds.), *Applying qualitative methods to marketing management research* (pp. 141–156). Basingstoke, UK: Palgrave Macmillan.

Big Buck Bunny. (2008). Retrieved from http://www.bigbuckbunny.org/

Buchinger, S., Kriglstein, S., Brandt, S., & Hlavacs, H. (2011). A survey on user studies and technical aspects of mobile multimedia applications. *Entertainment Computing*, *2*(3), 175–190. doi:10.1016/j.entcom.2011.02.001.

Buchinger, S., Kriglstein, S., & Hlavacs, H. (2009). *A comprehensive view on user studies: Survey and open issues for mobile TV*. Paper presented at 7th European Conference on Interactive Television EuroITV'09. Leuven, Belgium.

Burger, A. (2012). *Report: Mobile video growing, fueling demand for multiscreen options*. Retrieved July 22, 2012, from http://www.telecompetitor.com/report-mobile-video-growing-fueling-demand-for-multiscreen-options/

Carlsson, C., Carlsson, J., Puhakainen, J., & Walden, P. (2006). *Nice mobile services do not fly: Observations of mobile services and the Finnish consumers*. Paper presented at 19th Bled eCommerce Conference. Bled, Slovenia.

Carlsson, C., & Walden, P. (2007). *Mobile TV-to live or die by content*. Paper presented at 40th Annual Hawaii International Conference on System Sciences. Hawaii, HI.

Chipchase, J., Yanqing, C., & Jung, Y. (2006). *Personal television: A qualitative study of mobile TV users in South Korea*. Paper presented at Mobile Human Computer Interaction 2006. Espoo, Finland.

Cohen, J. (1960). A coefficient of agreement for nominal scales. *Educational and Psychological Measurement*, *20*(1), 37–46. doi:10.1177/001316446002000104.

Creswell, J. W. (2003). *Research design: Qualitative, quantitative, and mixed method approaches*. Thousand Oaks, CA: Sage Publications.

De Moor, K., Ketyko, I., Joseph, W., Deryckere, T., De Marez, L., Martens, L., & Verleye, G. (2010). Proposed framework for evaluating quality of experience in a mobile, testbed-oriented living lab setting. *Mobile Networks and Applications*, *15*(3), 378–391. doi:10.1007/s11036-010-0223-0.

Dellman, D. (2000). *Mail and Internet survey: The tailored design method*. New York: John Wiley & Sons Ltd..

Denzin, N. K., & Lincoln, Y. S. (Eds.). (2000). *Handbook of qualitative research*. Thousand Oaks, CA: Sage Publications.

Dey, I. (Ed.). (1993). *Qualitative data analysis - A user-friendly guide for social scientists*. New York, NY: Routledge. doi:10.4324/9780203412497.

Edmunds, H. (Ed.). (1999). *The focus group research handbook*. Chicago, IL: NTC/Contemporary Publishing Group, Inc..

Eronen, L. (2001). *Combining quantitative and qualitative data in user research on digital television*. Paper presented at 1st Panhallenic Conference on Human Computer Intteraction. Patras, Greece.

Eronen, L. (2006). Five qualitative research methods to make iTV applications universally accessible. *Universal Access in the Information Society*, *5*(2), 219–238. doi:10.1007/s10209-006-0031-2.

Frank, N. Magid Associates Inc. (2009). *The OMVC mobile TV study: Live, local programming will drive demand for mobile TV*. Retrieved July 22, 2010, from http://mobiletvworld.com/documents/OMVC%20Mobile%20TV%20Study%20December%202009.pdf

Froehlich, J., Chen, M., Consolvo, S., Harrison, B., & Landay, J. (2007). *My experience: A system for in situ tracing and capturing of user feedback on mobile phones*. Paper presented at the International Conference on Mobile Systems, Applications, and Services 2007. San Juan, Puerto Rico.

Hassenzahl, M., & Tractinsky, N. (2006). User experience - A research agenda. *Behaviour & Information Technology*, *25*(2), 91–97. doi:10.1080/01449290500330331.

Holst, O. R. (Ed.). (1969). *Content analysis for the social sciences and humanities*. Reading, MA: Addison-Wesley.

Huber, J., Steimle, J., & Mühlhäuser, M. (2010). *Toward more efficient user interfaces for mobile video browsing: An in-depth exploration of the design space*. Paper presented at 16th international conference on Multimedia. Firenze, Italy.

ITU-R. (2004). *Methodology for the subjective assessment of quality for television pictures*. Recommendation BT. 500-11. Retrieved from http://www.itu.int/rec/R-REC-BT.500/en

ITU-T. (1999). *Subjective video quality assessment methods for multimedia applications*. Retrieved from http://www.itu.int/md/T01-SG09-040510-D-0108/en

ITU-T. (2001). *Tutorial - Objective perceptual assessment of video quality*. Retrieved from http://www.itu.int/ITU-T/studygroups/com09/docs/tutorial_opavc.pdf

Jumisko-Pyykkö, S., & Häkkinen, J. (2005). *Evaluation of subjective video quality on mobile devices*. Paper presented at 13th ACM International Conference on Multimedia. Singapore.

Jumisko-Pyykkö, S., & Häkkinen, J. (2006). *I would like see the face and at least hear the voice: Effects of screen size and audio-video bitrate ratio on perception of quality in mobile television.* Paper presented at 4th European Interactive TV Conference. London, UK.

Jumisko-Pyykkö, S., Häkkinen, J., & Nyman, G. (2007). *Experienced quality factors – Qualitative evaluation approach to audiovisual quality.* Paper presented at SPIE Multimedia on Mobile Device. New York, NY.

Jumisko-Pyykkö, S., & Hannuksela, M. M. (2008). *Does context matter in quality evaluation of mobile television?* Paper presented at 10th International Conference on Human Computer Interaction with Mobile Devices and Services. Amsterdam, The Netherlands.

Jumisko-Pyykkö, S., Strohmeier, D., Utriainen, T., & Kunze, K. (2010). *Descriptive quality of experience for mobile 3D video.* Paper presented at Nordic Conference on Human Computer Interaction 2010. Reykyavik, Iceland.

Jumisko-Pyykkö, S., Vadakital, V. K. M., & Hannuksela, M. M. (2008). Acceptance threshold: A bidimensional research method for user-oriented quality evaluation studies. *International Journal of Digital Multimedia Broadcasting*, 20.

Jumisko-Pyykkö, S., Weitzel, M., & Strohmeier, D. (2008). *Designing for user experience: What to expect from mobile 3D TV and video?* Paper presented at 1st International Conference on Designing Interactive User Experiences for TV and Video. Silicon Valley, CA.

Ketyko, I. N., De Moor, K., Joseph, W., Martens, L., & De Marez, L. (2010). *Performing QoE-measurements in an actual 3G network.* Paper presented at 2010 IEEE International Symposium on Broadband Multimedia Systems and Broadcasting (BMSB). Shanghai, China.

Knoche, H., & McCarthy, J. D. (2004). *Mobile users' needs and expectations of future multimedia services.* Paper presented at Wireless World Research Forum (WWRF)12. Toronto, Canada.

Knoche, H., & McCarthy, J. D. (2005). *Design requirements for mobile TV.* Paper presented at Mobile Human Computer Interaction 2005. Salzburg, Austria.

Knoche, H., McCarthy, J. D., & Sasse, M. A. (2005). *Can small be beautiful? Assessing image resolution requirements for mobile TV.* Paper presented at 13th Annual ACM International Conference on Multimedia. Singapore.

Knoche, H., McCarthy, J. D., & Sasse, M. A. (2008). How low can you go? The effect of low resolutions on shot types in mobile TV. *Multimedia Tools and Applications*, *36*(1), 145–166. doi:10.1007/s11042-006-0076-5.

Knoche, H., & Sasse, M. A. (2006). *Breaking the news on mobile TV: User requirements of a popular mobile content.* Paper presented at Multimedia on Mobile Devices II. New York, NY.

Knoche, H., & Sasse, M. A. (2008a). Getting the big picture on small screens: Quality of experience in mobile TV. In Ahmad, A. M. A., & Ibrahim, I. K. (Eds.), *Multimedia Transcoding in Mobile and Wireless Networks* (pp. 31–46). Hershey, PA: Information Science Reference. doi:10.4018/978-1-59904-984-7.ch003.

Knoche, H., & Sasse, M. A. (2008b). *The sweet spot: How people trade off size and definition on mobile devices.* Paper presented at 16th ACM International Conference on Multimedia. Vancouver, Canada.

Knoche, H., & Sasse, M. A. (2009). The big picture on small screens: Delivering acceptable video quality in Mobile TV. *ACM Transactions on Multimedia Computing. Communications and Applications*, *5*(3), 27.

Krippendorff, K. H. (Ed.). (2004). *Content analysis: An introduction to its methodology*. Thousand Oaks, CA: Sage Publications.

Lawson, S. (2011). *Video dominates mobile traffic, survey shows*. Retrieved July 22, 2012, from http://www.computerworld.com/s/article/9218660/Video_dominates_mobile_traffic_survey_shows

Lazar, J., Feng, J. H., & Hochheiser, H. (2010). *Research methods in human-computer interation*. Chichester, UK: John Wiley and Sons Ltd..

Lindlof, T. R., & Taylor, B. C. (2002). *Qualitative communication research methods* (2nd ed.). Thousand Oaks, CA: SAGE.

Mäki, J. (2005). *Finnish mobile tv pilot*.

Manning, M., & Munro, D. (Eds.). (2007). *The survey researcher's SPSS cookbook* (2nd ed.). Sydney, Australia: Pearson Education Australia.

McCarthy, J. D., Sasse, M. A., & Miras, D. (2004). *Sharp or smooth? Comparing the effects of quantization vs. frame rate for streamed video*. Paper presented at SIGCHI Conference on Human Factors in Computing Systems. Vienna, Austria.

Menkovski, V., Exarchakos, G., & Liotta, A. (2011). The value of relative quality in video delivery. *Journal of Mobile Multimedia*, *7*(3), 151–162.

Menkovski, V., & Liotta, A. (2012). Adaptive psychometric scaling for video quality assessment. *Signal Processing Image Communication*, *27*(8), 788–799. doi:10.1016/j.image.2012.01.004.

Miyauchi, K., Sugahara, T., & Oda, H. (2008). *Relax or study? A qualitative user study on the usage of mobile TV and video*. Paper presented at 6th European Interactive TV Conference. Salzburg, Austria.

Morgan, D. L. (Ed.). (1998). *The focus group guidebook*. Thousand Oaks, CA: Sage Publications, Inc..

O'Hara, K., Black, A., & Lipson, M. (2006). *Everyday practices with mobile video telephony*. Paper presented at SIGCHI Conference on Human Factors in Computing Systems. Montréal, Canada.

O'Hara, K., Mitchell, A. S., & Vorbau, A. (2007). *Consuming video on mobile devices*. Paper presented at SIGCHI on Human Factors in Computing Systems. San Jose, CA.

Obrist, M., Meschtscherjakov, A., & Tscheligi, M. (2010). User experience evaluation in the mobile context. In Mobile, T. V. (Ed.), *Customizing Content and Experience* (pp. 195–204). London: Springer-Verlag. doi:10.1007/978-1-84882-701-1_15.

Orgad, S. (2006). *This box was made for walking*. Retrieved Sep 20, 2009, from http://www.nokia.com/NOKIA_COM_1/Press/Press_Events/mobile_tv_report,_november_10,_2006/Mobil_TV_Report.pdf

Petrovic, O., Fallenböck, M., Kittl, C., & Langl, A. (2006). Mobile TV in Austria. *Schriftenreihe der Rundfunk und Telekom Regulierungs-GmbH, 2*.

Ries, M., Nemethova, O., & Rupp, M. (2008). *On the willingness to pay in relation to delivered quality of mobile video streaming*. Paper presented at International Conference on Consumer Electronics. New York, NY.

Robitza, W., Buchinger, S., Hummelbrunner, P., & Hlavacs, H. (2010). *Acceptance of mobile TV channel switching delays*. Paper presented at Workshop on Quality of Multimedia Experience. Trondheim, Norway.

Sasse, M. A., & Knoche, H. (2006). *Quality in context-An ecological approach to assessing QoS for mobile TV*. Paper presented at 2nd ISCA/DEGA Tutorial & Research Workshop on Perceptual Quality of System. Berlin, Germany.

Schatz, R., & Egger, S. (2008). *Social interaction features for mobile TV services*. Paper presented at 2008 IEEE International Symposium on Broadband Multimedia Systems and Broadcasting. Las Vegas, NV.

Schatz, R., Egger, S., & Platzer, A. (2011). *Poor, good enough or even better? Bridging the gap between acceptability and QoE of mobile broadband data services*. Paper presented at IEEE International Conference on Communications ICC 2011. Kyoto, Japan.

Schilling, J. (2006). On the pragmatics of qualitative assessment: Designing the process for content analysis. *European Journal of Psychological Assessment*, *22*(1), 28–37. doi:10.1027/1015-5759.22.1.28.

Södergård, C. (2003). *Mobile television-technology and user experiences. Report on the Mobile-TV project*. Finland: VTT.

Song, W., & Tjondronegoro, D. (2010). *A survey on usage of mobile video in Australia*. Paper presented at Australian Human-Computer Interaction Conference 2010. Brisbane, Australia.

Song, W., Tjondronegoro, D., & Docherty, M. (2010). Exploration and optimisation of user experience in viewing videos on a mobile phone. *International Journal of Software Engineering and Knowledge Engineering*, *8*(20), 1045–1075. doi:10.1142/S0218194010005067.

Song, W., Tjondronegoro, D., & Docherty, M. (2011a). *Quality delivery of mobile video: In-depth understanding of user requirements*. Paper presented at Australian Human-Computer Interaction Conference 2011. Canberra, Australia.

Song, W., Tjondronegoro, D., & Docherty, M. (2011b). *Saving bitrate vs. pleasing users: Where is the break-even point in mobile video quality?* Paper presented at ACM Multimedia 2011. Scottscale, AZ.

Strauss, A., & Corbin, J. (Eds.). (1990). *Basis of qualitative research: Grounded theory procedures and techniques*. Thousand Oaks, CA: Sage Publications.

Sveriges Television AB (SVT). (n.d.). Retrieved Feb 2010, from ftp://vqeg.its.bldrdoc.gov/HDTV/SVT_MultiFormat/

Tasaka, S., Yoshimi, H., & Hirashima, A. (2008). *The effectiveness of a QoE-based video output scheme for audio-video IP transmission*. Paper presented at ACM Multimedia Information System. Vancouver, Canada.

Technical University of Munich. (n.d.). Retrieved Feb 2010 ftp://ftp.ldv.e-technik.tu-muenchen.de/pub/test_sequences/

The Nielsen Company. (2011a). *The cross-platform report, Quarter 1 2011*. Retrieved Feb 11, 2012, from http://www.tvb.org/media/file/nielsen_cross-platform_report_Q1-2011.pdf

The Nielsen Company. (2011b). *Telstra smartphone index*. Retrieved Jul 22, 2011, from http://sensisdigitalmedia.com.au/Files/Mobile/Nielsen_Telstra_Smartphone_Index_June2011_Presentation.pdf

The Nielsen Company. (2012). *Australian multi-screen report (Quarter 1, 2012) - Trends in video viewership beyond conventional television sets*. Retrieved Feb 11, 2012, from http://www.nielsen.com/content/dam/corporate/au/en/reports/2012/MultiScreenReportQ12012_FINAL.pdf

Tominaga, T., Hayashi, T., Okamoto, J., & Takahashi, A. (2010). *Performance comparisions of subjective quality assessment methods for mobile video*. Paper presented at Second International Workshop on Quality of Multimedia Experience 2010. Trondheim, Norway.

Trefzger, J. (2005). *Mobile TV launch in Germany: Challenges and implications*. Cologne, Germany: Institute for Broadcasting Economics Cologne University.

Vangenck, M., Jacobs, A., Lievens, B., Vanhengel, E., & Pierson, J. (2008). *Does mobile television challenge the dimension of viewing television? An explorative research on time, place and social context of the use of mobile television content*. Paper presented at European Interaction TV Conference 2008. Salzburg, Austria.

Winder, J. (2001). *Net content: From free to fee*. Retrieved Aug 19, 2009, from http://hbr.harvardbusiness.org/2001/07/net-content-from-free-to-fee/ar/1

Zhang, Y., & Wildemuth, B. M. (2009). Qualitative analysis of content. In Wildemuth, B. (Ed.), *Applications of social research methods to questions in information and library science* (pp. 308–319). Westport, CT: Libraries Unlimited.

ADDITIONAL READING

Aron, A., Aron, E., & Coups, E. J. (2009). *Statistics for psychology* (6th ed.). Harlow, UK: Pearson Education.

Bazeley, P. (2004). Issues in mixing qualitative and quantitative approaches to research. In Buber, R., Gadner, J., & Richards, L. (Eds.), *Applying qualitative methods to marketing management research* (pp. 141–156). Basingstoke, UK: Palgrave Macmillan.

Buchinger, S., Kriglstein, S., Brandt, S., & Hlavacs, H. (2011). A survey on user studies and technical aspects of mobile multimedia applications. *Entertainment Computing*, *2*(3), 175–190. doi:10.1016/j.entcom.2011.02.001.

Denzin, N. K., & Lincoln, Y. S. (Eds.). (2000). *Handbook of qualitative research*. Thousand Oaks, CA: Sage Publications.

Edmunds, H. (Ed.). (1999). *The focus group research handbook*. Chicago, IL: NTC/Contemporary Publishing Group, Inc..

Eronen, L. (2006). Five qualitative research methods to make iTV applications universally accessible. *Universal Access in the Information Society*, *5*(2), 219–238. doi:10.1007/s10209-006-0031-2.

Holst, O. R. (Ed.). (1969). *Content analysis for the social sciences and humanities*. Reading, MA: Addison-Wesley.

ITU-R. (2004). *Methodology for the subjective assessment of quality for television pictures*. Recommendation BT. 500-11. Retrieved from http://www.itu.int/rec/R-REC-BT.500/en

Jumisko-Pyykkö, S., Vadakital, V. K. M., & Hannuksela, M. M. (2008). Acceptance threshold: A bidimensional research method for user-oriented quality evaluation studies. *International Journal of Digital Multimedia Broadcasting*, 20.

Jumisko-Pyykkö, S., Weitzel, M., & Strohmeier, D. (2008). *Designing for user experience: What to expect from mobile 3D TV and video?* Paper presented at 1st International Conference on Designing Interactive User Experiences for TV and Video. Silicon Valley, CA.

Knoche, H., & Sasse, M. A. (2009). The big picture on small screens: Delivering acceptable video quality in mobile TV. *ACM Transactions on Multimedia Computing. Communications and Applications*, *5*(3), 27.

Krippendorff, K. H. (Ed.). (2004). *Content analysis: An introduction to its methodology*. Thousand Oaks, CA: Sage Publications.

Kumar, A. (2010). *Implementing mobile TV*. Focal Press.

Law, E. L. C., & van Schaik, P. (2010). Modelling user experience - An agenda for research and practice. *Interacting with Computers*, *22*(5), 313–322. doi:10.1016/j.intcom.2010.04.006.

Lazar, J., Feng, J. H., & Hochheiser, H. (2010). *Research methods in human-computer interation*. Chichester, UK: John Wiley and Sons Ltd..

Menkovski, V., Exarchakos, G., & Liotta, A. (2011). The value of relative quality in video delivery. *Journal of Mobile Multimedia*, *7*(3), 151–162.

Menkovski, V., & Liotta, A. (2012). Adaptive psychometric scaling for video quality assessment. *Signal Processing Image Communication*, *27*(8), 788–799. doi:10.1016/j.image.2012.01.004.

Obrist, M., Meschtscherjakov, A., & Tscheligi, M. (2010). User experience evaluation in the mobile context. In Mobile, T. V. (Ed.), *Customizing Content and Experience* (pp. 195–204). London: Springer-Verlag. doi:10.1007/978-1-84882-701-1_15.

Roto, V. (2009). Demarcating user experience. In Gross, T., Gulliksen, J., Kotzé, P., Oestreicher, L., Palanque, P., Prates, R., & Winckler, M. (Eds.), *Human-Computer Interaction – INTERACT 2009* (*Vol. 5727*, pp. 922–923). Berlin: Springer. doi:10.1007/978-3-642-03658-3_112.

Song, W., Tjondronegoro, D., & Docherty, M. (2010). Exploration and optimisation of user experience in viewing videos on a mobile phone. *International Journal of Software Engineering and Knowledge Engineering*, *8*(20), 1045–1075. doi:10.1142/S0218194010005067.

Song, W., Tjondronegoro, D., & Docherty, M. (2011b). *Saving bitrate vs. pleasing users: Where is the break-even point in mobile video quality?* Paper presented at ACM Multimedia 2011. Scottscale, AZ.

Zhang, Y., & Wildemuth, B. M. (2009). Qualitative analysis of content. In Wildemuth, B. (Ed.), *Applications of social research methods to questions in information and library science* (pp. 308–319). Westport, CT: Libraries Unlimited.

Chapter 3
Modeling and Documenting Aspect-Oriented Mobile Product Lines

Camila Nunes
Pontifical Catholic University of Rio de Janeiro, Brazil

Uirá Kulesza
Federal University of Rio Grande do Norte, Brazil

Roberta Coelho
Federal University of Rio Grande do Norte, Brazil

Carlos Lucena
Pontifical Catholic University of Rio de Janeiro, Brazil

Flávia Delicato
Federal University of Rio de Janeiro, Brazil

Paulo Pires
Federal University of Rio de Janeiro, Brazil

Thais Batista
Federal University of Rio Grande do Norte, Brazil

ABSTRACT

Aspect-Oriented Software Development (AOSD) has evolved as a software development paradigm over the last decade. Recent research work has explored the use of Aspect-Oriented Programming (AOP) to modularize variations in product lines. This chapter presents a strategy for modeling and documenting aspect-oriented variations by integrating two existing approaches: (1) use cases are used to express the crosscutting nature of the variations of a mobile product line; and (2) crosscutting interfaces help the definition of the relevant variation join points that are raised by the mobile product line core and are extended by its respective variations. The synergy and benefits of the integration between these approaches are demonstrated by modeling and documenting MobileMedia, a software product line that provides support to manage different media (photo, music, and video) on mobile devices. Evolution scenarios of the MobileMedia are used to illustrate the benefits of the integrated usage of use cases and crosscutting interfaces in order to identify and analyze the change impact on the mobile product line.

DOI: 10.4018/978-1-4666-4054-2.ch003

INTRODUCTION

Aspect-Oriented Software Development (AOSD) (Kiczales et al., 1997) is a promising paradigm to improve the separation of concerns. Some concerns have impact in several modules of a software system and bring problems such as scattering, tangling and replicated code. Such concerns are known as *crosscutting concerns*. The existing object-oriented mechanisms and techniques are not able to suitably modularize these crosscutting concerns. Examples of typical crosscutting concerns are: exception handling, persistence, distribution, monitoring and security. The goal of AOSD is to support advanced separation of concerns by modularizing crosscutting concerns through smaller units, called *aspects*. The aspects introduce new abstractions and composition mechanisms, such as join point, pointcut, advice and inter-type declarations. AOSD has been used as a technique for improving the system reusability and maintainability.

Over the last years, AOSD techniques have been widely explored in both research and industry. Recent research work has explored the use of AOSD at the design and implementation levels in different contexts, such as, object-oriented application frameworks (Kulesza et al., 2006), design patterns (Hannemann & Kiczales, 2002), multi-agent systems (Nunes et al., 2009), model-driven development (Alvez et al., 2008; Greenfield, 2004; Sánches et al., 2008) and software product lines (Alves et al, 2005). Many of these research works indicate that the synergy between Software Product Lines (SPLs) and AOSD is very promising to boost the reuse in software systems development. Software Product Lines (SPLs) (Clements & Northrop, 2001) aim to improve the software reuse through the modeling and implementation of the commonalities and variabilities of artifacts (Clements & Northrop, 2001). In the context of SPLs, there are some works using AOSD to modularize variations in the domain of mobile devices applications (Figueiredo et al., 2008; Alves et al., 2005). Examples of known variations in the mobile domain are: screen size, number of colors, available memory, and types of devices. In addition, there are non-functional features such as portability, performance, security. Most of these kinds of variabilities have a fine-grained and usually tend to exert a crosscutting behavior. In order to overcome this problem, AOSD techniques have been a suitable mechanism for modularizing these variations features.

As previously mentioned, there are many recent works that evaluate the (dis)advantages of AOSD in the implementation level. Although the literature have focused at the implementation level, the problem of using aspects in the context of product lines should be handled in the earliest stages of the software development. Current modeling approaches (Gomaa, 2004; Anthonysamy & Somé, 2008; Lopez-Herrejon & Batory, 2006) do not offer a complete solution to address the modeling of crosscutting variations in both domain analysis and design development stages. Besides, these approaches do not deal with change impact analysis during the evolution of SPLs. In other words, they do not maintain a synergy between the requirements and design models, and the respective artifacts responsible for the implementation of these variations. For example, the approach proposed by (Gomaa, 2004) does not take into account crosscutting variations that are part of the mobile SPL. In addition, when this approach is used alone, it does not allow an adequate evolution and management of the models.

This work proposes an integrated approach to allow the modeling and documentation of crosscutting variations in aspect-oriented mobile product lines. Our approach integrates: (1) the product line development approach based on Extension Joint Points (EJPs) (Kulesza et al., 2006) and Crosscutting Interfaces (XPIs) (Sullivan et al., 2005) – which addresses the modularization of crosscutting and non-crosscutting features in SPL architectures; and (2) the AO use cases (Jacobson & P.-W, 2004) in conjunction with the notations proposed by (Gomaa, 2004) – which has the potential to specify and model crosscutting variations at the domain analysis stage. Our work emphasizes the strong synergy between these approaches demonstrated through the modeling and documentation of MobileMedia, a software product line that provides support to manage different medias (photo, music and video) on mobile devices. In addition, some discussions and lessons learned of the integrated use of these approaches through the change impact analysis performed in the MobileMedia are also presented. Several benefits and advantages can be observed from the integrated use of these approaches, such as: (1) early identification and modeling of (non)-crosscutting features during the requirements phase; (2) seamless mapping of extension points in AO use cases to EJPs, thus aiding SPL engineers to design and implement the crosscutting extension points of the SPL architecture; and (3) better support to traceability of non-(crosscutting) features along the different artifacts produced in domain engineering.

The main contributions of this work are: (1) to use existing approaches to model and document crosscutting variations on mobile product lines (AO use cases and crosscutting interfaces/extension join points); and (2) to report discussions and lessons learned about the integration between the two approaches, emphasizing the benefits and drawbacks. The integrated approach is generic enough to be applied to different SPL domains, although the focus of this work is on the mobile domain.

BACKGROUND

This section presents an overview of the underpinning techniques used to build the proposed approach, which are: (1) the AO use case approach; (2) the crosscutting interfaces (XPIs); and (3) the extension join points (EJPs).

Aspect-Oriented Use Case Approach

The use case approach is a well-known specification technique to model systems during the requirements analysis and specification activities of software development process. Jacobson and Ng (Jacobson, 2004) proposed an approach that combines use cases and AOSD motivated by the fact that use cases crosscut a set of components, thus being itself characterized as a crosscutting concern. During the elaboration of an use case, there are many classes involved that contain pieces of several use cases. This fact raises two problems in the generated artifacts: scattering (many classes involved) and tangling (pieces of several use cases). The adoption of AOSD techniques aims at helping the resolution of these problems by allowing a better separation of concerns. Jacobson and Ng (Jacobson, 2004) discuss that AOSD provides an important mechanism to keep the use case modularity in the design and implementation of software systems. Basically, they propose to map extension use cases to aspects. Extension use cases behave as alternative flows and take place in a specific condition. Extension use case adds a new behavior to any other use case, which can be performed by an advice in a aspect-oriented programming language at the implementation level. An advice is executed when a specified join point from a pointcut is reached. Advices are similar to methods and add behavior to be executed at extension points designated by pointcuts. This approach can be used to model all the relationships – extend, include and generalization – between use cases.

Crosscutting Interfaces (XPI)

Crosscutting interfaces (XPIs) are used to abstract an existing crosscutting behavior in the base code. They were proposed to specify design rules (Sullivan et al., 2005) between core classes of a system (or SPL) and a set of aspects that extend these core classes. The use of XPIs appears as an alternative to mediate the relationships and interactions between classes and aspects thus contributing to improve the reuse and modularity of a system (or SPL). XPIs (Griswold et al., 2006) were originally proposed on the top of AspectJ language. The main idea behind the use of them is to abstract advised joinpoints. An XPI has four elements: (1) the XPI's name; (2) a scope over which the XPI abstracts join points; (3) one or more sets of abstract join points; and (4) a partial implementation. Basically, the set of abstract joint points is represented by: (1) a pointcut descriptor signature declaring a name and exposed parameters, which constitutes the abstract interface; and (2) a semantic part where the conditions of the exposed join points (pre and post-conditions) are specified to be verified statically or dynamically. By using XPIs, it is possible to specify aspects that define constraints for the exposed join points and include behavioral constraints on aspects. The XPI's syntactic part exposes two named pointcuts descriptor: *joinpoint()* and *topLevelJoinpoint()*. To implement and design them, it is not necessary to know or implement the aspects in advance. In other words, we design general XPIs as a mechanism to completely separate the detailed design of the core classes and the aspects. Finally, XPIs provide a better comprehensibility, concurrent development, and evolution.

Extension Join Points (EJP)

In (Kulesza et al., 2006, 2007), a framework extension approach is proposed based on the use of Extension Join Points (EJPs). EJPs enable systematic extension of object-oriented frameworks or software product line architectures by means of aspects. In this way, EJPs are adopted to facilitate the implementation of crosscutting variability. The EJP (Extension Join Point) consists on a unified way of designing and documenting existing crosscutting extension points. In the EJP approach, aspects are developed using an aspect-oriented programming language to modularize and to document variabilities. Object-oriented application frameworks define extension points (hotspots) by means of abstract classes and interfaces, which need to be implemented during the framework instantiation. EJPs enable us to expose crosscutting extension points that can be used to implement the crosscutting variabilities through aspects. Thus, EJPs can be used to modularize not only crosscutting extension points, but also some kinds of non-crosscutting optional and alternative features.

The concept of EJPs extends the XPI approach proposed by (Griswold et al., 2006) to the context of object-oriented framework and product line development. EJPs can be used in two ways: (1) to expose a set of framework events that can be used to implement a crosscutting integration with other application modules; and (2) to enable a set of join points that are spread and/or tangled in the framework and that are candidates to be extended with aspects. The main concepts that comprise an EJP in the AspectJ language are: (1) a name that is represented by the aspect's name; (2) a scope that defines all the framework elements that are encapsulated by the EJP; (3) a set of crosscutting extension points, which specifies the framework join points that represent relevant events or transition states occurring during the execution of the framework functionalities; and (4) a set of internal and extension contracts – which are used to guarantee the independent evolution of the framework core or extension aspects.

The EJP contracts are classified in the following way: (1) *framework internal contracts* – define constraints to be complied by the framework. These contracts have the purpose of assuring that framework refactoring and evolution do not affect the functionality of its extension aspects; and (2) *framework extension contracts* – define constraints to be complied by the extension aspects. These contracts are used to assure that each extension aspect respects constraints and invariants of the framework.

AN INTEGRATED APPROACH TO MODEL AND DOCUMENT CROSSCUTTING VARIATIONS

This section describes how AO approaches previously described can be used together to address the modeling and documentation of crosscutting variations in SPL, especially in the mobile applications domain. It gives an overview of our integrated approach to model and document crosscutting variations along different SPL stages, highlighting the benefits and advantages of adopting the proposed approach.

Approach Overview

SPL development processes involve several activities (Clements & Northrop, 2001), which are typically organized in three main phases: domain engineering, application engineering, and management. Domain engineering is the process in which the commonality and variability of the SPL are identified, modeled and realized. It produces as main result a set of code assets to be reused in the application engineering. The application engineering is the process of synthesis of products from the SPL, which reuses the code assets produced in the domain engineering to improve the quality and productivity of the development of products. Finally, the management is concerned with the production strategy of the SPL by guaranteeing the optimal allocation of resources to the different development activities. Our approach is mainly concerned with the domain-engineering phase, as it provides support to model non-crosscutting and crosscutting features throughout the domain analysis, design and implementation activities. Figure 1 gives an overview of our integrated approach by illustrating thc mapping from features to uses cases in the design and implementation elements, using AO use cases and EJP/XPIs.

Our approach focuses mainly in the following activities:

1. **Domain Analysis:** Identify and model the common and variable features of the SPL products using adequate modeling techniques. This phase encompasses the domain analysis activities, where the scope of the SPL is analyzed and defined. During this phase, we focus on the use case modeling to collect and express the functional and non-functional requirements. However, in this approach, we are concerned about

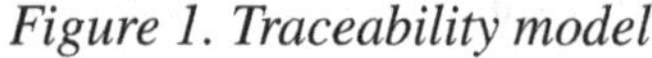
Figure 1. Traceability model

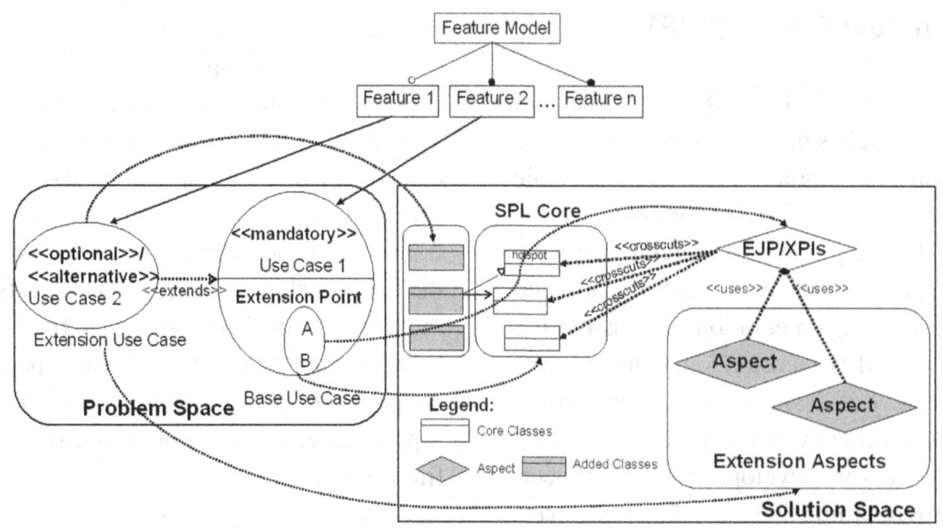

crosscutting use cases. In this stage, we used the approach proposed by (Jacobson, 2004), which makes use of aspects to model the crosscutting variations through extension points.

2. **Design and Implementation of the SPL Core Architecture:**. This activity comprises the implementation of the reusable software assets that address the common and variable features of the SPL. It is a creative activity that requires technical experience from the software engineers involved with them. It uses the artifacts produced during the domain analysis as a base to define the SPL core architecture.
3. **Design and Implementation of EJP/XPIs:** This activity consists of exposing a set of join points for the inclusion of crosscutting extensions. During this activity, we are going to model potential EJPs (optional/alternative features) that are relevant to be implemented by extension aspects. In this activity, we can use the information contained in use cases to design and implement the EJPs. As the crosscutting variations were already captured in the use case modeling, it is easier to design and implement the EJPs. However, during this activity it is possible to identify new crosscutting variations and update the use case modeling.
4. **Design and Implementation of the Extension Aspects:** In this activity, extension aspects are implemented to address the optional and alternative crosscutting features. Each extension aspect must be applied to the exposed EJP/XPIs from the SPL architecture. In addition, they can also contain variability in their implementation.

Our approach promotes the integration between the AO use cases and EJP/XPIs techniques with the aim to support the modeling and documentation of complex AO SPL architectures. It covers all the activities of domain engineering by making explicit the modeling, design and implementation of crosscutting variability and their intrinsic relationships with the SPL core.

The benefits of the adoption of our approach are: (1) early identification and modeling of crosscutting features during the requirements phase; (2) seamless mapping of extension points in AO use cases to EJP/XPIs, thus helping SPL engineers to design and implement the crosscutting extension points of the SPL architecture; and (3) support to traceability of non-crosscutting and crosscutting features along the different artifacts produced in domain engineering.

An Illustrative Example

In this section, we present an illustrative example showing how a crosscutting variation can be modeled and documented using the AO use case approach along with XPI/EJPs specifications. A well-known example of crosscutting variation is logging. For example, let us suppose a banking system, where the user can withdraw or deposit money. Figure 5(a) illustrates the use case modeling, showing base use cases and Logging extension use case. In the base use cases, the extension points are specified, which are: Make Withdraw and Make Deposit. When these extension points are reached, the Logging extension use case inserts a new behavior within them. Logging is represented by an aspect, which affects the extension points of base use cases.

Figure 2(b) illustrates the design of the EJPs. Basically, we have the Account class that is affected by the Logging aspect, which is an EJP. This EJP is composed of variability aspects – an extension aspect that implements and introduces a variability in the SPL architecture – to implement the way to store the logging. We extend this EJP through an aspect called LogStream. This aspect is responsible for storing the information into a file. Another possible implementation of

Figure 2. Use case and EJPs modeling. (a) AO Use Case. (b) EJPs.

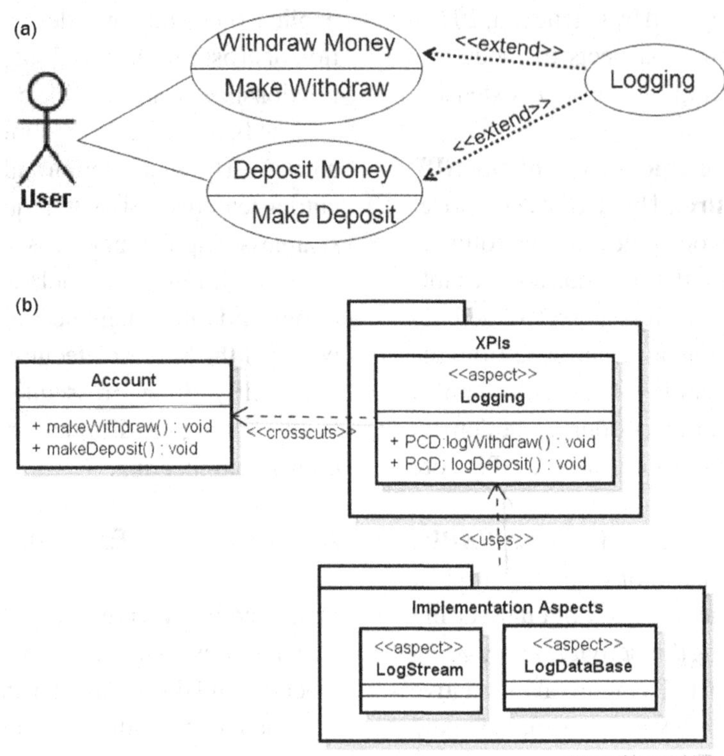

variability aspects could be an aspect to store the *logging* in a database. When the makeWithdraw() and makeDeposit() methods are completed, the logWithdraw() and logDeposit() advices of the LogStream aspect are invoked to register the information performed by the user. Another variability aspect is the LogDataBase aspect, which is responsible for storing the log information into a database. Code 1 illustrates the Logging aspect, which defines the pointcuts to affect the Account class. Code 2 illustrates the extension aspect implementation, the LogStream aspect.

MODELING THE MOBILEMEDIA VARIATIONS

This Section details how our approach can be used to model and document the crosscutting variations encountered in the MobileMedia software product line. We illustrate how the combination of AO use cases and XPIs/EJPs specifications allows a better modeling and documentation of different kinds of crosscutting variations.

Code 1. The AspectJ code of the logging EJP

```
public aspect Logging {
   public pointcut logWithdraw (): call (Account.makeWithdraw());
        public pointcut logDeposit(): call (Account.makeDeposit());
}
```

Code 2. The code of the LogStream aspect

```
public aspect LogStream {
        after(): Logging.logWithdraw() {
                File lfile = new File(TransactionConstants.LOG_FILE);
                BufferedReader input = new BufferedReader(new
FileReader(lfile));
...
    }
         after(): Logging.logDeposit() {
        }
}
```

The Target SPL: MobileMedia

MobileMedia (MM) (Figueiredo et al., 2008) is a Software Product Line (SPL) that provides support to manage (create, delete, visualize, play, send) different medias (photo, music and video) on mobile devices. It was developed based on a previous software product line called Mobile Photo (Young, 2005) conceived at the University of British Columbia. During the development and evolution of the mobile product line, mandatory, optional, and alternative features were included in the MM core architecture. The core features are: to create/delete media (photo, music or video), to label media, and to view/play media. Some optional features are, among others: to transfer photos via SMS, to count and sort media, to copy media and to set favorites. Nine releases (products) of the MM product line were generated. In each release, the core artifacts (base model) and non-crosscutting features were implemented in Java, using the conditional compilation technique, and aspects were implemented in AspectJ. These variability implementation techniques (conditional compilation and aspect-oriented programming) are used to improve the management of the SPLs variability and allow the product derivation process. MM was developed for a family of 4 brands of devices, namely Nokia, Motorola, Siemens, and RIM.

Our Approach in Action

We selected some crosscutting features/variation in MM to present the proposed approach for modeling and documenting using AO use cases and EJP/XPIs. In the use case modeling of the MM we used the notations proposed by (Jacobson & P.-W, 2004) and (Gomaa, 2004). The notations proposed by (Gomaa, 2004) indicate that a use case: (1) is part of the SPL kernel being (<<mandatory>>); (2) is present in some products (<<optional>>); or (3) varies among the SPL products (<<alternative>>).

Figure 3 illustrates the AO use cases and EJPs modeling of the exception handling crosscutting feature. Figure 3(a) illustrates the use case modeling. In this Figure the mandatory use cases of the MobileMedia and the use case extension Handle Exception are presented. The Handle Exception use case inserts a new behavior in the extension points specified in base use cases.

Figure 3(b) shows the ControllerException, DataException and ImageException EJPs, which crosscuts MobileMedia core components and are used by the extension aspects. According to the Figure 3, it is possible to see the similarities between the AO use cases and EJPs. Basically, the extension points captured during the requirements phase are translated to the EJPs approach. In other words, these extension points represent the point cuts specified in the EJPs.

Figure 3. Use case and EJPs modeling of the exception handling feature. (a) AO Use Case. (b) EJPs.

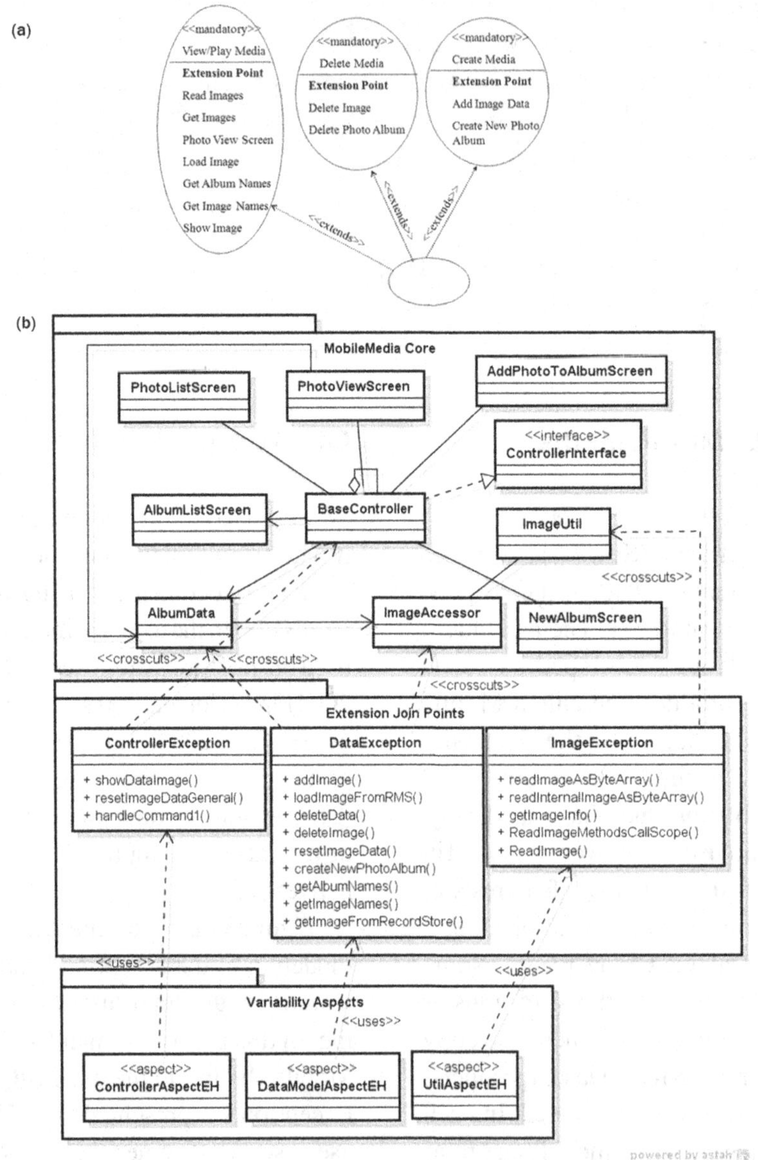

Figure 4 illustrates the AO use cases and EJP modeling of the optional feature used to count the number of times a photo has been viewed. This optional feature extends the base use cases in following extension points (Figure 4(a)). In terms of EJP modeling, we created the CountViews EJP and two variability aspects: BaseControllerCV and ImageCountViews. The CountViews EJP affects three components in the MobileMedia core: BaseController, ImageUtil, and PhotoViewScreen (Figure 4(b)). In this modeling, we reused two EJPs: ImageException and ControllerException.

Figure 5 illustrates the AO use cases and the EJP modeling of the optional feature to view the favorite photos. This optional feature extends the base use cases at the same extension join points

Figure 4. Use case and EJPs modeling of the count photo views optional feature. (a) AO Use Case. (b) EJPs.

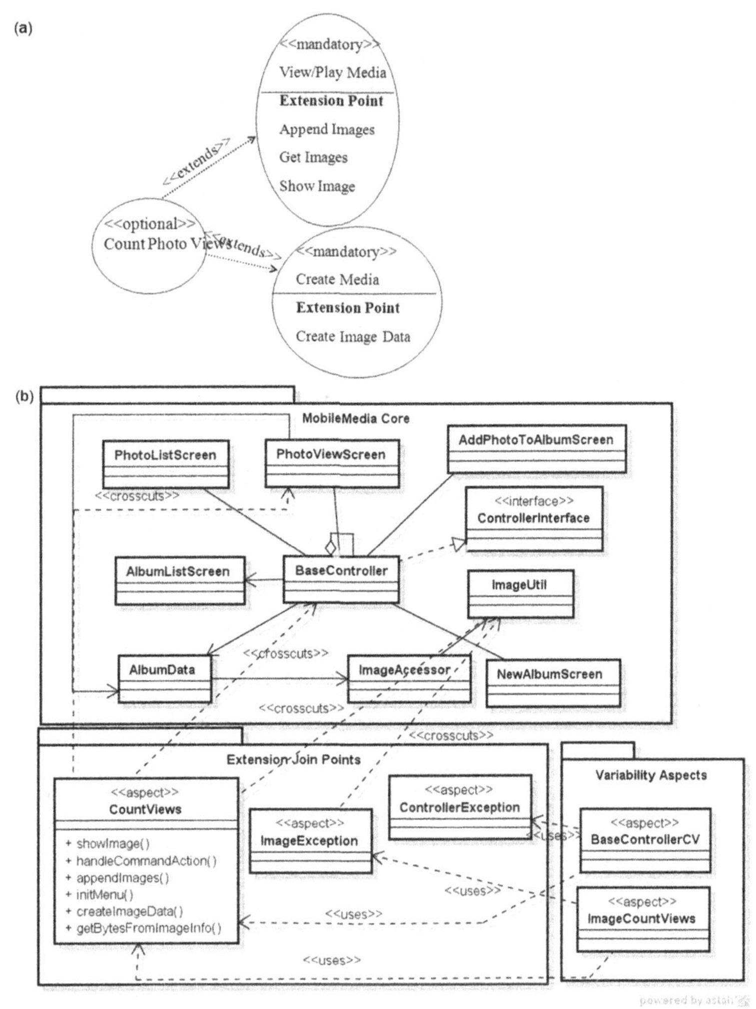

than the count photo views feature. Due to that, the EJP defined during the inclusion of the count photo view feature was reused in this feature to expose the join points. For the modeling of the view favorite photo feature, it was necessary to create the Favorites EJP with only an extension point. After that, we created two variability aspects: BaseControlerFavorites and ImageFavorites. In this feature, we reused the EJPs previously defined: CountViews and ImageExceptions.

The integration between the AO use cases with the notations proposed by (Gomaa, 2004) and XPI/EJPs allows a better management of variabilities. This improvement is achieved since with such approach details of the crosscutting variations are presented both in the requirements level and in the design and implementation of the SPL.

DISCUSSIONS, LESSONS LEARNED, AND OPEN ISSUES

In this section, we discuss how the use of our integrated modeling approach brings several benefits to the development and evolution of the MobileMedia product line. Our discussions

Figure 5. Use case and EJPs modeling of the favorite photos optional feature. (a) AO Use Case. (b) EJPs.

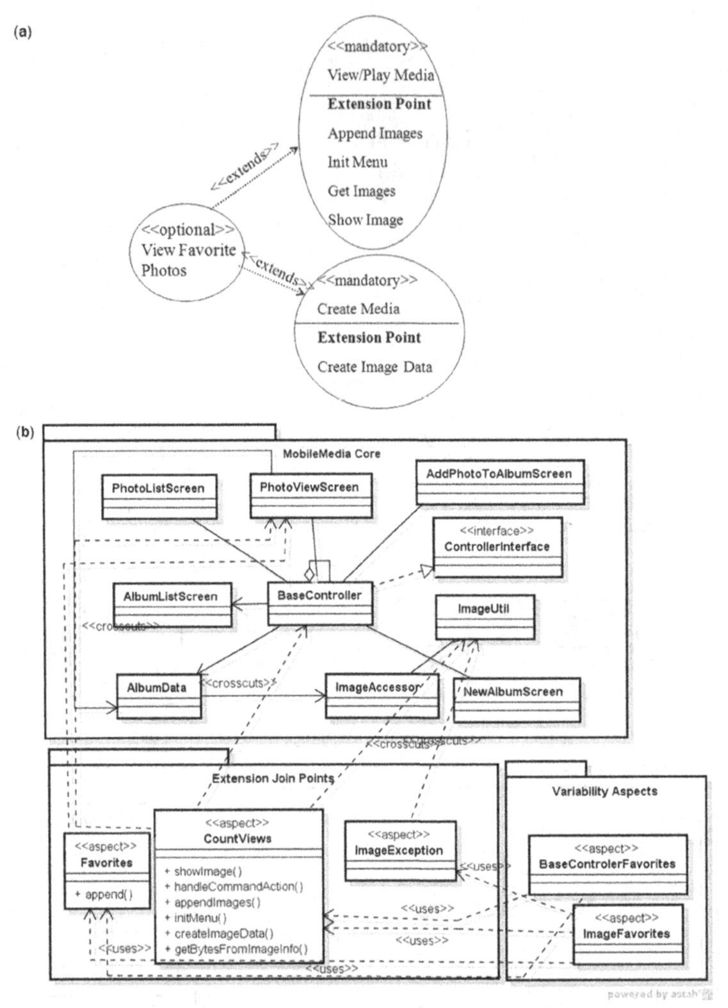

emphasize the benefits of the approach mainly in terms of documentation of crosscutting variations, traceability of variations, change impact analysis, and model-driven tool supporting.

Documentation of Crosscutting Variations

Our approach allows the documentation and modeling of crosscutting variations through development of SPLs. Initially, that documentation is performed during the requirements phase, using the AO use case modeling, where the crosscutting variations are identified. Those crosscutting variations are inserted in specific extension points of the base use cases, modifying their behavior during the choice of a determined optional or alternative feature. After that, we consider the design and implementation of the XPI/EJPs, which represent the interfaces between the core SPL and the extension aspects. This design and implementation is performed based on the AO use case modeling. Thus, during the design and implementation of the XPI/EJPs, it is possible to evolve the AO use cases and also the XPI/EJPs, identifying new requirements required by the extension aspects to implement specific variability in the SPL architecture. This integrated approach allows a better understanding and evolution of the crosscutting variations of the SPL, because the documentation does not get spread and tangled along different system modules. It allows visualizing which specific extension join points in the SPL architecture/design are affected and extended by existing extension aspects.

Traceability Model

Our approach facilitates the management, implementation, and evolution of the features through a traceability model. In the domain analysis stage, during the conception of the SPL features, the variations points are mapped to extension points into the use cases. After that, in the domain design phase, all these extension points are then mapped to XPIs, which are explicitly extended by implementation aspects. Finally, all these elements can be implemented using appropriate mechanisms to modularize the SPL core, EJPs and extension aspects. The use of AO techniques during the phases of domain analysis, design and implementation allows the homogenous addressing and a better understanding of both crosscutting and non-crosscutting features throughout the whole development process. As a result, our approach improves the traceability between the different elements produced in SPL development thus contributing to the change impact analysis.

Change Impact Analysis

The techniques for modeling and documenting SPL can be used for the three kinds of SPL development: proactive, extractive and reactive (Krueger, 2003). MobileMedia was developed following the reactive approach, where new features are incrementally included. Therefore, the models incrementally evolved during the conception of each MobileMedia release. The effectiveness and modularization of a SPL depends on how the features are managed and modeled during the domain analysis, design and implementation. When new features were included, it was observed that models were not extensively modified to incorporate the new variation added. Our approach shows the synergy between AO use cases with UML notations and XPI/EJP to model (non-) crosscutting variations. It addresses the seamless modularization of features from requirements to implementation through the use of AO techniques. These improvements in the modularity are important to allow a good evolution and maintenance of these features. The mappings from feature models to AO use cases, and AO use cases to design and implementation elements improve the understanding of the SPL features and also contribute to the flexibility to add new features or evolve the existing ones.

Model-Driven Development

Although our approach can bring benefits to the management and traceability of crosscutting and non-crosscutting features in SPLs, it is fundamental that it provides automatic tool support to facilitate the navigation along the different artifacts developed in order to support an effective change impact analysis. Different implementation strategies can be adopted to support our approach using model-driven development techniques. One possible strategy is only using information provided by the mapping between the artifacts in our approach to produce a set of trace links which can be analyzed and processed by a more elaborated traceability tool (Anquetil, 2009). The specification of these different trace links should be supported using the Variability Modeling Language (Alferez, 2009). Another interesting strategy to explore is to allow the systematic refinement of models from domain analysis to domain design and implementation. In this strategy, model-driven tools could provide the necessary support to allow the semi-transformation of models from different abstraction levels.

ALTERNATIVE AO APPROACHES: A BRIEF REPORT

In this section, we present other existing alternative approaches to model crosscutting variations. We also briefly discuss how our integrated approach can bring several benefits compared to existing ones. Eventual drawbacks will be also depicted.

Several extensions of UML to model product lines have been proposed. One example is Product Line UML-based Software engineering (PLUS) (Gomaa, 2004). This method proposes the use of use cases with stereotypes indicating the kind of feature: mandatory, optional and alternative. PLUS also contains the same information of traditional feature models. The PLUS approach defines a customization of the use case model to specify and document the SPL requirements. This approach does not provide clear support to deal with the crosscutting variations. Besides, it does not allow an easy evolution of the models due to SPL evolution scenarios.

(Anthonysamy & Somé, 2008) applies the AO use case with some adaptations to model software product lines. Basically, they introduced the terms *variability* and *commonalities*. They model variability and crosscutting commonalities as use cases and link them with a <<variability>> relation. The <<variability>> relation is a specialization of the <<aspect>> relation. This work is based on a previous one, which is a use case modeling tool, named Use Case Editor (UCEd), which takes a set of related use cases written in a restricted natural language and automatically generates executable State Charts. This work focuses only on the requirements phase to generate the possible products of the SPL.

(Lopez-Herrejon & Batory, 2006) explores the use case slices proposed by Jacobson and Ng (Jacobson & P.-W, 2004) and FOP (Feature Oriented Programming) to model and modularize features in aspect-based product lines. In this paper, Lopez-Herrejon & Batory presented a case study to illustrate the benefits and similarities between use cases slices and FOP program. They observed that current proposals lack from a composition model to map use case slides to concrete working implementations. Thereby, our approach proposes the mapping of extension points elements to implementation elements through XPI/EJPs, becoming clear how crosscutting variations can be modularized and implemented.

As it was aforementioned, there are some works using AOP to model variability in SPL. Our work differs from the other ones because it presents an integrated approach, which allows performing the traceability from requirements (AO use cases) to the design (XPI/EJPs) level. Using these two techniques, we can anticipate the crosscutting variations using AO use cases and facilitate the design and development of the XPI/EJPs. Besides, we explore the synergy among the models thus facilitating the evolution and management.

CONCLUSION AND FUTURE WORK

This work presented an integrated approach for modeling and documenting crosscutting and non-crosscutting features along with the SPL development process (domain analysis, design and implementation). The approach was illustrated by using the MobileMedia SPL. Some evolution scenarios were modeled in order to show the benefits of the integrated use of AO use cases joined with the PLUS method and crosscutting interfaces.

We have found in our study that our approach brings benefits to the management (modeling, documentation and traceability) of SPL commonalities and variability by making explicit the specification of (non-)crosscutting features. Also, it suitably captures the relationships and interactions between them through the use of extension points in use cases and XPIs in detailed design and implementation. Besides, our approach allows a better evolution of the SPL features by using AO techniques in the modularization process throughout the SPL phases. Due to the use of structured models, it is possible to easily map the elements from the problem space to the solution space.

As future work, we intend to apply our approach to more complex AO SPLs in order to validate them qualitatively and quantitatively. We also intend to develop a model-driven tool to provide automatic support to our approach and contributing to help maintenance of the models during SPL evolution. This tool will provide the traceability model the composition mechanisms associated to modeling and implementation elements. Besides, these models can be used to process transformations and generate code automatically.

REFERENCES

Alferez, M., Santos, J., Moreira, A., Garcia, A., Kulesza, U., Araújo, J., & Amaral, V. (2009). Multi-view composition language for software product line requirements. In *Proceedings of the 2nd International Conference on Software Language Engineering (SLE'2009)*. Denver, CO.

Alves, M., Pires, P., Delicato, F., & Campos, M. (2008). CrossMDA: A model-driven approach for aspect management. *Journal of Universal Computer Science*, *14*(8), 1314–1343.

Alves, V., et al. (2005). Extracting and evolving mobile games product lines. In *Proceedings of SPLC'05* (LNCS), (vol. 3714, pp. 70-81). Berlin: Springer.

Anquetil, N., Kulesza, U., Mitschke, R., Moreira, A., Royer, J., Rummler, A., & Sousa, A. (2009). A model-driven traceability framework for software product lines. *Software & Systems Modeling*.

Anthonysamy, P., & Somé, S. S. (2008). Aspect-oriented use case modeling for software product lines. In *Proceedings of the AOSD Workshop on Early Aspects*, (pp. 1-8). New York, NY: ACM.

Aspect, J. Team. (2009). *The AspectJ programming guide*. Retrieved from http://eclipse.org/aspectj/

Clements, P., & Northrop, L. (2001). *Software product lines: Practices and patterns*. Reading, MA: Addison-Wesley Professional.

Czarnecki, K., & Eisenecker, W. (2000). *Generative programming: Methods, tools, and applications*. Reading, MA: Addison-Wesley Publishing.

Figueiredo, E., et al. (2008). Evolving software product lines with aspects: An empirical study on design stability. In *Proceedings of the 30th ICSE'08*, (pp. 261-270). ICSE.

Gomaa, H. (2004). *Designing software product lines with UML: From use cases to pattern-based software architectures*. Redwood City, CA: Addison Wesley Longman Publishing.

Greenfield, J., Short, K., Cook, S., & Kent, S. (2004). *Software factories: Assembling applications with patterns, models, frameworks, and tools*. New York: Wiley.

Griswold, W. et al. (2006). Modular software design with crosscutting interfaces. *IEEE Software*, *23*(1), 51–60. doi:10.1109/MS.2006.24.

Hannemann, J., & Kiczales, G. (2002). Design pattern implementation in Java and aspectJ. In *Proceedings of the 17th ACM SIGPLAN Conference on Object-Oriented Programming, Systems, Languages, and Applications*, (pp. 161-173). New York, NY: ACM.

Jacobson, I., & Ng, P.-W. (2004). *Aspect-oriented software development with use cases*. Reading, MA: Addison-Wesley Professional.

JBOSS AOP. (2009). Retrieved from http://www.jboss.org/community/docs/DOC-10201

Kiczales, G., Irwin, J., Lamping, J., Loingtier, J., Lopes, C. V., Maeda, C., & Mendhekar, A. (1997). Aspect-oriented programming. [). Berlin: Springer-Verlag.]. *Proceedings of the ECOOP*, *1241*, 220–242.

Krueger, C. (2003). Easing the transition to software mass customization. In *Proceedings of the 4th International Workshop on Software Product-Family Engineering* (LNCS), (vol. 2290, pp. 282-293). Berlin: Springer-Verlag.

Kulesza, U., Alves, V., Garcia, A., Lucena, C., & Borba, P. (2006a). Improving extensibility of object-oriented frameworks with aspect-oriented programming. In *Proceedings of the 9th International Conference on Software Reuse (ICSR)*, (pp. 231-245). Torino, Italy: ICSR.

Kulesza, U., Alves, V., Garcia, A., Neto, A., Cirilo, E., Lucena, C., & Borba, P. (2007). Mapping features to aspects: A model-based generative approach. In A. Moreira & J. Grundy (Eds.), *Early Aspects: Current Challenges and Future Directions, 10th International Workshop* (LNCS), (vol. 4765, pp. 155-174). Berlin: Springer-Verlag.

Kulesza, U., Coelho, R., Alves, V., Neto, A., Garcia, A., Lucena, C., & Borba, P. (2006b). Implementing framework crosscutting extensions with EJPs and AspectJ. In *Proceedings of Brazilian Symposium on Software Engineering*, (pp. 117-192). IEEE.

Lopez-Herrejon, R. E., & Batory, D. (2006). Modeling features in aspect-based product lines with use case slices: An exploratory case study. [MoDELS.]. *Proceedings of MoDELS Workshops, 2006*, 6–16.

Nunes, C., et al. (2009). Comparing stability of implementation techniques for multi-agent system product lines. In *Proceedings of the 3th European Conference on Software Maintenance and Reengineering (CSMR'09)*. Kaiserslautern, Germany: CSMR.

Sánchez, P., Fuentes, L., Stein, D., Hanenberg, S., & Unland, R. (2008). Aspect-oriented model weaving beyond model composition and model transformation. *Proceedings of MoDELS, 2008*, 766–781.

Spring AOP Aspect Library. (2009). Retrieved from http://www.springframework.org

Sullivan, K. et al. (2005). Information hiding interfaces for aspect-oriented design. In *Proceedings of ESEC/FSE'2005* (pp. 166–175). Lisbon, Portugal: ESEC.

Young, T. (2005). *Using AspectJ to build a software product line for mobile devices*. (MSc Dissertation). University of British Columbia, Vancouver, Canada.

KEY TERMS AND DEFINITIONS

Application Engineering: Application Engineering is one of the stage of the SPL development process. This stage is responsible to develop products from the core assets produced during domain engineering.

Aspect-Oriented Software Development (AOSD): AOSD has been proposed as a technique for improving the separation of concerns in the construction of OO software, supporting improved reusability and ease of evolution.

Crosscutting Concern (CC): CC is a concern is spreading among several modules and we want modularize in order to treat as a single unit.

Crosscutting Interfaces (XPI): XPIs are used to help the definition of the relevant variation join points that are raised by the SPL core and can be extended by a set of implementation aspects.

Domain Engineering: Domain Engineering is one of the stages of the SPL development process. This stage encompasses the following stages: domain analysis, design and implementation.

Extension Join Points (EJP): EJPs establish a contract between the framework classes and a set of aspects extending the framework functionality. EJPs aim at increasing the framework variability and integrability.

Software Product Lines (SPL): SPL is a well-known approach to allow the systematic development of program families for a particular domain. They enable modular, large-scale reuse through a core software architecture and a set of common and variable features of a product family.

Use Cases: Use cases are used to capture the SPL requirements during the domain analysis.

Chapter 4
Developing a Multimodal Application for a Scientific Experiment on Smartphones:
Case Study, Tools, and Results

José Rouillard
University of Lille, France

ABSTRACT

Designing and developing multimodal mobile applications is an important knowledge for researchers and industrial engineers. It is crucial to be able to rapidly develop prototypes for smartphones and tablet devices in order to test and evaluate mobile multimedia solutions, without necessarily being an expert in signal processing (image processing, objects recognition, sensors processing, etc.). This chapter proposes to follow the development process of a scientific experiment, in which a mobile application will be used to determine which modality (touch, voice, QRcode) is preferred for entering expiration dates of alimentary products. For the conception and the generation of the mobile application, the AppInventor framework is used. Benefits and limitations of this visual tool are presented across the "Pervasive Fridge" case study, and the obtained final prototype is discussed.

1. INTRODUCTION

According to the global research and advisory firm Forrester, "by 2016, smartphones and tablets will put power in the pockets of a billion global consumers" (Schadler et al. 2012). In 2011, sales of this kind of devices have exceeded sales of personal computers. Yet very few Web sites have been optimized to take advantage of these new devices. The introduction of many types of mobile devices and the rise of their capabilities has revealed a fundamental challenge in the needs

DOI: 10.4018/978-1-4666-4054-2.ch004

for designing and deploying mobile interactive software that optimize the human-computer interaction. Indeed, modern terminals are natively equipped with many input and output resources required for multimodal interactions, such as camera, vibration, accelerometer, etc. This creates the urgent necessity for easier access to information, whether at the office, home, or on the train, etc. This need is felt all the more with the constant new arrival of soft/hardware materials, the success of the pocket computers and mobile telephones. Moreover, pervasive and ubiquitous computing are new principles that developers have to keep in mind, in order to provide satisfying experiences to end-users while using smartphones, especially in mobile situations. However, it is not straight forward to conceive, develop, test and evaluate real applications for such smartphones and tablets devices. Designing for mobile is not exactly the same as designing for desktop or laptop, and it is not only a matter of screen size.

This chapter presents some aspects of this scientific problem and is structured as follows: section two explains the background and motivation for understanding our objective while designing and deploying a mobile application for a scientific purpose. Section three gives an overview of the AppInventor visual tool used to achieve this goal. Section four is describing the case study and the obtained results and future trends for further work are presented in the conclusion of the chapter.

2. BACKGROUND AND MOTIVATION

Currently, according to the Food and Agriculture Organization of the United Nations, consumers worldwide waste about 1.3 billion tons of food annually. Consumers in rich countries waste about 222 million tons of food products (FAO, 2011). People buy food items that are kept in a fridge or cupboard and they are often unaware when products are at their consumable deadline, until the food needs to be thrown away. One part of this work is related to the field of pervasive and ubiquitous computing, whereby mobile application is to be used to help consumers avoid wasting food. In this context, we are developing multimodal mobile applications in order to conduct experiments with end-users.

The existing development tools available are often associated with the kind of device and OS targeted. For instance, Visual Studio Microsoft's integrated development environment (IDE) software package allows developing applications that can be run in Windows, Windows Mobile, CE, dot NET Framework, and Silverlight. The iOS Software Development Kit (formerly known as iPhone SDK) is a software development kit developed by Apple Inc. and released in February 2008 to develop native applications for iOS. The Android SDK provides the API libraries and developer tools necessary to build, test, and debug apps for Android. All the IDEs are used by programmers through various programming languages (C# or Visual Basic, Objective-C, Java, etc.), using both the Eclipse IDE and the SDK Android. PhoneGap, an open-source mobile development framework (PhoneGap 2012) currently supports development for multiple operating systems, including Apple iOS, Google Android, HP WebOS, Microsoft Windows Phone, Nokia Symbian OS and RIM BlackBerry. It requires Eclipse (version 3.4+) and/or other tools like Java Development Kit (JDK) in order to be used. Some tools are dedicated to more specific devices. For instance, MotoDev (MotoDev 2012) is particularly adapted for development of mobile application running on Android Motorola smartphones. AgilePlatform (AgilePlatform 2012) is considered more as a tool to develop all aspects of Web and mobile applications using a fully integrated visual modeling environment (user interfaces, business logic, databases, workflows, business processes, security rules, asynchronous jobs, etc.). Despite the advancements of tools that assist developers, it is still not a straightforward process to develop interactive multimodal applications that will be used in mobile situations while employing various components of the devices.

In this chapter, we will describe a case study in which a rapid development of a mobile and multimodal application is needed, in order to test and experiment scientific hypotheses about management of foods and fight against alimentary waste.

2.1. A Scientific Experiment on Uneaten Food Loss

Mobile computing is a generic term describing the ability to use computer technology while moving. This situation of mobility is stronger than the notion of portable computing, where the users often use their laptop, but not necessarily while they are moving. Many types of mobile computers have been introduced since the 1990s, including Pocket PC, Wearable computer, PDA, Smartphone, Ultra Mobile PC (UMPC), etc.

The use of various ways of communication inputs (keyboard, mouse, speech recognition, eye tracking, gesture recognition, etc.) and outputs (screen, sounds, speech synthesis, force feedback, etc.) are the so-called modalities of a multimodal system (Rouillard, 2009). Some of these are inherently bidirectional, such as touch-screens or haptic interfaces. Multimodal interfaces offer the possibility of combining the advantages of natural (but ambiguous) inputs, such as speech, and less natural (but unambiguous) inputs by way of direct manipulation. In the domain of mobility, particularly, software and hardware are becoming more and more suitable for multimodal uses. It is now possible to have multiple interfaces on the same traditional screen and keyboard device that include touch and speech interface capabilities. Various kinds of connections and technologies are also, sometimes natively available on modern mobile devices, such as RFID (Radio-Frequency IDentification), NFC (Near Field Communication), or QR (Quick Response) Codes.

In situation where a user is mobile, it is very important to provide the right information at the right time in the right place. For instance, there could be a way to easily retrieve nutrition information based upon a QRCode on a burger package, paying for a can of drink at a vending machine, getting bus information in real time, obtaining in-store information about a product, speeding up flight check-in and boarding etc. Other areas where this QRCode technology could be beneficial for mobile persons may include cultural, entertainment and advertising of relevant information. Embedded camera, GPS, Wifi, Bluetooth, barcode detection, accelerometers, and other mobile devices features are primarily used individually, without possessing the ability to optimize other applications capable of adapting to a given context. For example, it is not easy to implement voice, stylus, or keyboard applications on a phone, where the user can choose to act and speak freely, due to the lack of standards available to achieve this goal.

Mobiles devices and phones, originally created for vocal interaction between humans are nowadays being used to interact with machines. Some mobile phones devices are adapted with user interface features that are dedicated to aiding the partially sighted and blind persons, while others are used to smoothly interact with everyday smart objects such as refrigerators or interactive TV using a ubiquitous and multimodal adaptive interaction. A person using an iPhone with one finger pressure can triggers an application in charge of decoding 2D barcodes simply by aiming the camera of his phone. The barcode displayed on a poster against the wall for example, can be instantly mapped and displayed on the phone screen that shows the location of the restaurant requested. The user can zoom in and navigate within this map. A link between the physical and the digital world is made.

In this chapter, we will follow the development process of a scientific experiment, in which a mobile application will be used in order to determine which modality (touch, voice, QRcode) is preferred for entering expiration dates of alimentary products. The "Pervasive Fridge" project (Rouillard, 2012) will be our case study context towards the design and deployment of a prototype application. In this project, we argue that mobile smartphones could be helpful in helping users to avoid wasting food. The idea relies on two key moments: (1) people use their own smartphone to scan product and enter the expiry dates as they store them in a fridge or cupboard, and (2) later, the system notifies the user just before the product expiration date occurs. In this paper, we are working on the first key moment, and we are searching what is the best way to enter expiration date, when employing a multimedia smartphone.

2.2. Needs for our Project

For our project, we need to provide to users an application running on various kind of smartphone under Android and able to manage various modalities. As illustrated on Figure 1, we have several different devices available such as Galaxy Tab, Acer Liquid, Acer stream and ZTE Link Android, in order to try different size and kinds of smartphone and tablets.

The technological needs for our project are important, and the final application has to manage many features such as barcode scanner, voice recognition, calendar widget, connection to distant resources, timer, and area hidden for end-users, but available for experimenters.

Figure 1. Android smartphones, tablets, and QR-codes for expiration dates of products

2.2.1. Barcode Scanner

In our project, we need to manage smartphone's camera in order to obtain the barcode of each product, encoded with EAN13 barcodes, and the expiration date of the product, encoded with QRcodes, especially by us for our experiment. On android smartphone, one of the most used barcode recognition tool is Zebra Crossing, from Google (ZXing, 2009), a multi-format 1D/2D barcode image processing library.

EAN13 is a one-dimensional type of barcode. EAN stands for "European Article Numbering." It is employed to access data related to billions of products carrying EAN-13 barcodes (ISO/IEC 15420:2000), and was invented by George J. Laurer in 1973 (EAN13, 1973) (See Figure 2).

QRcodes are used to encode and decode data at a rapid rate. QR stands for "Quick Response" as the creator intended the code to allow its contents to be decoded at high speed. A QRcode is a two-dimensional barcode introduced by the Japanese company Denso-Wave in 1994 (Denso-Wave 1994). This kind of barcode was initially used for tracking inventory in vehicle parts manufacturing and is now used in a variety of industries. The use of camera phones to read two-dimensional barcodes for various purposes is currently a popular topic in both research and in practical applications.

The QRcode represented in Figure 3 is encoding the date "17/04/2012", and was generated with the Kaywa online generator (Kaywa, 2012).

Figure 2. Scanning EAN code of a product (milk here)

Figure 3. Expiration date of a product (QRcode: 17/04/2012)

Figure 4. Calendar date picker

2.2.2. Voice Recognition

In our project, we want users to (try to) pronounce expiration dates of products, thanks to the smartphone's microphone. The Automatic Speech Recognition (ASR) is needed for this kind of service, which uses the Internet network to send (to a particular server) the raw speech file and receive back the best voice recognition result in a textual format. The advantage of this "cloud" approach is that (1) it gives very good results, (2) without the need of a training phase, and (3) it is not necessary to declare a grammar before using that kind of speech recognition. The drawbacks of this solution are that: (1) as circumscribed grammars are not declared, ASR's results are not restricted to a specific type or domain, such as currency, date, integer, etc., and (2) obviously, the system must be connected to the Internet network with a good bandwidth, otherwise the results will not be satisfying.

2.2.3. Calendar Date Picker

In our project, we also want to display a calendar and provide the ability to choose a date, by scrolling and picking it from a calendar widget, as shown of Figure 4.

Because App Inventor does not provide calendar tools for the moment, we choose to use an external application provided by Jefferson Software (Activity Starter DatePicker, 2011) for this purpose. This will be a nice illustration of activity starter managed by App Inventor (called here "asDP" for "activity starter Date Picker").

2.2.4. Connection to Distant Website

As our application will request information from other services on the Internet network, we need a connection mechanism able to send and receive data. Prixing services (Prixing, 2011) will be reached across their API, in order to get product details according to a barcode and a personal PHP Web site (000Webhost.com, for instance) (000Webhost, 1997) will be used to store our metric data (durations, statistic data, etc.) that will be analyzed later.

2.2.5. Timer

In our project, we need to measure elapsed time and to determine how long it takes to enter expiration dates of products with calendar, voice, or QRcode. This information will not be displayed on the smartphones and users will not be aware that timers are measuring their interactions.

2.2.6. Shake Detection

During the experiments, at the beginning of each user session, the experimenter would have to enter a secret zone, that is not directly reachable for end-users, and he/she will type in the number of the session, and then hide this area, before giving the smartphone to the next user. We have searched a smart way to have a private zone, which is not available for other users during normal interaction, and reachable with a specific interaction, only known by experimenters, in this case by shaking the smartphone for a while.

3. DEVELOPING WITH APP INVENTOR

We could have used HTML5 for developing our application, because it's a W3C standard for cross platform applications, but HTML5 is not yet a stable version and some features using media and physical components (like vibration, camera, video, etc.) are not treated in the same way, according to the used platform and browser. Other tools such as WinDev are available but sometimes very expensive and not always completely suitable for development on mobile devices.

In a fast prototyping approach, we choose App Inventor (App Inventor 2010) as a mobile application development tool that could be used by students and researchers not completely fluent in Java (Wolber et al., 2011). Originally unveiled by Google in July 2010 and now supported by MIT since early 2012, Google's App Inventor for Android (AIA) is a visual "blocks" programming language that offers relatively simple tools for conceiving, developing, testing, deploying and sharing Android mobile applications (App Inventor, 2010). AIA is still in beta version, and there is already a list of known issues that will be fixed in the final release.

This section will talk about App Inventor; how we can design mobile application with it, using visual blocks instead of classical textual computing language.

3.1. Presentation of App Inventor

From a simple Web interface, it is possible to design some applications with a WYSIWYG editor (See Figure 5), just by dragging some elements, adding buttons, menus, icons or placing some media on the interface area. Then, for the programming part, the developer is able to manage visual blocks (See Figure 9, Figure 8, Figure 10), representing the expected behaviors according to

Figure 5. AppInventor online editor

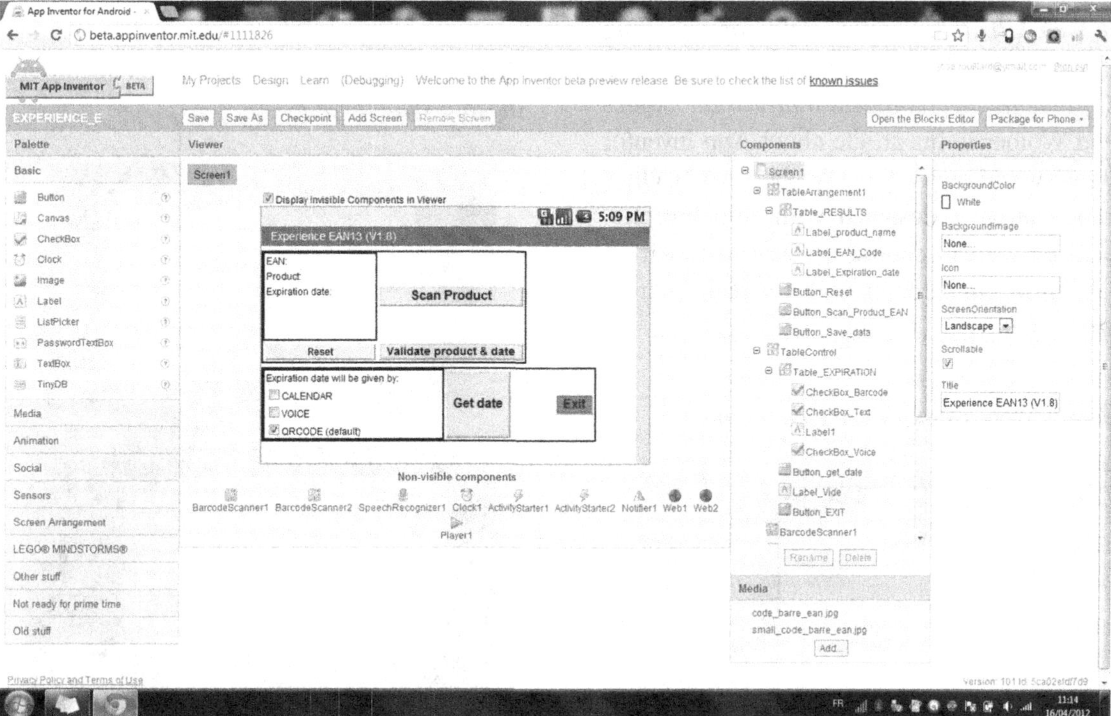

various events. The resulting application can be seen under the App Inventor Android emulator (See Figure 6), but also, on a real Android smartphone connected to the PC via USB. With this last solution, it is very interesting to see, in real time, that every modification made in the editor is directly reflected on the application running on the smartphone, without going through a phase of compilation, for example. It is also possible to export the final project via an Android application package file (APK) format, to download the package to the computer or to show the barcode associated to the generated package. This last feature is only working with a developer Gmail account.

3.2. Designing with App Inventor

Designing with App Inventor is used since a few years for introducing students and developers, to the creation of mobile applications though a visual programming approach. "*App Inventor provides high-level components for process incoming SMS texts, interfacing with the GPS location sensor of the phone, scanning barcodes, and communicating with Web APIs. The mobile world was exploding and here was this visual language the let beginning students be creative in that world.*" argues David Wolber in his article about App inventor and real-world motivation (Wolber, 2011). Once an application is designed with App Inventor, the developer can choose to see and test it on an Android emulator or under a real Android smartphone.

Figure 6. AppInventor device emulator

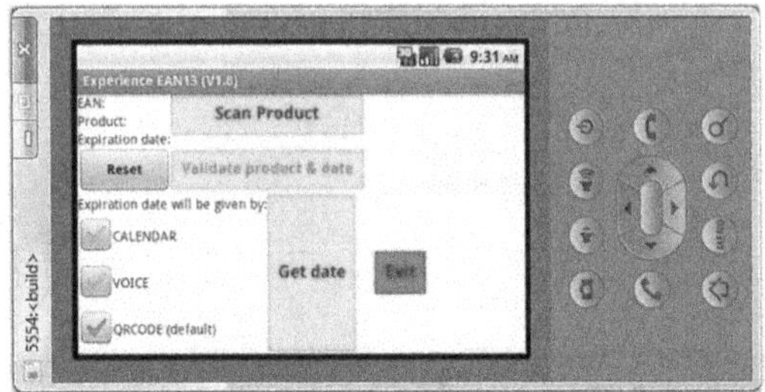

3.2.1. On Emulator

Figure 6 is a capture of the application developed for our experiment, tested with the emulator provided by App Inventor. With this tool, it is possible to click with the PC mouse on the buttons and other widgets presented on the emulator.

Actions are triggered and executed normally, except those depending on physical resources, such as camera or microphone, for instance. It is possible to choose the orientation of the application screen or to let it being controlled dynamically by the smartphone. For our project, we have fixed this orientation to the "landscape" mode.

3.2.2. On Real Connected Android Device

Figure 7 is a photo of the same application running on two different Android ACER smartphones. As previously illustrated on Figure 1, we have deployed this application on various different devices (Galaxy Tab, Acer Liquid, Acer stream, and ZTE Link Android), without additional changes.

Figure 7. Our application running on two different Android smartphones

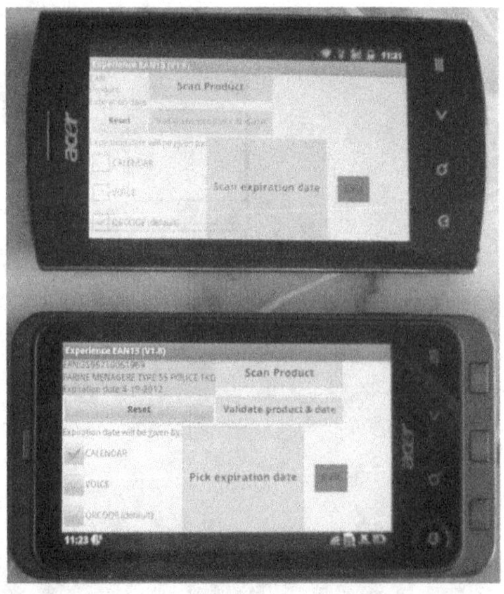

Figure 8. Details of a block (button click)

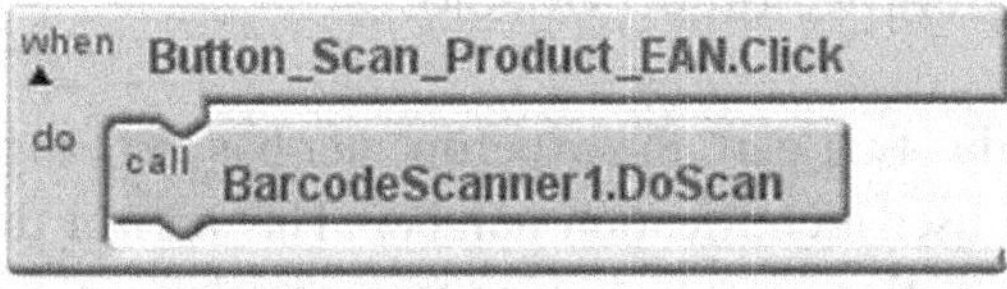

Figure 9. Details of a block (barcode scanner)

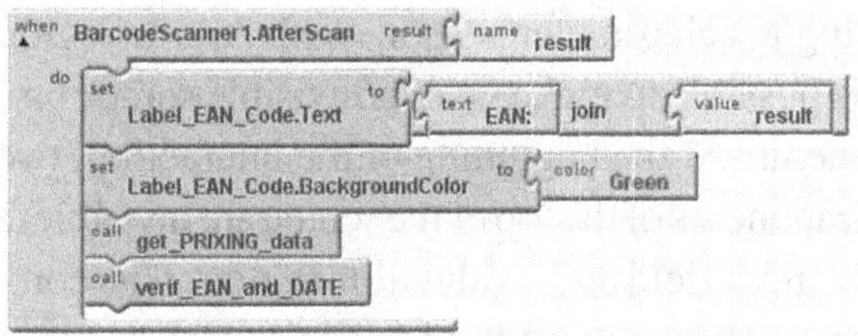

On the top of the Figure 7, we can see that the QRcode choice is checked, and that the large main button is labeled as "Scan expiration date." On the bottom of the Figure 7, the same button's label changed for "Pick expiration date," because the user selected the calendar option. We also can see that the three main fields (EAN, Product, Expiration date) are filled or not, according the data provided by the user during the interaction.

A step between designing and deploying applications is needed, in order to give the rules to follow and actions to do, for each event. In App Inventor, this programming phase is not done with a textual language, but with a visual block editor tool, as explained in the following.

3.3. App Inventor Blocks

As easy as assembling pieces of Lego™, it is possible, thanks to App Inventor, to build an application entirely in a visual manner, and without writing a single line of code (Java in this case). Figure 8 is an example of elementary blocks that can be joined in order to obtain a more complex one. In this particular case, a click on the "But-

Figure 10. App inventor global view of our application

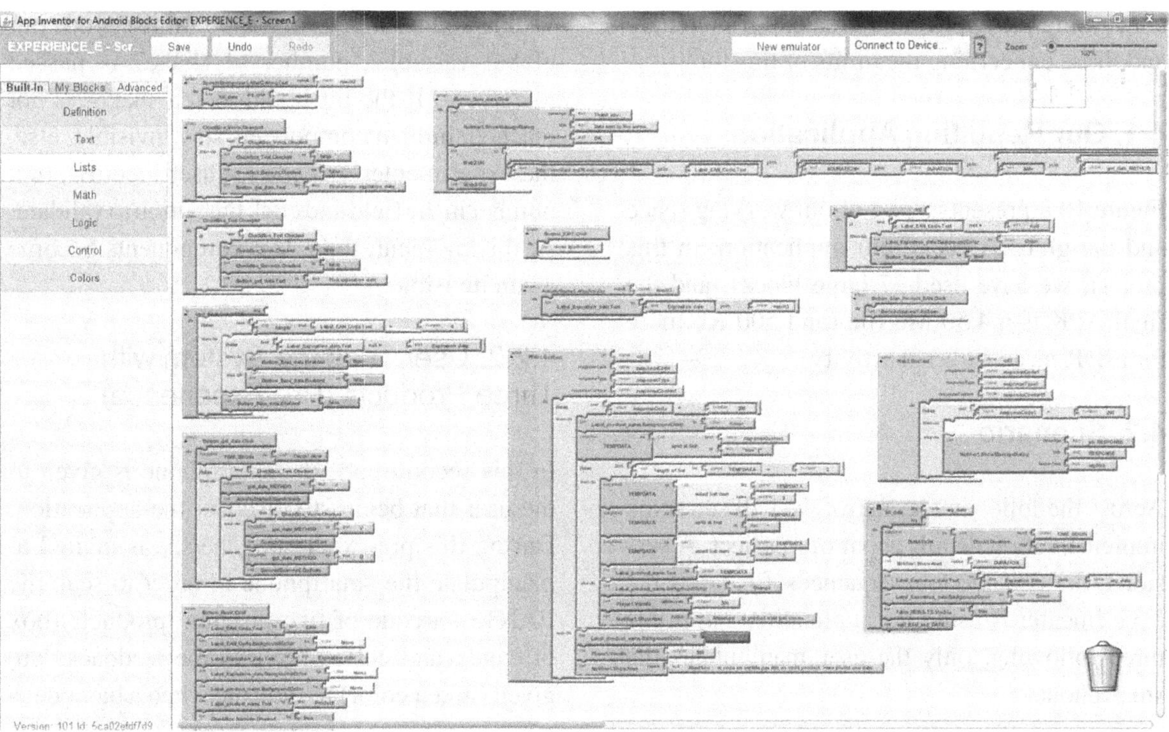

ton_Scan_Product_EAN" component will call the "DoScan" routine of the "BarcodeScanner1" component.

On Figure 9, we can see how the result, obtained after the scan of a barcode, will be employed, as an argument (named "result" here), and joined to the text "EAN:" in order to change dynamically the label of a component named "Label_EAN_Code". The color label will then be changed in green, and two subroutines will be called sequentially: "get_PRIXING_data" and "verif_EAN_and_DATE."

4. RESULTS

Thanks to App Inventor, it was relatively easy to design and test different versions of our application, within a few hours. What is really interesting with that tool is the fact, that, unlike with compiled languages, here, each change on the designer Web side is directly visible and usable on the connected smartphone side.

We are now going to talk about the results concerning the application created for our study and those concerning the limits of this tool.

4.1. Our Resulting Application

Figure 10 represents a part of our working space and the global view of our application. In this project, we have used 27 large blocks, and the final APK is a 4.50 Mo file (and 500 Ko more for asDP, as an external activity).

4.2. Scenario

Across the following scenario, we will see some implementation details about our project. A session is divided into four sequences: first, only the experimenter is using the application and in the three following, only the user manipulates the smartphone.

4.2.1. Preparation of the Experimentation Session

In this first part, the experimenter has to change the user identification number. This part of the interaction must be hidden from the classical users and we need a mechanism in order to show and hide some elements of the application. To do so, we have decided to use a shake method using accelerometer sensor. With App Inventor, it is possible to make visible/invisible some components. At the beginning of the interaction, two components of the top of the screen are invisible: a text input field and a validation button. This component senses the Android device's accelerometer, which detects shaking and measures acceleration in three dimensions. A clock component is also used in this mechanism. In the blocks editor, we are measuring if the user (experimenter) is shaking the smartphone for at least 5 seconds. This is done to prevent involuntary shakes. An integer variable is increased each time a shake is detected. If this variable exceeds the threshold of ten, then, the input text field and the validate button become visible and usable. Every 5000 milliseconds our global variable "number_of_shake" is passed to zero, so if the threshold of ten shakes is not reached, hidden components stay invisible, else, the experimenter can enter a user identification number in the field and click the button to validate. At this moment, these two components become again invisible.

4.2.2. User Tries the System with Three Products (Pre-Experience)

In this second part, the smartphone is given to the user that begins to discover the application. During this pre-experience, he/she is invited to manipulate the smartphone in order to scan the EAN13 barcode of two different products (box of candy and coffee box). Some feedbacks are given (green color) to indicate when a barcode is correctly identified. Then, thanks to a non-visible component (called "Web") that provides functions for HTTP GET and POST requests, the application calls the Prixing API with a parameter (the barcode number) in order to get several details about the product (name, price, composition...). If the product is identified, then its name is indicated in a label with a green background.

Then, the user has to try the three different manners to enter the limit date of the product: by a touch modality, with the calendar widget shown of Figure 4; by pronouncing the date and by scanning a QRcode encapsulating the deadline date.

The Figure 11 shows a capture of the block used for the management of the "button_get_date. Click" component. First, this method records, in a global variable, the time (Clock1.Now), captured when the user clicks the button. After that, the global variable "get_DATE_METHOD" is set to P, V or Q and respectively, an ActivityStarter, a speech recognition or a barcode scanning is launched, depending on the results of the "if... else" sub-blocks.

Once a date is obtained by one of these modalities, the button "validate product and date" becomes clickable and the user can press on it to send the appropriate data (measures) to the server. The user has to validate three modalities for a product, then, he/she can click and the button "reset" before entering data the next product. After that stage, the real experiment begins.

Figure 11. The button_get_date.click block

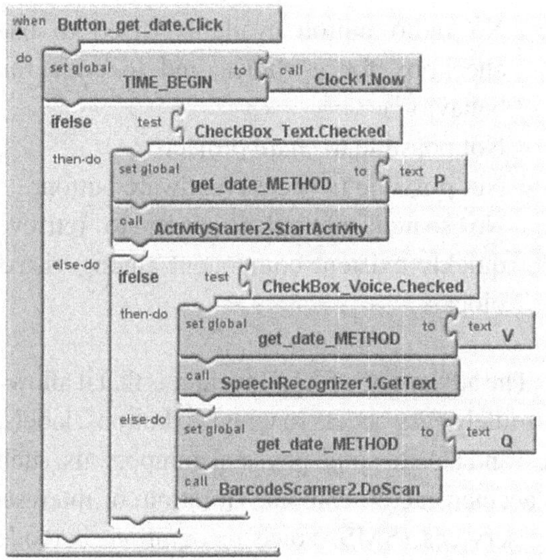

4.2.3. Kernel of our Experiment with Eight Products

In this part of the experiment, people use the system with eight different products and the three modalities. Each time a user clicks on the button that allows entering the date (calendar, voice, QRcode), a timer is initialized; and stopped automatically, after the end of this interaction. Elapsed times (in milliseconds) are sent to the server at the end of each interaction.

4.2.4. Preferred Modality (Post-Experience)

In the third and last part of the experiment, the users are asked, after scanning the barcode of the products, to employ a single modality, at their convenience, in order to enter the deadline of three other products.

4.2.5. Stored Data

The scientific results obtained thanks to this application have been submitted to a conference in a paper that presents a comparative study of products expiration dates entered on smartphone, according to three interaction possibilities: touch (calendar), voice and QRcodes encapsulating the relevant deadline. 87% of users prefer to use QRcodes for entering expiration dates. The measurements collected during our experiments confirm these figures: QRcode is faster than voice and calendar (6.8 seconds against 7.3 and 15.2 seconds). This indicates that mobile applications could help users avoid wasting food by facilitating the entry of product data via QRcodes specially designed for this purpose.

4.3. Outcome

The experiment involved 11 men and 4 women (mean age 30.6 years, SD = 7.68) with 67% of them use a smartphone daily or often, while the rest use it occasionally, rarely or never. Respondents have a very similar frequency of use of speech recognition and QRcodes. For speech recognition 47% never use it, versus 53% rarely or occasionally. For QRCodes 53% never use it, 40% rarely or occasionally, and 7% often or daily; 60% of people have never scanned product with their phone. Concerning input modalities, 93% of users considered that the calendar input technique was not fast (versus 7% of "fast" or "very fast"). Oppositely, 73% perceived the use of voice as fast or very fast (27% somewhat or not at all fast) against 93% for QRcodes (7% somewhat or not at all fast). Moreover, QRcodes were considered the fastest way to enter the date (80% for QRcodes, 13% for voice and 7% for the calendar). In contrast, the calendar was elected slower with 67% against 33% for voice and 0% for QRcodes.

QRcodes are also given as the most efficient (in terms of error) for date input (73% against 27% for the calendar and 0% for voice). In contrast, the voice is perceived as the least effective with 60% against 33% for the calendar and 7% for QRcodes.

After the experiment, 60% of people say they are ready to use such an application to prevent food waste, 27% maybe and 13% not. 40% think that this kind of application can avoid food waste, 40% maybe and 20% not. For these kinds of applications, 87% of users believe that QRcodes are the best way to enter dates (7% for voice, 0% for the calendar, and 7% by other means not specified in the questionnaire). Average speeds measured for input deadline (of 8 products in the kernel of the experience) are as follows (in milliseconds) for the methods P (Picker Calendar), V (Voice) and Q (QRcode) (See Table 1).

All deadlines obtained by QRcodes or calendar are exploitable in the context of a system to fight against waste with proactive notification.

Table 1. Average speeds measured for input deadline (P = picker calendar; V = voice; Q = QRcode)

Method	P	V	Q
Mean	15241	7353	6841
SD	8170	5146	3967
Min	2736	3453	2948
Max	53748	38896	30445

However, some of the dates obtained by speech recognition would be difficult to use without a proper format transformation.

4.4. Limitations

As we have shown, App Inventor is not a toy language and it can help in building sophisticated apps for Android smartphones. It really can be used to build complex applications with classical loops and tests. There are mechanisms for communicating with Web services and databases.

Obviously, there are also some limitations. Krishnendu has noticed in his paper some things that work and some other that need improvement (Krishnendu, 2012). App Inventor main capabilities and limitations are known and documented (Wolber, 2012). During our project, we have particularly observed the following:

- No radio button available (need to use checkbox for a choice and to code the behavior[1]).
- Not possible to share projects.
- Not possible to access hardware button.
- No search method in order to retrieve quickly existent component among visual blocks.

The advantage of App Inventor is that it allows extremely easy access to widgets (buttons, labels, etc.) and also to some physical components, such as accelerometer, camera, vibration, or interesting services (SMS, voice recognition, barcode scanner, etc.).

The main drawback for us was the limitation of the workspace. When the blocks become important and numerous, it is very difficult to navigate among them, and the reorganizing function does not really helps. Moreover, even if we have the opportunity to comment the blocks, this feature adds some graphic content and ultimately the illegibility remains. It is also difficult to reuse existing projects and not possible to export projects into Java code, for instance. Another limitation exists for applications developed with App Inventor: the impossibility to be published into the Android Market, limiting the distribution of the app.

5. CONCLUSION

We have demonstrated how App Inventor can be used to create a mobile application for Android smartphones in a green IT approach, and for the needs of a scientific experiment. This application was really simple to conceive and deploy. We tested it by changing quickly and simply various components (with/without vibration, portrait or landscape orientation, with/without color feedbacks, etc.).

Our main requirements were oriented toward the multimodality and the connectivity of the mobile application needed in our case study, guided by the desire to understand what kind of modalities are used by people in the context of this project against uneaten food loss.

Many features have been incorporated smoothly in our mobile application, such as barcode scanner, voice recognition, calendar date picker, Web connections, timers, and shake detection. We have conducted some tests on the emulator of App Inventor and also on real connected smartphones and tablets, and it was very pleasant to see dynamically and concretely, the look and feel on the real time manipulated application.

In conclusion, we have seen more benefits than drawbacks in using App Inventor for our scientific project. The obtained final prototype has been used for an experiment with 15 users and the results will be soon proposed for publication in a conference or a journal, around pervasive and ubiquitous computing. Some limitations of this framework are known, but it still remains a useful tool to rapidly prototype small mobile applications. In future, a component development kit is being considered. This will allow programmers to build App Inventor components with Java and expand the functionality of App Inventor.

ACKNOWLEDGMENT

We are grateful to the French National Agency ANR MOANO project for providing support for this research, and to the Prixing Company for special API provided for this study.

REFERENCES

Activity Starter DatePicker. (2011). *Jefferson software, version 0.2.1*. Retrieved October 2012 from https://play.google.com/store/apps/details?id=com.jsoft.android.util

AgilePlatform. (2012). Retrieved October 2012 from http://www.outsystems.com/agile-platform/

App. Inventor. (2010). Retrieved October 2012 from http://www.appinventor.mit.edu/

Denso-Wave. (1994). *Creator of QRcode*. Retrieved October 2012 from http://www.densowave.com/qrcode/aboutqr-e.html

EAN13. (1973). *George J. Laurer, ISO/IEC 15420:2000 - Information technology – Automatic identification and data capture techniques - Bar code symbology*. EAN/UPC.

FA0. (2011). *Global food losses and food waste*. Düsseldorf, Germany: Interpack2011. Retrieved October 2012 from http://www.fao.org/docrep/014/mb060e/mb060e00.pdf

Kaywa. (2012). Retrieved October 2012 from http://qrcode.kaywa.com/

Krishnendu, R. (2012). App. inventor for android: report from a summer camp. [Raleigh, NC: ACM.]. *Proceedings of SIGCSE, 12*, 283–288.

MotoDev. (2012). Retrieved October 2012 from http://developer.motorola.com/

PhoneGap. (2012). Retrieved October 2012 from http://www.phonegap.com/

Prixing. (2011). Retrieved October 2012 from http://www.prixing.fr

Rouillard, J. (2009). Multimodal and multichannel issues in pervasive and ubiquitous computing. In Kurkovsky, S. (Ed.), *Multimodality in Mobile Computing and Mobile Devices: Methods for Adaptable Usability* (pp. 1–23). Hershey, PA: Idea Group. Inc. doi:10.4018/978-1-60566-978-6.ch001.

Rouillard, J. (2012). *The pervasive fridge: A smart computer system against uneaten food loss*. Paper presented at the Seventh International Conference on Systems, ICONS 2012. Saint Gilles, Reunion Island.

Schadler, T., McCarthy, J. C., Brown, M., Martyn, H., & Brown, R. (2012). *Mobile is the new face of engagement – An information workplace report*. Retrieved October 2012, from http://www.forrester.com/Mobile+Is+The+New+Face+Of+Engagement/fulltext/-/E-RES60544?objectid=RES60544

000. *Webhost*. (1997). Retrieved October 2012 from http://www.000Webhost.com

Wolber, D. (2011). App. inventor and real-world motivation. In *Proceedings of the 42nd ACM Technical Symposium on Computer Science Education*, (pp. 601–606) Dallas, TX: ACM.

Wolber, D. (2012). *App. inventor capabilities and limitations*. Retrieved June 2012 from http://www.appinventor.org/capabilities-limitations

Wolber, D., Abelson, H., Spertus, E., & Looney, L. (2011). *App. inventor: Create your own android apps*. O'Reilly Media.

ZXing. (2009). *Zebra crossing*. Retrieved October 2012 from http://code.google.com/p/zxing

ENDNOTES

1 http://code.google.com/p/app-inventor-for-android/issues/detail?id=113

Section 2
Sensors-Based Interactivity

This section focuses on achieving an engaging and intuitive interaction by leveraging mobile devices' multimodal sensors.

Chapter 5
Contextual and Personalized Mobile Recommendation Systems

Jitao Sang
Chinese Academy of Sciences, China

Changsheng Xu
Chinese Academy of Sciences, China

Tao Mei
Microsoft Research Asia, China

Shipeng Li
Microsoft Research Asia, China

ABSTRACT

Mobile devices are becoming ubiquitous. People are getting used to using their phones as a personal concierge to discover what is around and decide what to do. Mobile recommendation therefore becomes important to understand user intent and simplify task completion on the go. Since user intents essentially vary with users and sensor contexts (time and geo-location, for example), mobile recommendation needs to be both contextual and personalized. While rich user mobile data is available, such as mobile query, click-through, and check-in record, there exist two challenges in utilizing them to design a contextual and personalized mobile recommendation system: exploring characteristics from large-scale and heterogeneous mobile data and employing the uncovered characteristics for recommendation. In this chapter, the authors talk about two mobile recommendation techniques that well address the two challenges. (1) One exploits mobile query data for local business recommendation, and (2) one exploits mobile check-in record to assist activity planning.

INTRODUCTION

Mobile devices are becoming ubiquitous these days, which revolutionizes the way people surf information and make decisions. While on the go, people are using their phones as a personal concierge discovering what is around and deciding what to do. Therefore, mobile device has become a recommendation terminal, making it important to understand user intent to simplify task completion on the go.

DOI: 10.4018/978-1-4666-4054-2.ch005

However, understanding user intent on the go is not trivial. On one hand, user intent is implicit and complicated – sometimes user does not have a very explicit intent in mind, but just wants some suggestions. In this case, user modeling is needed by analyzing user history activities and summarizing their behavior patterns. On the other hand, user intent changes with users and sensor contexts (time and geo-location, for example). It is easy to understand that mobile recommendation should be customized to individuals to meet the personalized requirements. More importantly, user intent is not static but sensitive to contexts – users have different information needs under different circumstances. Therefore, mobile recommendation needs to be both contextual and personalized.

The widespread use of mobile devices has offered opportunity to gain insights on user mobile activity and behavior patterns. Rich mobile data is available, including cellular data (Gonzalez, Hidalgo, & Barabasi, 2008), mobile query log (Zhuang, et al., 2011), click-through history (Nicholas, et al., 2010), check-in record (Anastasios, et al., 2011), etc. There involved two challenges in understanding the complicated user intent to provide contextual and personalized recommendation services. (1) How to explore meaningful characteristics of user behavior patterns from the large-scale, noisy and heterogeneous mobile data, and (2) How to employ the discovered characteristic to perform efficient recommendation customized to contextual and personalized information needs.

In this chapter, we will introduce two mobile recommendation techniques that well address the prior two challenges: one exploits mobile query data for local business recommendation and one exploits mobile check-in record to assist activity planning.

1. **Mobile Query Log-Oriented Local Business Recommendation:** From data analysis on a large-scale real-world mobile query database, we found that much query is related to local business and mobile search is usually personalized and context-aware. The observations motivate a contextual and personalized POI (Point of Interest, i.e., local business) recommendation approach, which consists of three key components: (1) POI crawler which collects POIs with attributes (e.g., cuisine for restaurant) from the Web, (2) POI extraction which detects and recognizes POIs from a query or click-through, and (3) POI ranking which ranks the POIs in a context-sensitive and personalized way without requiring any input. Specifically, a probabilistic entity ranking algorithm is proposed to model the generating probability of a POI by the user conditioned on the mobile context.
2. **Check-in Record-Oriented Activity Planning Assistant:** User activity usually consists of a series of actions, where each action can be referred to one POI. Compared with recommending single POI, when user planning activities, it is more desired to recommend a series of consecutive POIs. In the second work, we move one step beyond to address the problem of serial POI recommendation for activity planning assistant. The rise of online social networks has provided a useful mobile data source - check-in record, where location data, user activity and relationships are available. Based on the analysis of a real-world check-in database, we found significant POI category transition characteristic in consecutive check-in actions. The proposed approach estimates the transition probability from one POI to another, conditioned on current context and check-in history in a Markov chain. To alleviate the context discritization error and sparsity problem, context collaboration and prior information are integrated into the probabilistic formulation.

Based on the two proposed techniques, real applications have been developed on Window Phone 7 devices. The interfaces and application scenarios are illustrated in Figure 1.

MOBILE QUERY LOG-ORIENTED LOCAL BUSINESS RECOMMENDATION

As shown in Figure 1(top), the local business recommendation system works as follows. A user named "clark" gets the recommendation of POI categories in (b) when he was in the context of (a), and then can check the ranked POIs in (c) by clicking a specific POI category (e.g., nearby restaurants that favor his taste in the POI category of restaurant). The POI category and the POIs within each category are ranked according to the relevance to user and sensory context. In this section, we will introduce inspiring mobile query log analysis and the implementation of the system.

Mobile Query Log Analysis

We collected a large-scale query log data from a commercial mobile search engine. The time range of query database is from 2009-09-30 to 2010-03-

Figure 1. The interfaces of two contextual and personalized mobile recommendation systems: (top) local business recommendation, (bottom) activity planning assistant

28. All these queries were conducted in United States. During the six months, the number of raw queries is up to 75,221,037, which were issued by a total of 13,711,497 users, corresponding to 417,895 queries per day. Each query log consists of "user id," "time," "GPS location," "query," and "URL." We show the results of data analysis in Figure 2 and conclude the data characteristics in the following.

Figure 2(a) shows the detailed #query in eight cities. We can see that *mobile search is becoming pervasive, especially in big cities.* Figure 2(b) shows the number of queries (#words) with different lengths. We observe that #query decreases sharply with #word. About 62.3% of the queries contain less than three words (three excluded), and 76.8% queries contain less than four words. On average, each query contains 2.52 words. Figure 3(c) shows the distribution of queries with different length measured by #letter. Each query contains 18.76 letters on average. These results confirm the assumption that *search queries on mobile platform are usually short.* Users are not willing to type long sentences on the keyboard of very limited size. Therefore, it would reduce user's effort significantly if we can design systems to avoid time consuming user-phone interaction.

Figure 2(d) shows #query in different time slot (of the day). The highest peak occurs near to 5--6 pm, when the search on mobiles tends to be active. One possible reason is that people are about to get off work. The lowest point occurs at about 2-3 am. This is reasonable as few people are active in the midnight. The previous observation implies that *mobile search is usually context-aware.* Using POI extraction technique, we can detect the POI query, i.e., the queries target at searching POIs. The ratio of #POI query to #raw query is 15.28% (the ratio depends on the recall of an POI extraction algorithm). This verifies that search on *mobile is local and POI-oriented.*

The prior characteristics of mobile query motivate the design of a contextual and personalized recommendation system. Due to the abundant attributes of extracted POIs (as shown and Figure 1), a user can get a quick glance of his information need without typing a query explicitly.

Figure 2. The statistics of data analysis in the collected mobile query database

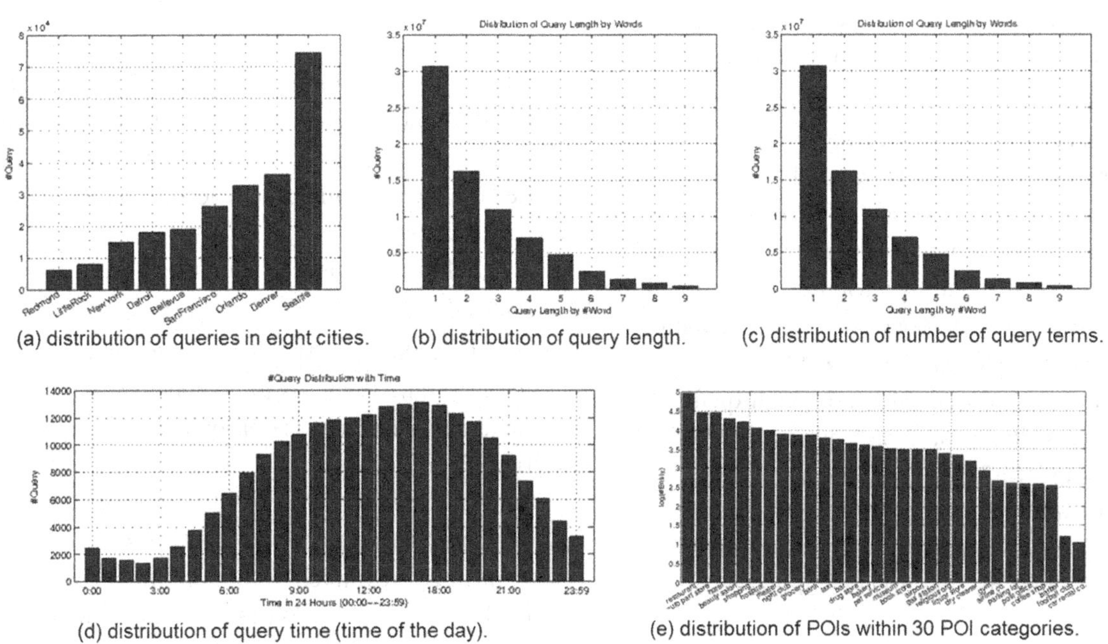

(a) distribution of queries in eight cities. (b) distribution of query length. (c) distribution of number of query terms.

(d) distribution of query time (time of the day). (e) distribution of POIs within 30 POI categories.

Figure 3. The accuracy of the examined methods on POI and POI category ranking

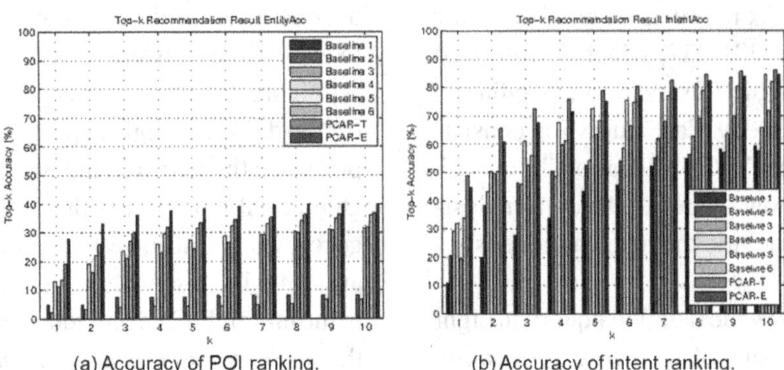

(a) Accuracy of POI ranking.

(b) Accuracy of intent ranking.

Approach

In this section, we introduce the proposed recommendation approach, which consists of two key components: (1) POI extraction which detects and recognizes POIs from a textual query log, and (2) POI ranking which ranks a candidate set of POIs and the corresponding POI categories to the user. We pay more attention to the POI ranking algorithm.

We build a database of extracted POIs from several major local business sites on the Web. The extraction algorithms are similar to the previous POI extraction system *KnowItAll* (Oren, et al., 2005), where the key techniques are rule-based. That is, the first step of POI extractor is to automatically create a collection of extraction rules for each kind of POI category. Beyond of the extraction rule of POIs, we make a further step to extract the attributes of POIs as well. This step aims to get a high recall of POIs and the corresponding attributes. After obtaining the initial extracted ones, we can pass them to a search engine to retrieve more POIs. Then a pattern learner is employed to filter out high-quality POIs for expansion. In summary, we extract 5,457,192 POIs under 37 POI categories.

Besides the common features, each POI category has a unique set of attributes. For example, a restaurant has the price level and cuisine type as the major features. Due to the limited space, we do not introduce the detailed features of each category here. After obtaining the results by POI extractor, we can design the query parser which maps a query to a specific POI. This can be implemented by supervised classification techniques or topic models (Guo, et al., 2009).

By using POI extractor and query parser, we can obtain a query database $Q = \{q_1, \ldots, q_m\}$, where m is the number of mobile queries containing POIs in our database. Each mobile query is a 5-dimensional tuple:

$$Q = < O, C, U, L, T >$$

where $O \in \mathcal{O}$ is the POI searched in Q, C indicates the POI category of O, $U \in \mathcal{U}$ is the user who conducted the query, L and T are the context, i.e., the location and time when Q was generated, respectively.

Each mobile user U consists of a history of queries: $U := \mathcal{Q}_u = \{q_1, \ldots, q_u\}$. Given a user $u \in U$ in the context $< l, t >$, the task POI ranking is to rank the POIs in $\mathcal{O}$ such that the higher ranked POI has a larger probability of being queried by u.

We consider the probability of generating the POI o_j by user u_i under the context $< l, t >$. The POIs in $\mathcal{O}$ are ranked by the conditional probability $P(O \mid U : L, T)$. We assume the final ranking score of a specific entity is proportional to this conditional probability. To this end, we introduce a variable C to indicate the user's intent. For example, if C indicates "eat" for the time being, probably the user intends to go to a restaurant. With this intuition, the possible choice of C is determined by the number of POI categories N_C. We deem $P(O \mid C)$ as the popularity of some O under the category C, and $P(C \mid U)$ proportional to the frequency of querying C by U. We have the conditional probability.

$$P\left(o_j|u_i;l,t\right) = \sum_{k=1}^{N_C} P\left(e_j|c_k;l,t\right) P\left(c_k|u_i;l,t\right)$$

Due to the limited query history of U, it is common that some POI categories have never been queried by U. Assigning a zero conditional probability of these POI categories is apparently not proper. To overcome this problem, we leverage the idea of collaborative recommendation techniques (Gediminas, A. & Alexander, T., 2005). That is, we introduce the query record of other users to help estimate the probability. If other users similar to u_i have queried an POI before, u_i also has the possibility of searching for the POI. Thus we have:

$$P\left(o_j|u_i;l,t\right) = \sum_{k=1}^{N_C} P(o_j \mid c_k;l,t) \sum_{n=1}^{N_u} P\left(c_k \mid u_n;l,t\right) P(u_n \mid u_i)$$

The transition probability $P(u_n \mid u_i)$ is proportional to the similarity $s(u_n, u_i)$ between them, where $s(\cdot,\cdot) : \mathcal{U} \times \mathcal{U} \rightarrow \mathbb{R}^+$ measures the user similarities. It is intuitive to model the probability of search intent using $P\left(c_k \mid u_n;l,t\right)$ and POI popularity $P(o_j \mid c_k;l,t)$ directly by counting frequencies in the collected dataset. To instantiate the user similarity function, we represent each as a query history record. Depending on how to represent users with the query records, two types similarity function can be adopted here: *POI-based* and *POI-category-based*. Each user is converted into a fixed-length vector according to term frequency-inverse document frequency (Fabrizio, S., 2002). Here 'document' implies user U, 'term' indicates either POI or POI category.

As shown in Figure 1, the returned POIs are organized by POI categories. Besides ranking POIs, we also need to rank the categories as the left diagram in Figure 1 to facilitate user's browsing purpose. To this end, we consider the scheme akin to the probabilistic framework of POI ranking:

$$P\left(c|u;l,t\right) \propto P(c \mid u;l,t)s(u_n, u)$$

We organize the POIs by ranked categories with ranking score $P(C \mid U; L, T)$. The POIs under each category are sorted by $P(O \mid U; L, T)$.

Experiment

We build the user similarity graph based on the mobile query data from the first five months, while adopting the data in the last one month for test. We randomly selected 2,000 users and use their queries in the March of 2010 as test data, which contains a total of 58,111 query records. We denote this test set by $\mathcal{Q}_t$.

For each test query record $q \in \mathcal{Q}_t$, we use its location l and time t as the search context and call the system to recommend the POIs to its user u. We split the time into 7 intervals: 0:00-6:00, 6:00-8:00, 8:00-12:00, 12:00-13:00, 13:00-18:00, 18:00-20:00, 20:00-24:00. We extract a set of queries from the query database with the context $< l,t >$. Then we sort the POIs contained in these queries by POI ranker. To this end, we extract the queries conducted within five kilometers to u and fallen into the same time interval as t.

Let $\mathcal{O}_t$ denote the set of the ranked POIs, we can measure the top-*k* accuracy of $\mathcal{O}_t$ by counting the position $\grave{\lambda}(q.O)$ of q.O in $\mathcal{O}_t$. Formally the top-*k* accuracy of the whole test set $\mathcal{Q}_t$ is computed by:

$$\text{Accuracy}\left(\mathcal{Q}_t, \text{k}\right) = \frac{\sum_{q \in \mathcal{Q}_t} \mathbb{I}(\grave{\lambda}(q.O) \leq k)}{|\, \mathcal{Q}_t \,|}$$

where $\mathbb{I}(c)$ returns 1 if *c* is true and 0 otherwise.

We measure accuracy of two kinds of recommendation:

- **EntityAcc:** the top-*k* accuracy of the recommended POIs.
- **IntentAcc:** the top-*k* accuracy of the recommended POI categories. It reflects the intent of the users at the query location.

The recommendation results have a two-level organization (refer to Figure 1). According to the definition, *EntityAcc* should always be less than *IntentAcc*. The abbreviation of the examined schemes are summarized as follows:

Baseline 1 $< l >$**:** Recommend POIs according to their distance to the user's current location, and rank them by distance to the user.

Baseline 2 $< l, p >$**:** Recommend POIs as *Baseline 1*, but rank them by the popularity pp.

Baseline 3 $< l, p, t >$**:** Recommend POIs as *Baseline 1*, but rank them by their popularity in current time slot.

Baseline 4 $< l >$**:** Recommended POIs that have been queried in current location, and rank them by their distance to the user.

Baseline 5 $< l, p >$**:** Recommend POIs as *Baseline 4*, but rank by popularity.

Baseline 6 $< l, p, t >$**:** Recommend POIs that have been queried in both current location and time slot, then rank by popularity.

PCAR-T $< u, l, t >$**:** The Personalized Context-Aware POI Ranking (PCAR) algorithm proposed in this paper. It is essentially a hybrid recommendation algorithm. The tf-idf representation of users for building user similarity graph is POI-category based.

PCAR-E $< u, l, t >$**:** The same as PCAR-T except the tf-idf representation of users are POI-based.

It is worth noticing that *Baseline 1-3* and *Baseline 4-6* can be categorized into two groups: the first group only considers the POIs whose physical locations are close to user's current location, while thc second only considers the POIs which are queried in user's current location.

Figure 3 shows the recommendation accuracy of the evaluated schemes. We can draw several observations from this figure. First, among all the baseline methods 1-6, baseline 4-6 reports significantly better results than baseline 1-3. Note that baseline 1-3 deem all the existing POIs as candidates when recommending to users, while baseline 4-6 only use the POIs that have been queried in current context as candidates. Therefore, we conclude POIs used to be searched are more likely to be searched again. Baseline 5-6 are slightly better than baseline 4, which implies: (1) popularity is more important than geometric distance, and (2) time is an effective context for recommendation.

Second, PCAR produces significantly higher accuracy than all the Baseline methods. PCAR not only considers the popularity of POIs through $\text{P}(\text{O} \mid \text{C})$, but leverages the information of each individual user through the generating probability $\text{P}(\text{C} \mid \text{U})$ and the transition probability $\text{P}(\text{C} \mid \text{C}')$. Besides the context.

The entity-oriented search depends on the user in nature. For example, when searching a restaurant, different users may have different cuisine preference. People who like Chinese food may conduct totally different queries comparing with people who favor Western food. Therefore, it is important to make the preference personalized. The proposed PCAR system yields the accuracy of 10% higher than that of baselines. Thus we conclude that personalization exists in POI-oriented search.

Third, comparing the two user representation schemes, *PCAR-E* is better than *PCAR-T* for POI recommendation, while *PCAR-T* is better than *PCAR-E* for category recommendation. It implies that the tf-idf representation for users when computing user similarity graph is task-dependent. The top-10 *EntityAcc* is 40% and *IntentAcc* approaches to 90%. This high accuracy confirms the efficacy of the proposed probabilistic entity ranking algorithm. It also verifies the importance of taking context and personalization into account.

CHECK-IN RECORD-ORIENTED ACTIVITY PLANNING ASSISTANT

As shown in Figure 1(bottom), the activity planning assistant system works as follows. A user named "Emily" checks in a shopping mall at the context of (1), and gets recommendation of (2) sequential POI categories (e.g., café → restaurant) and (3) the corresponding packages of sequential POIs when selecting a specific POI category sequence in (4). The recommendation of sequential POI categories and POIs rather a single POI is a metaphor of "plan activities in real life" and thus more natural to mobile users. Users will have much less interactions with the phone to complete their tasks.

Check-In Record Analysis

We resort to exploit the real-world check-in data for activity planning assistant, as it provides a concise recording and rich context of people status. Compared with traditional mobile data like query log, and click-through history, check-in record differentiates itself in two-fold. (1) Instead of requiring a POI extraction step, the venue where a user chooses to check-in usually corresponds to a specific POI. The check-in POI is a good indicator of the activity being conducted. For example, user's check-in at a Chinese restaurant implies his/her food preference. Therefore, check-in data naturally bridge user intent and the behaviors in the physical world. (2) Different from mobile browsing log data in which users have no intent to broadcast, in check-in services, users voluntarily share their status and make them publicly available. These two characteristics of check-in data make it a promising source for user activity modeling. Recently, check-in data have been used in various applications, including popular POI identification (Alves, Rodrigues, & Pereira, 2011), user activity recognition (Lian, & Xie, 2011), friendship prediction (Adam, Henry, & Jeffrey, 2012), and user localization (Eunjoon, Seth, & Jure, 2011).

We collected a check-in dataset from Dianping[1], the check-in Website which cooperates with China's biggest microblog, Sina Weibo[2]. The collected check-in records are located at Beijing and Shanghai, from Jan 7 to June 11 in 2011. Each check-in record R consists of user identification U, check-in time T, GPS latitude and longitude L, check-in POI O, and the POI category C. The overall number of check-in records is 152,154, issued by 2,342 unique users. For the duplicate records where two consecutive check-in POIs are identical, we removed the latter. After filtering, we obtained 128,937 records in total, with 55 average records per user. 32,891 unique POIs were identified from these records, resulting in 24 POI categories with public facilities, entertainments, life services, and so on. The distribution of check-in POI categories is shown in Figure 4(d). In the following, we first analyze the properties of check-in data, and then discuss key observations on these data.

We first investigate the temporal interval between two consecutive check-in records, i.e., the inter check-in time (Anastasios, et al., 2011). Figure 4(a) shows the distribution of inter check-in time. It can be seen that some users intended to submit consecutive check-ins within a short period. Since it is unusual for one user conducting

Figure 4. The statistics of data analysis in the collected check-in record database

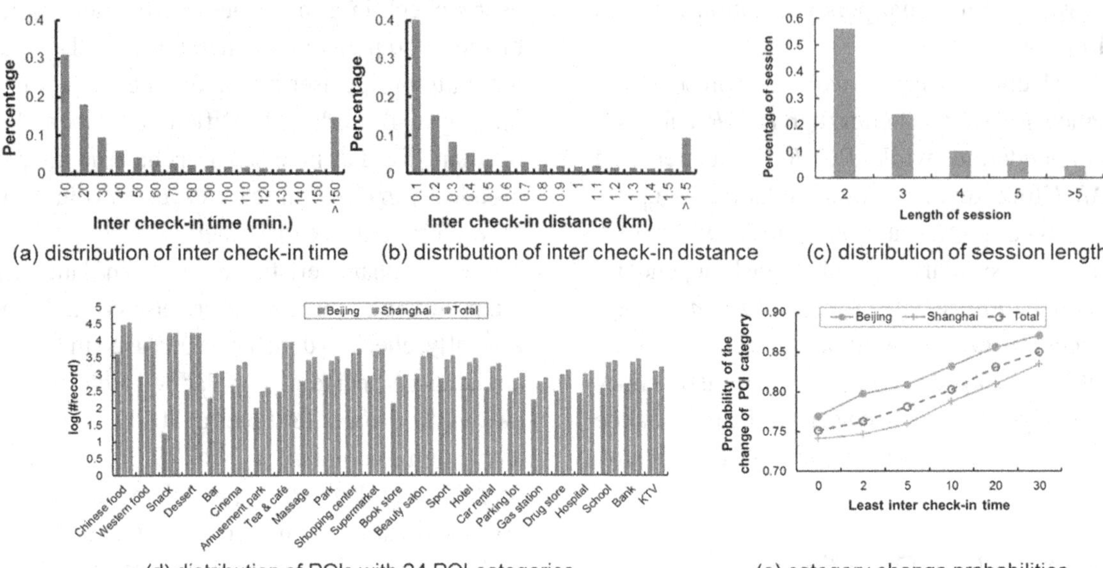

(a) distribution of inter check-in time (b) distribution of inter check-in distance (c) distribution of session length

(d) distribution of POIs with 24 POI categories (e) category change probabilities

different actions in a very short time, most of these check-ins are for fun or for the sake of benefit, and thus do not accord with real user behaviors (Lian, & Xie, 2011). On the other hand, if inter check-in time between two records is very long, it is assumed that the latter check-in action has little relation with the former. In this work, we use *session* to represent an activity with a series of consistent check-in actions. Based on the previous discussions, we empirically set the range of inter check-in time as [10 min, 120 min]. Each session can be represented as:

$$S := < U, T_1, L_1, O_1, C_1, \cdots, T_{N_s}, L_{N_s}, O_{N_s}, C_{N_s} >$$

where N_S is the length of session, T_k, L_k, O_k and C_k denote the time, location, POI name, and POI category of the k-th record in the session S. As a result, 13,078 sessions are extracted. The distribution of session lengths is shown in Figure 4(c)[3]. We can observed that around 80% sessions containing two and three check-in records, which demonstrates that a typical "activity planning" consists of more than one but less than four activities. For example, activity of gathering friends may include the succeeding actions of eating, massage, and going to a bar. Since the task of this work is to predict POIs within a session, we will focus on data analysis of sessions in the following. By aggregating the inter check-in time within individual sessions, the estimated average inter check-in time $\overline{T}_{inter} = 38.9$ min.

We introduce inter check-in distance to investigate the geographic distance between two consecutive check-in records from unique user. Figure 4(b) plots the distribution of inter check-in distance within a session in Shanghai. It shows that the inter check-in distances span a long range from 0.1 km to 1.5 km. The average inter check-in distance $\overline{D}_{inter} = 0.33$ km. The estimated POI density in Shanghai is about 361 per km^2, which means if we attempt to recommend the next POI within the range with the radius $\overline{D}_{inter}$ (centered at the current POI), the number of next POI candidates will be 108 $\left(\text{À} \times 0.33^2 \times 261\right)$. This provides a good understanding for the challenge of sequential POIs recommendation problem. However, note that this is only a coarse estimation. In practice, the inter distance between consecutive actions is related to moving speed and travel vehicles. It would be more reasonable not to restrict the POI candidates within a fixed area. Therefore, the actual candidate number could be much larger.

To understand the user behavior associated with the check-in records, the category of the check-in POI is very important as POI category carries more abstract activity information. For sequential POIs recommendation problem, it is interesting to investigate how different POI categories succeed to each other in a session. We first investigate the probability of the change of POI category. The results for Beijing, Shanghai and the overall collected dataset are shown in Figure 4(e). We can see that all the three curves ascent with the increase of inter check-in time. If we set the least inter check-in time to 10 min, the overall category change probability is around 80%, indicating that four out of five check-in records have different POI categories with the previous one. This observation indicates *significant transition patterns at the POI category level*, which much motivates the proposed solution.

Typically users will continually conduct a series of activities, e.g., visiting a snack after shopping, going to a gas station or parking lot after visiting a car rental. It is easy to understand that category transition within a single session follows certain patterns. We examine the transition from category i to j as the probability that, within a session s, the POI category of former and latter records are i and j.

In particular, we first investigate the category transition patterns in different time intervals. We split the time into 14 time intervals in a weekday and weekend[4]. The category transition matrices for four selected time intervals are shown in Figure 5(a), where the depth bar on the right shows the corresponding probability values. We can see that time intervals follow very different category transition patterns and implies that *category transition is time-sensitive*. We further investigate the category transition patterns in different locations. The category transition matrices for Beijing and Shanghai are shown in Figure 5(b). We can see that the transition patterns are much different for these two cities. For example, in Beijing, the

Figure 5. The category-transition patterns for different contexts and users

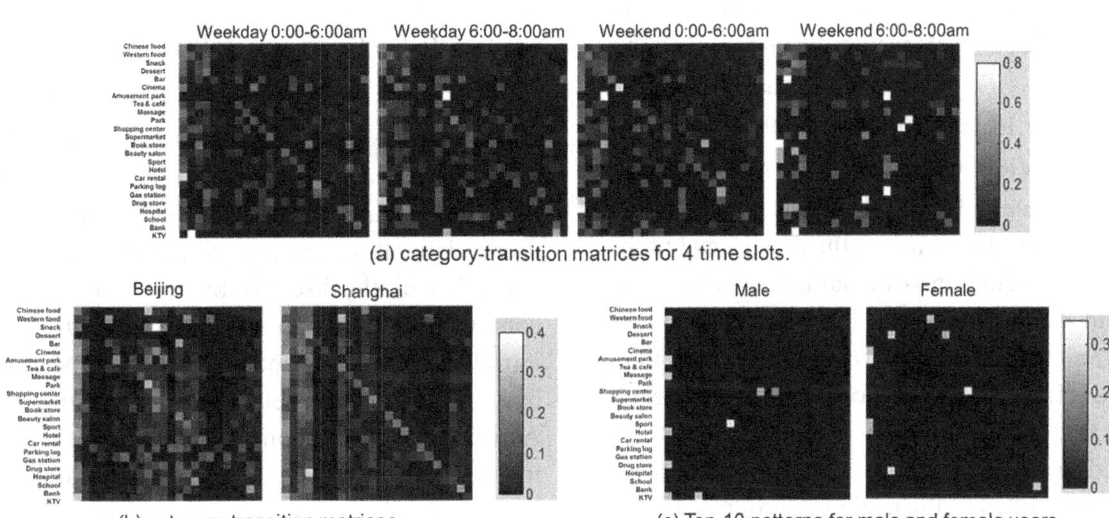

(a) category-transition matrices for 4 time slots.

(b) category-transition matrices.

(c) Top 10 patterns for male and female users.

transition pattern of *western food* → *bar* (2, 5) is significant, while in Shanghai, preferred POI category after *western food* is replaced by *tea & café* (2, 8). Therefore, we can see that *category transition is also location-sensitive*.

Top transition patterns indicate the typical sequential activities of users. Figure 5(c) plots the 10 most popular category-transition patterns for female users and male users. The differences of popular transition patterns do not surprise us as female and male users usually conduct very different sequential actions. This observation implies that *category transition is gender-sensitive* or in a more general concept - *personalized*.

The previous observations of check-in records inspire the design of a contextual and personalized sequential POIs recommendation. In the following, we introduce a category-transition centric approach by explicitly considering context as well as user collaboration.

Approach

In this section we present the details of the proposed probabilistic Markov recommendation approach. We first introduce the probabilistic ranking formulation, and then provide the implementation for context and user collaboration.

We consider the recommendation problem from a ranking perspective. Specifically, the recommended POI sequences are ranked by $P(O_2, O_3, \cdots, O_N \mid O_1; U, L, T)$, where U, L, T, O_1 denote user, location, time, and POI of the current check-in action, $O_2, \cdots, O_N$ is the recommended POI sequence including (N-1) POIs. Three reasons are considered to compute the conditional probability based on Markov assumption: (1) according to the session length distribution in Figure 4(c), most user activities contain no more than three actions; (2) in practical applications, user's real action can be incorporated into the recommendation cycle; (3) investigating 1-order transition patterns can fully exploit the collected dataset. Following the Markov assumption, the conditional probability is formulated as:

$$P\left(o_2, \cdots, o_N \mid o_1; u, l, t\right)$$
$$= P(o_2 \mid o_1; u, l, t) \prod_{k=2}^{N-1}$$
$$P(o_{k+1} \mid o_k; u, l\left(o_k\right), t + (k-1)\overline{T}_{inter})$$

Note the new location is estimated by location of the candidate POI and the new time is estimated by current timestamp plus the average inter check-in time $\overline{T}_{inter}$.

Our data analysis shows distinct category transition patterns in sequential check-in actions. Inspired by this, we propose a category transition-centric formulation and assume the probability of visiting next-POI o_j is determined by two components: (1) transition probability - the probability of conducting action denoted by o_j's category $c(o_j)$ after the current check-in POI o_i, and (2) POI popularity - the probability of choosing o_j given its category $c(o_j)$. Therefore, the target 1-order POI transition probability can be decomposed as:

$$P\left(o_j \mid o_i; u, l, t\right)$$
$$= P\left(c\left(o_j\right) \mid o_{;u}, l, t\right) P\left(o_j \mid c\left(o_j\right); u, l, t\right)$$

$$\propto P\left(c\left(o_j\right) \mid c\left(o_i\right); u, l, t\right) \; P(o_j \mid c(o_j); u, l, t)$$

We can see the approximated transition probability in the second step is actually a category-transition probability. We also utilize the method of user collaboration for calculation here. Moreover, to tackle with the discretization error along with context granularity (Cristiana, et al., 2007), we introduce context collaborative recommendation by considering the correlations between contexts. Assuming the generation of users and contexts are independent, we have the following formulation for the category-transition probability and POI popularity:

$$P\left(c\left(o_j\right) | c\left(o_i\right); u, l, t\right) \propto \sum_{u_k; l_m; t_n} P\left(c\left(o_j\right) | c\left(o_i\right); u_k, l_m, t_n\right) a(u_k, u) a(l_m, l) a(t_n, t)$$

$$P\left(o_j | c\left(o_j\right); u, l, t\right) \propto \sum_{u_k; l_m; t_n} P\left(o_j | c\left(o_j\right); u_k, l_m, t_n\right) a(u_k, u) a(l_m, l) a(t_n, t)$$

where $a\left(\cdot, \cdot\right)$ is a measurement function for the similarity between users or contexts. The conditional probability $P(c(o_j) \mid c(o_i); u_k, l_m, t_n)$ and $P(o_j \mid c(o_j); u_k, l_m, t_n)$ can be computed by directly counting frequencies in the collected check-in dataset.

Two correlations are considered to formulate the similarity function: prior correlation and history correlation. History correlation is easy to understand that the users' or contexts' similarity can be represented by the history check-in records of the users or within the contexts. We use the prior correlation to model the prior information for users and contexts, where user prior information includes their gender and residence, time prior information indicates the temporal partition for different time intervals and location prior means the geometrical information.

We utilize heuristic methods for user and context prior similarity calculation. We define that two users have high prior similarity if they are of the same gender or from the same residence. The prior similarity between two locations is inversely proportional to their distance, while two time intervals have high prior similarity if they are adjacent within day or both from weekend or weekday. Note that alternative complex measurement of prior information can also be explored.

Since transition probability and POI popularity focus on category transition and POI probability inside category, we construct different functions for the computation of their history similarities. Specifically, for transition probability history similarity, we represent each user or context by category-transition history for transition probability similarity, and check-in record history for POI popularity similarity. Each user or context is then converted into a fixed-length vector according to tf-idf. Here "document" indicates user, location area or time interval, while "term" refers to either a category transition pattern $< C_i, C_j >$ or a POI O.

Similar to the local business recommendation, besides recommending sequential POIs, another related problem is to recommend sequential POI categories, which will narrow the POI candidates and facilitate users with a hierarchical recommendation service. Sequential POI categories recommendation is relatively easy as it can be viewed as a sub-problem of sequential POIs recommendation. Based on similar Markov assumption, the contextual and personalized POI category transition probability is calculated as:

$$P(c_k \mid o; u, l, t) \propto P(c_k \mid c(o); u, l, t)$$

where c_k is the recommended POI category in the k-th step, o is current check-in POI, and c(o) is the category of POI o.

Experiment

We conducted experiments on the collected check-in data of Shanghai, which include \$105,463\$ unique check-in records, 11,483 check-in sessions from 1,114 users. The number of POIs used in evaluation is 25,503, which were used as candidates for recommendation in test set. We randomly selected 311 users and their check-in sessions from May 21 to June 11 in 2011 as test data. The test set contains 1,265 check-in sessions, denoted by S_t.

For each test check-in session s, we use the location $s.L_1$ and time $s.T_1$ of its first check-in record as the context and employ the recommendation approach to recommend POIs based on user identification s.U, current check-in POI $s.O_1$ and POI category $s.C_1$. We utilize top-*k* accuracy as the evaluation metric. We consider evaluating the performance on situations of recommending one and two POIs. All these 1,265 test sessions $\left(N_S \geq 2\right)$ constitute the test set S_{t1} for recommending one POI. 317 sessions out of the 1,265 test sessions contain no less than three check-in actions $\left(N_S \geq 3\right)$, which constitute the test set S_{t2} for recommending two POIs. Denoting the candidate POI set for recommending two POIs as $\mathcal{O}_{t2} = \mathcal{O}^2$, $\grave{A}(s.O_2,\ s.O_3)$ indicates the position of $s.O_2$, $s.O_3$ in $\mathcal{O}_{t2}$ according to probabilistic POIs ranking. Then, the top-k accuracy of recommending two POIs is computed by

$$\text{Accuracy}\left(\mathcal{S}_{t2}, \mathcal{O}_{t2}, \text{k}\right) = \frac{\sum_{s \in \mathcal{S}_{t2}} \mathbb{I}(\grave{A}(s.O_2,\ s.O_3) \leq k)}{|\,\mathcal{S}_{t2}\,|}$$

We compared the performance among the following recommendation schemes:

- **Co-Occurrence Based (*CO*):** recommending candidates that most co-occur with current POI, which is implemented according to (James, et al., 2010).
- **POI-Transition Based (*PT*):** POI transition centric recommendation without considering user or context information, which can be regarded as a one-order implementation of (Jose, & Mark, 2010).
- **Context and Personalized POI Transition Based (*CPPT*):** contextual and personalized POI transition centric recommendation, which directly computes the conditional probability $P(O_2 \mid O_1; U, L, T)$.
- **Context and Personalized POI Category Transition Based (*CPCT*):** POI category transition centric recommendation, which is our proposed approach in this work.
- ***CPCT_context*:** POI category transition centric recommendation without considering context collaboration.
- ***CPCT_prior*:** POI category transition centric recommendation without considering prior information in constructing similarity functions.
- ***CPCT_checkin*:** POI category transition centric recommendation without considering current check-in record, which ranks based on $P(O_2, \cdots, O_N \mid U, L, T)$.

Figure 6(a) shows the top-k accuracy for recommending one POI. Several observations can be drawn as follows. (1) Top-9 accuracy of two baseline methods, *CO* and *PT* is only around 25%. *CO* and *PT* recommend POI based on general inter-POI relationship and ignore the sensitivity to context and user,

which results in inferior performances. This result demonstrates the need for contextual and personalized methods. (2) *CPPT* has the worst performance among the evaluated methods. This can be explained by two reasons: First, due to the large number of POI, the POI transition matrix is extreme sparse and it is impractical to explicitly calculate POI-level transitions. (3) *CPCT_context* produces lower accuracy than *CPCT*. *CPCT_context* only considers collaboration for the user factor and does not leverage context collaboration, which subjects to the sparsity issue and fails to make full use of the data. (4) Another alternative, *CPCT_prior*, also achieves inferior performance than CPCT. Employing the prior information can reduce the impact of context discretization. (5) The fact that CPCT significantly outperforms *CPCT_checkin* validates the advantage of utilizing the current check-in record. User's current action indicates the activity he/she is conducting or will conduct, which is important for user intent understanding and activity plan.

Figure 6. The recommendation accuracy for the examined approaches

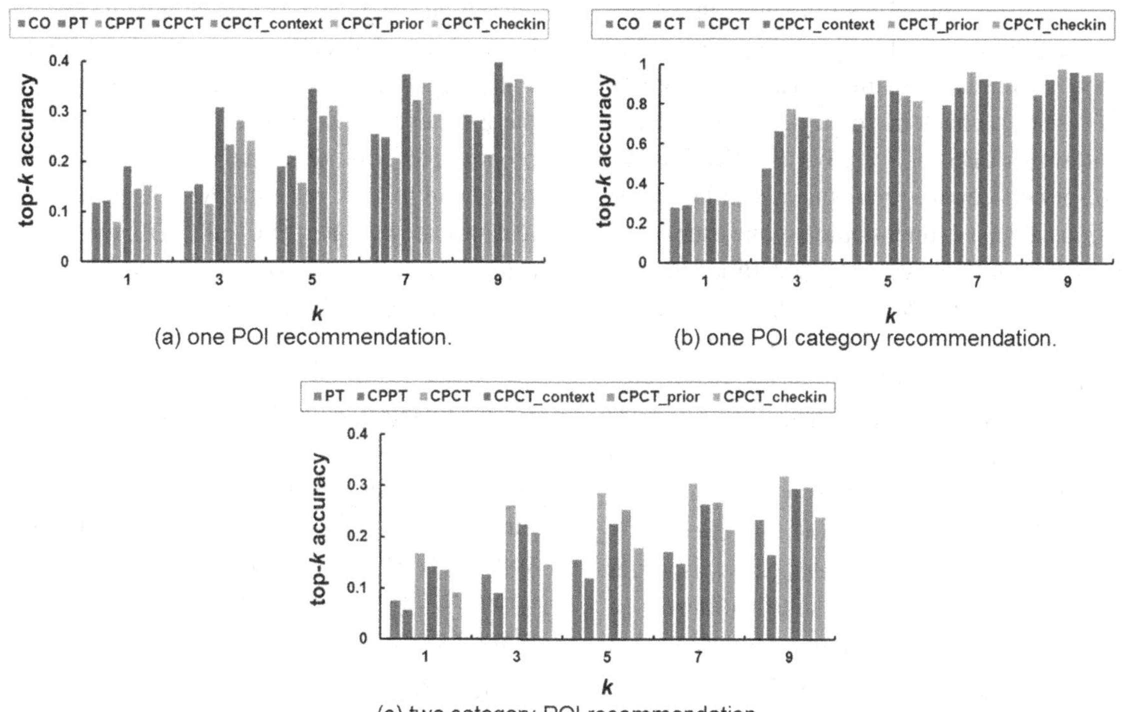

(a) one POI recommendation.

(b) one POI category recommendation.

(c) two category-POI recommendation.

We also evaluate the performance of POI category recommendation. Since category recommendation involves no POI transition, we remove the evaluated method of *CPPT* and replace *PT* with category-transition centric recommendation method (*CT*). The results are shown in Figure 6(b). It is observed that the top-k accuracy of POI category recommendation follows similar patterns with that of POI recommendation. The proposed method consistently outperforms the baseline methods. The top-3 accuracy of *CPCT* approaches 77%, which demonstrates its efficacy.

Along the two-level organization as shown in Figure 1(b), we are interested to the accuracy of recommending POI sequence when user chooses the desired category sequence $s.C_2$, $s.C_3$ in the first level. We denote its top-k accuracy as category-POI accuracy, which is shown in Figure 6(c). Note the number of candidate POI sequences $|\mathcal{O}^2|$ dramatically reduces to $N_{s.C_2} \times N_{s.C_3}$. The top-9 of *CPCT* is around 30%, which means for one out of three trials, the nine top recommended POI sequences may include the users' actual choices for the following two actions. To summarize, the experimental results validate the effectiveness of the proposed approach for recommending one and two POIs and leaves room for improvement in the future work.

DISCUSSION

From mobile data analysis, we also found some phenomenons not easy to explain. For example, the category transition probabilities for *Chinese food* and *Western food* are only around 75%. This indicates that 1/4 of users eat at consecutive Chinese or Western restaurants within a short time, which looks not reasonable. Two interesting discussions are from these findings. (1) *We need to detect check-in spams*. Where there is reward, there is spam. Due to the limited positioning accuracy and the large number of POI, current check-in service cannot differentiate a spam from a normal check-in. Therefore, check-in spam detection is very important for LBS services. We note that investigation on consecutive check-ins will initiate some solutions, e.g., it is impossible for one user going to very distant POIs within a short time interval. (2) *We need to separate visit and pass-by check-in behaviors*. Assuming one user did go to a POI, there are still two types of check-in behaviors: visit and pass-by. Visit check-in indicates the user paid a real visit to the POIs, while pass-by means the user just went through the POIs without entering. It is easy to understand only the visit check-in is valuable. Observations from this paper also offer some clues for visit check-in identification, e.g., it is not common eating at different restaurants, but it is possible passing by different restaurants in a short time. Besides, the visit check-in identification is category-dependent. For example, visiting dozens of shopping centers succeedingly is very common for most shopaholics.

The future research directions for developing contextual and personalized mobile recommendation systems include: The future works include: (1) considering more attributes of the recommended items (e.g., price range, ratings, comments of POI, etc.), (2) leveraging social signals for better understanding user preference and thus improving recommendation performance, and (3) integrating more context for recommendation, e.g., weather, user status (walking, driving, etc.).

CONCLUSION

Mobile data analysis plays important roles in designing practical mobile systems. In this paper, we first conduct data analysis on large-scale mobile query data and check-in records collected from commercial mobile application platforms. Our observations indicate that the (1) mobile queries are typically short, context-aware, and local; (2) check-in category transition is common, context-sensitive, and user-sensitive. Motivated by these observations, probabilistic POI ranking is proposed for solutions. The proposed approaches is capable to rank relevant POI categories (e.g., restaurant, hotel, bar, etc.) and POIs within each category (e.g., "I love sushi" and "MacDonalds" in the category of"restaurant").

REFERENCES

Adam, S., Henry, A. K., & Jeffrey, P. B. (2012). Finding your friends and following them to where you are. In *Proceedings of the Fifth ACM International Conference on Web Search and Data Mining* (pp. 723-732). ACM.

Alves, A. O., Rodrigues, F., & Pereira, F. C. (2011). Tagging space from information extraction and popularity of points of interest. In *Proceedings of the 2nd International Conference on Ambient Intelligence* (pp. 115-125). IEEE.

Anastasios, N., Salvatore, S., Cecilia, M., & Massimiliano, P. (2011). An empirical study of geographic user activity patterns in foursquare. In *Proceedings of the 5th International AAAI Conference on Weblogs and Social Media* (pp. 570-573). AAAI.

Cristiana, B., Carlo, C., Elisa, Q., Fabio, A. S., & Letizia, T. (2007). A data-oriented survey of context models. *SIGMOD Record*, *36*(4), 19–26. doi:10.1145/1361348.1361353.

Eunjoon, C., Seth, A. M., & Jure, L. (2011). Friendship and mobility: user movement in location-based social networks. In *Proceedings of the 17th ACM SIGKDD International Conference on Knowledge Discovery and Data mining* (pp. 1082-1090). ACM.

Fabrizio, S. (2002). Machine learning in automated text categorization. *ACM Computing Surveys*, *34*(1), 1–47. doi:10.1145/505282.505283.

Gediminas, A., & Alexander, T. (2005). Toward the next generation of recommender systems: A survey of the state-of-the-art and possible extensions. *IEEE Transactions on Knowledge and Data Engineering*, *17*(6), 734–749. doi:10.1109/TKDE.2005.99.

Gonzalez, M. C., Hidalgo, C. A., & Barabasi, A. (2008). Understanding individual human mobility patterns. *Nature*, *453*(7196), 779–782. doi:10.1038/nature06958 PMID:18528393.

Guo, J. F., Xu, G., Cheng, X. Q., & Li, H. (2009). Named entity recognition in query. In *Proceedings of the 32th Annual International ACM SIGIR Conference* (pp. 267-274). ACM.

James, D., Benjamin, L., Liu, J., Palash, N., Taylor, V. V., & Ullas, G. Dasarathi, S. (2010). The YouTube video recommendation system. In *Proceedings of the 4th ACM Conference on Recommender Systems* (pp. 293-296). ACM.

Jose, B., & Mark, L. (2010). A comparison of scoring metrics for predicting the next navigation step with Markov model-based systems. *International Journal of Information Technology and Decision Making*, *9*(4), 547–573. doi:10.1142/S0219622010003956.

Lian, D. F., & Xie, X. (2011). Collaborative activity recognition via check-in history. In *Proceedings of 3rd ACM SIGSPATIAL International Workshop on Location-Based Social Networks* (pp. 45-48). ACM.

Nicholas, D. L., Dimitrios, L., Zhao, F., & Andrew, T. C. (2010). Hapori: Context-based local search for mobile phones using community behavioral modeling and similarity. In *Proceedings of the 12th International Conference on Ubiquitous Computing* (pp. 109-118). IEEE.

Oren, E., Michael, J. C., Doug, D., Ana-Maria, P., Tal, S., & Stephen, S. et al. (2005). Unsupervised named-entity extraction from the Web: An experimental study. *Artificial Intelligence*, *165*(1), 91–134. doi:10.1016/j.artint.2005.03.001.

Sawyer, S., & Tapia, A. (2005). The sociotechnical nature of mobile computing work: Evidence from a study of policing in the United States. *International Journal of Technology and Human Interaction*, *1*(3), 1–14. doi:10.4018/jthi.2005070101.

Zhuang, J. F., Mei, T., Steven, C. H., & Li, S. P. (2011). When recommendation meets mobile: Contextual and personalized recommendation on the go. In *Proceedings of the 13th International Conference on Ubiquitous Computing* (pp. 153-162). IEEE.

ENDNOTES

1. http://www.dianping.com/
2. http://weibo.com/
3. We only consider sessions with no less than two check-in records.
4. We treated weekday and weekend separately, as the activities on weekday and weekend are different.

Chapter 6
Gamifying Everyday Activities using Mobile Sensing

Zachary Fitz-Walter
Queensland University of Technology, Australia

Dian Tjondronegoro
Queensland University of Technology, Australia

Peta Wyeth
Queensland University of Technology, Australia

ABSTRACT

The addition of game design elements to non-game contexts has become known as gamification. Previous research has suggested that framing tedious and non-motivating tasks as game-like can make them enjoyable and motivating (e.g., de Oliveira, et al., 2010; Fujiki, et al., 2007; Chiu, et al., 2009). Smartphone applications lend themselves to being gamified as the underlying mobile technology has the ability to sense user activities and their surrounding environment. These sensed activities can be used to implement and enforce game-like rules based around many physical activities (e.g., exercise, travel, or eating). If researchers wish to investigate this area, they first need an existing gamified application to study. However if an appropriate application does not exist then the researcher may need to create their own gamified prototype to study. Unfortunately, there is little previous research that details or explains the design and integration of game elements to non-game mobile applications. This chapter explores this gap and shares a framework that was used to add videogame-like achievements to an orientation mobile application developed for new university students. The framework proved useful and initial results are discussed from two studies. However, further development of the framework is needed, including further consideration of what makes an effective gamified experience.

DOI: 10.4018/978-1-4666-4054-2.ch006

INTRODUCTION

Over the last few years, a growing interest in both research and industry has explored how the engaging aspects of video games can be translated to non-game contexts to make them more motivating. This has become known as *gamification* (Deterding et al., 2011). An increasing number of software and Web applications are integrating elements of games, for example; achievements, points, and competitive leaderboards have been added to the location-sharing application *foursquare*[1] to reward players for visiting physical locations; game aesthetics and language have been used to style the to-do list application *Epic Win!* [2] to look and feel like a role-playing game; and in the online application *Chore Wars*[3], a dungeons and dragons style game has been created to make housework more motivating (See Figure 1).

These applications have included game elements as a way to engage users who may not find an activity incredibly motivating to begin with, such as completing housework. In these applications, various activities are being captured and used as input for the game elements. The game rules in these applications may be enforced in a number of different ways. Rules could be enforced by the user, like in *Epic Win!*, where the user adds and completes tasks themselves to gain experience points. Rules may also be enforced by another user, such as a dungeon master in *Chore Wars*, or they may be enforced automatically using the technology available, like in *foursquare* which uses a Global Positioning System (GPS) sensor to determine a user's location automatically.

Using technology like this to capture different interactions and activities means game rules can be enforced automatically. This relieves the user of the burden of implementing the rules themselves, like in many digital games, and can also help to prevent cheating. As more sensors continue to be included in commodity mobile technology, more and more activities can be recognized and used as input for game elements. The persuasive use of game elements combined with mobile technology provides a number of interesting research opportunities for exploring the gamification of mobile activities, such as exercise. If researchers wish to investigate the effect of gamification in these areas then an existing gamified application is required to study. However if, for various reasons, an existing application doesn't exist, then the researcher may need to create their own prototype to study, and therefore will need to know how to integrate

Figure 1. Screenshot of the chore wars website (©2012 Kevan Davis. Used with permission)

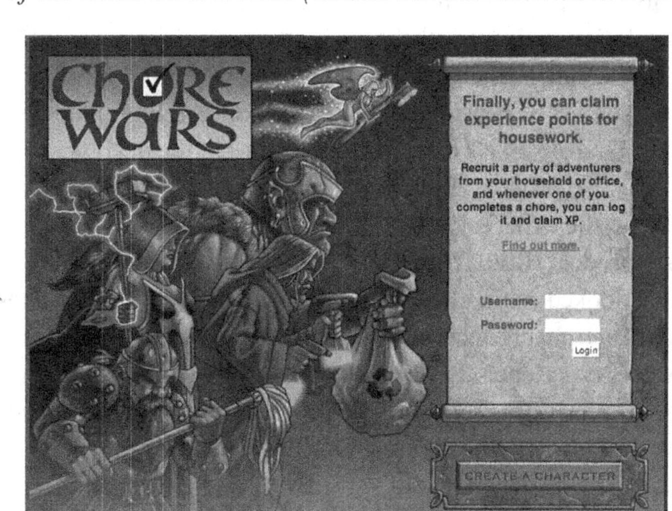

game elements appropriately with this prototype. Unfortunately, there are few explanations and processes that detail the design and integration of game elements in non-game contexts.

This chapter addresses this gap by discussing the addition of game elements to software applications, focusing in particular on the opportunities of using mobile sensing. First an *overview of gamification and related research* is provided in which gamification is discussed, and research gaps and opportunities are identified. The advantages of *using mobile sensors to capture user behaviour* for gamification are then evaluated. A *gamification-sensing framework* is then presented to provide a way in which to design gamified experiences. This framework focuses on using sensors available to the designer to link activity goals with game elements. A case study is then presented that *discusses how the framework was used* to integrate an achievement system into a mobile application prototype for new students attending university orientation. Initial study results are provided, along with a discussion on the usefulness of the framework in this particular scenario and its applicability to other areas.

BACKGROUND

The function and usability of a product plays a very important part in its design, however other aspects such as the role of fun and play in the design of application software can also be considered important (Malone, 1981; Draper, 1999). Over the last decade or so, interactive product design has focused more closely on the experience users have with, and from, a product (Roto, 2006). Concurrently, research has looked more closely at how to elicit feelings of play from products and interfaces to improve a user's experience (Blythe et al., 2004). This includes exploring the role that games can play in these designs as a means to make non-game applications more engaging and motivating to use. Games in general have had a history of being used primarily for purposes other than entertainment, such as to instruct and inform (Abt, 1987) and for education and teaching (Michael & Chen, 2006; Prensky, 2001). However instead of creating fully-fledged serious games, recent research has looked at how the addition of *elements of games* might be used in non-game applications. For example video game achievement systems, like those found in all Microsoft Xbox games, were introduced to a photo sharing service to study their effect on the user experience and motivation to use the service (Montola et al., 2009). Another example is the addition of leaderboards and points to a medication reminder system to encourage medication to be taken on time (de Oliveira et al., 2010). The use of game elements in this way has become known as *gamification* (Deterding et al., 2011).

The term gamification has been defined as "the use of game design elements in non-game contexts" (Deterding et al., 2011). The act of playing a game is generally considered an enjoyable and intrinsically satisfying activity on its own (Ryan et al., 2006), therefore the general notion behind gamification can be seen as an attempt to harness this enjoyment and fun, and to use it to frame non-game activities as game-like in order to make them more motivating. Deterding et al. (2011) suggest that the term gamification denotes an artefact with *game design elements*, rather than a fully-fledged, serious game. However, the authors also discuss that the distinction between the serious game and gamification can be blurry, and it often relies on how people use the artefact or on what the designer intended.

Game design elements can include anything from game aesthetics such as graphics and progress bars, to complete games with an overarching goal, rules, story, levels, quests and achievements (Deterding et al., 2011). For the last few years though the concept of gamification has been tied to achievement and competition based game elements such as badges, points, and leaderboards, and these have been often applied to computer applications and social media contexts (Antin & Churchill, 2011). These elements can be implemented on top of existing systems like Websites by using *gamification platforms* such as Badgeville[4], Big Door[5], and Bunchball[6]. These platforms have been built to allow the integration of elements such as levels, points, achievements and tangible rewards into Websites to reward interactions, such as commenting, registering, social sharing and making purchases.

Research in the area of gamification has grown over the last few years, with more and more studies exploring the effect of game elements on encouraging adoption, or influencing behaviour, in a range of different domains. These primarily include health (e.g., de Oliveira et al., 2010; Fujiki et al., 2007; Chiu et al., 2009), education (e.g., Mieure, 2012), green living (e.g., Bång et al., 2009; Gustafsson et al., 2008), social and online communities (e.g., Vassileva, 2012) and software applications (e.g., Flatla et al., 2011). This research has generally explored the effect of game elements when added to a software application built to support a particular domain and activity. For example, de Oliveira et al. (2010) added game elements to a medication reminder system built to remind users to take their medication closer to the prescribed time. They found that even though traditional reminder systems were useful, they weren't having the desired effect on medication compliancy. Thus, two reminder systems were designed, developed, and compared - one with game elements and one without. The findings suggest that the added game elements created a more motivating experience for users. Previous research in this area has generally reported positive results, however further research could help to support early findings, or explore similar gamified solutions in different domains not yet studied.

Designing and Developing Gamified Applications

In order to research gamification in different domains, a process of integrating game elements into non-game contexts is needed. Generally, previous research has focused on reporting the effect of adding game elements, as opposed to explicitly detailing the process of how game elements were chosen, designed and integrated with each system studied. In one study (Flatla et al., 2011) the researchers did create a design framework that could be used for building different types of calibration games for interactive systems that required calibrating. This framework was based off isolating particular calibration tasks a user may need to undertake and then matching these with a suitable game mechanic (e.g., A "color space registration" calibration involves a "signal discrimination" task which maps to the game mechanic of "differentiating enemies or targets"). This provided the researchers with a way to match particular calibration tasks with related game mechanics and produce three different games to study further. The concept behind this framework, of matching activity goals to game elements using the input available could be generalised and used to apply game elements to other applications and domains. Early research by Malone (1981) has also provided heuristics for designing game elements for non-game contexts, however there has been little research describing the process of linking game elements with interactions and activities in non-game contexts.

SENSING AND GAMIFICATION

Previous research has used different ways to integrate and enforce game elements added to non-game applications. Sensors that automatically capture interactions have been used as game input, for example the medication reminder system *Movipill* uses a sensor that can detect if a pillbox has been opened or closed (de Oliveria et al., 2011). Crowd sourcing has also been used to enforce game elements, for example the *ESP Game* relies on matching an image descriptor tag given by one player with another player to make sure the tag is reliable (von Ahn & Dabbish, 2004). Other applications, such as *Epic Win!* the to-do list application, rely on the player enforcing the rules themselves. The advantage of using technology to automatically capture interactions is that players can be relieved of the burden of implementing the rules themselves. This is considered one of the most important benefits that technology brings to games (Adams, 2010). As technology continues to develop there are more and more opportunities for sensing new types of interactions. Commodity smart phones in particular have a range of different sensors available that can be used to capture a number of different activities. These sensors provide opportunities for both new types of software applications, and new ways to enforce gamification.

Sensors are an important facilitator for gamification, providing a way to integrate game elements and enforce game rules based on real activities. Sensors can be divided into three different groups; physical, virtual, and logical sensors (Indulska & Sutton, 2003):

- Physical sensors are *hardware* sensors in devices that can attain physical data regarding the user and their environment, such as location, movement or temperature.
- Virtual sensors source *context* data from software applications or services, such as current computer logins or search history.
- Logical sensors use multiple information sources and combine *physical* and *virtual* sensors to solve higher tasks – for example locating an employee by using his current login at a desktop PC and mapping that to device location information (Baldauf, Dustdar, & Rosenberg, 2007).

Taking values from a single sensor and providing a symbolic or sub-symbolic output for it can create different cues. Combining these cues together can help describe various contexts, as shown in Table 1.

Table 1. Describing contexts in terms of cues from Schmidt, Beigl, and Gellerson (1999)

Context	Cues
In the office	Artificial light, stationary or walking, room temperature, dry
Jogging	Natural light (cloudy or sunny), walking or running, dry or raining, high pulse

A combination of various cues can then be used to describe a range of available contexts that could be used for gamification input. For desktop and Web applications, virtual sensors would act as the primary source of available context, as few physical sensors are available on desktop machines. This would primarily include sensing interactions with software applications, for example on the *AusGamers* Website[7], user actions such as watching videos, downloading files and making forum posts are used as input to unlock various achievements on the Website. Current commodity smartphones include a range of cheap and powerful sensors such as an accelerometer, digital compass, gyroscope, location, microphone, and camera (Lane et al., 2010) and these provide an opportunity to sense a much larger range of contexts.

A range of mobile applications already use these various contexts to provide useful functionality, but these sensors can also be used to implement game elements. Take for example the popular location-sharing application *foursquare*, which combines the use of physical sensors (location and time) with virtual sensors (places of interest and their geographical coordinates) to allow users to check in to locations such as businesses, shops, parks, and buildings and share this with friends. Foursquare also uses this information, along with other virtual sensor information, such as the previous check in history of the user, friends and other users, to implement game elements – badges, points, a weekly leaderboard and mayorships - that are aimed at encouraging use of the service. Every time a user checks in to a location they generally receive points. If they check in to a new location, or a location they have visited three times in the same week then they receive bonus points. If they check in with a friend, check in to a new place before any other friends, or check in to a place a long distance from their last check in, then again they receive bonus points. If they check in to a specific location more than anyone else using the service then they become the foursquare *mayor* of that place.

By utilising the different contexts provided by physical sensors on smartphones, as well as other available virtual sensors, a wide range of activities can be recorded. These activities provide the opportunity for new types of software applications, and in turn can be used to power and enforce game elements added to encourage and motivate users.

A FRAMEWORK FOR LINKING SENSED ACTIVITIES TO GAME ELEMENTS

In order to design game elements for a gamified prototype a general, three-layered framework was devised based off the concept of matching goals to game elements using the contexts available (Flatla et al., 2011). This framework was built to aid in the gamification of software applications by breaking the design into three parts; identifying *goals* of a particular activity and linking these goals to game *elements* using available *sensing*. The framework is not meant to replace a design process (e.g., user-centered or iterative design), but instead complement it by providing additional activities.

Framework Overview

Gamification generally revolves around a serious goal, such as adopting the use of a Web service, or keeping fit and healthy. This goal may involve tasks that users might find tedious, boring or non-motivating. For example; encouraging more people to adopt a new service may be difficult because they already use another service that provides similar functionality; or encouraging people to exercise might be hard because they may not enjoy exercising in the first place. A software application may be built to make the task easier to complete, and to this game elements could be added to make the task more motivating. To map the goal successfully to the game elements chosen a thorough knowledge of how to sense the desired activities is needed.

Based on this interpretation, a gamification design could be divided into the three parts:

- **Goals:** The problems, motivation and underlying activities that the designers aim to address with gamification.
- **Sensing:** The triggers identified based on the available sensing and input to inform the design and implementation of game design elements.
- **Game:** The game design elements chosen to support the goals, driven by the sensing options available.

Presenting these three parts as layers provides a way to describe the gamification elements added to a software application (See Figure 2).

As an example, the *Movipill* system (de Oliveira, 2010) is based around addressing issues of medication compliance. Applications that reminded patients to take their pills at a certain time did "*not seem to be enough to motivate people to comply with their medication regimens*" (de Oliveira, 2010). Therefore the researchers designed and developed an application that used reward-based, game elements (specifically points

Figure 2. A framework for gamification

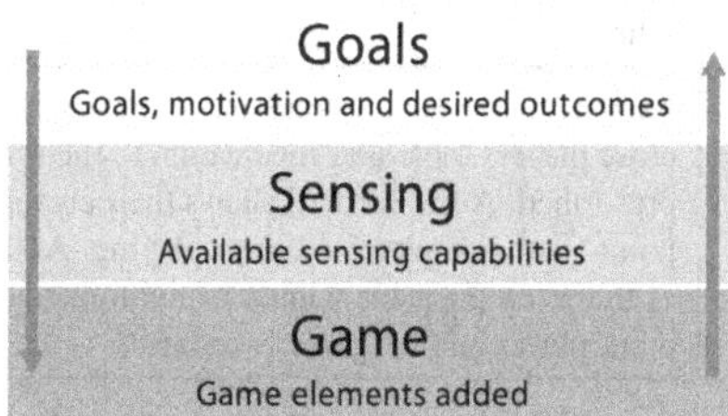

and a leaderboard) to encourage motivation and compliance. These were implemented through the use of a pillbox embedded with a sensor that registered *date* and *time* information when the pillbox was opened. This information was combined with *user* and *prescription* information to identify how close a user had taken their medication to the prescribed time. Using the framework before the gamified elements of the system could be described as shown in Table 2.

Describing gamification designs in this way can help to simplify and explain the implementation used and also identify further design questions. For example, *Movipill* rewards points based on when the pillbox was opened, but how does the system know that the pill was actually taken? Is this something that could be addressed with other sensing capabilities, or is it too difficult to enforce? If so, what is the likelihood of people cheating, and how would this affect the overall design and experience of the system? What happens when one person is prescribed to take more medication than someone else, do they gain more points and therefore receive an unfair advantage?

Using the Framework to Aid Gamification Design

More importantly, this framework could be used to help inform the design of new gamified systems, by helping to identify the goals to be addressed,

Table 2. Describing Movipill (de Oliveira, 2010) using the framework layers

Goals	Encourage better medication compliance (i.e., patients taking their pills at the prescribed time)
Sensing	Pillbox embedded with a sensor that registers two cues (date, time) when the pillbox is opened; combined with user information and prescription schedule
Game	Various points are awarded depending on how close players took their medication to the time prescribed. A leaderboard shows their current point score compared to others playing. At the end of the week the game winner is highlighted to each other player and the game is restarted.

and how available sensing can be used to implement the game elements chosen. These layers are discussed further later in terms of their use during the design process of a gamified application.

Goals

This layer should identify a *problem* to be addressed using gamification. The problem may have particular *goals,* such as encouraging a healthy lifestyle, or encourage adoption and use of an online social network. Based on these goals a number of specific *activities* can be outlined as key performance indicators that may address the problem (e.g., undertake thirty minutes of high-physical exercise everyday, or comment on a profile page). Problems, goals, and activities may be identified and gathered in a variety of ways, such as reviewing previous literature, or undertaking questionnaires, interviews, focus groups, workshops, observation and studying documentation (Rogers, et al., 2011). This might occur early in the design process, or after an issue with a current system is identified.

In summary, this layer specifies the problem and a number of goals and activities to be addressed using gamification. Based on these requirements, sensing options can be explored to link these activities to game elements.

Sensing

Once the goals and activities have been identified, finding the most automatic and accurate way to sense them is required to implement the game elements. If the desired activities are too difficult to sense automatically using the available technology, then other techniques can be employed instead, such as crowdsourcing. However, for a crowdsourcing approach, a willing community is needed, and cheating may also be potentially easier unless rigorous security measures are implemented. Game rules could also be self-enforced by the user; however, we propose that this is likely to be the least enjoyable experience, as users need to take on an additional role of referee for the game elements added. In this role, users could also choose to play unfairly, or cheat more easily than if the game elements were enforced automatically using technology.

This layer involves the following processes:

- Turning the goals and activities into specific contexts.
- Identifying if these contexts can be gathered from the cues available through logical and physical sensors (technology), or by other means available (crowdsourcing or self-enforcement).
- Using these contexts as input for the game elements chosen.

To summarize, the sensing layer aims to provide a link between game elements and goals by identifying available cues and contexts based on the underlying technology. This layer may indicate any issues with recognizing various activities or implementing particular game elements.

Game

The game layer is built based on the underlying goals and implemented using the available contexts. This layer involves choosing and designing game elements that aim to support the goals and activities identified. This may involve the addition of a range of different game design elements, such as graphics, points and progress bars, to more complete game experiences with an overarching goal, rules, story, levels, quests and achievements. Reward and competition based game elements such as badges, points, and leaderboards have been popular elements explored recently in a number of studies (e.g., Montola, 2009; de Oliveira, 2010). Flatla et al. (2011) designed gamified experiences by building on previous game design work (Hunicke et al., 2004; Malone, 1981), including features such as challenge, theme, reward and progress in their designs. Depending on what game design elements are chosen by the researcher to study, the design of these can be informed by the underlying goals and activities, and implemented using the available context sensing.

Advantages and Limitations of the Framework

As mentioned earlier this framework is not meant to replace a particular design process chosen (e.g., user-centered design), but rather support it by dividing the gamification design of an application into three layers, suggesting various activities and processes that may be useful. The advantage of this approach is that it provides a simple way to break down a gamification design, helping to identify the goals to be addressed by the game elements and how they can be enforced by the technology available. A limitation of this framework is that it may be too general and does not explicitly discuss the choice or design of the game elements to be used in the design. This choice depends on what the researcher intends to investigate and if they aim to study the affect of one particular game design element or design a more detailed gamification experience.

CASE STUDY: GAMIFYING UNIVERSITY ORIENTATION

This framework was created to aid in the design of achievements added to a university orientation event smartphone application. University orientation is an event that generally occurs before the first week of university and aids in a student's transition from a school to a university environment. The orientation event at the Queensland University of Technology (QUT) aims to introduce new students to the university campus, people and services. A review of two previous QUT student experience surveys, and a focus group with QUT orientation staff revealed three areas that needed improvement; encouraging *exploration* of the campus, *participation* in the event activities, and *social networking* between peers. Previously the organizers employed the use of a paper-based event list to provide event information to student. It was found that this could be improved as the list was reported as being easy to lose, and was out of date if a student changed their schedule online. To address these areas a mobile event application prototype was developed which provided a list of orientation events, a location-aware map of the university, a contact list to which students could add new friends, and important university information. The gamification framework previously described was then used as a basis to design and integrate different game elements.

The framework helped to link the goals identified to various game elements using the available sensing options, as outlined in Table 3.

The gamified design of the application is broken down further in the next section, with detail provided on how each layer of the framework was used to aid the design process undertaken.

Goals

A review of two past student orientation experience surveys, and a focus group with university engagement staff, revealed that new students could often feel lost, have trouble meeting new friends and trouble finding what services and events are available on campus while attending orientation. The mobile event application addressed some of these problems through the functionality it provided, but these functions did not necessarily engage or motivate the user. For example, the map helped the user to find their way around the university, but it did not necessarily encourage them to explore the campus. Therefore goals identified for the game elements to address included:

1. Encouraging exploration of the campus and services available.
2. Encouraging participation in the orientation events.
3. Encouraging social networking between students.

Table 3. Describing orientation passport using the gamification-sensing framework

Goals	Encourage exploration of the campus, participation in events, and social networking between new students
Sensing	A range of physical sensors provided by the smartphone technology combined with orientation information on events, activities and services
Game	Achievements are awarded for completing various orientation tasks, a leaderboard shows who has completed the most achievements, students can enter the draw for physical rewards if they complete a number of various challenges

These goals where then broken down into a number of specific activities that the student could undertake while at university orientation. For example, activities based on exploration included: visiting important university locations (e.g., the library – for borrowing books and for study spaces, Information Technology helpdesk – for help with any computer related issues), and learning about a particular university service (e.g., the university security phone number – so students know who to call in case of an emergency, campus shuttle bus – so students know there is a free shuttle bus available to take them between campuses). Activities based on participation included: attending a scheduled orientation event that the student had signed up to and collecting specific university related objects (e.g., student diary and student card). Activities based on social networking included: meeting other first year students as well as students from the same faculty and course.

Sensing

The prototype was developed for the iPhone 4, this technology provided a range of different sensors able to record a number of the activities previously listed. These were determined based on what could be obtained from relevant physical and virtual sensors available. In terms of physical sensors, location could be accessed through the use of the phone's Global Positioning System (GPS) sensor, cellular and Wi-Fi sensors, or by scanning a Quick Response (QR) code left at a specific location using the phone's camera. Barcodes on books and other items could be read using the phone's camera as well. Time could be obtained from the phone's internal clock and movement from the phone's accelerometer sensor. Students could also input data using phone's keyboard for answering questions. Virtual sensors available included an orientation event schedule with location and time information, a list of university related objects with barcode numbers (e.g., books and student identification cards), a list of important university services and their details (e.g., campus security phone number), a list of important places and the geographic co-ordinates, a list of QR codes used and their locations, and a list of contacts that the student has added to their contact list. These sensor cues were then combined to provide a set of activities that could be used as input for the game elements, as outlined in the Table 4.

Table 4. A list of student activities as identified using the available sensors

Student Activities	Sensing cues
Find the details of a university service	Keyboard input, a list of important university services (e.g., campus security) and their details (e.g., phone number).
Obtain an object with a barcode	Camera input, an object with a barcode (e.g., library book, student card), a list of orientation objects with barcodes and their barcode numbers.
Find a location on campus	Location sensor (GPS, Wi-Fi, cellular) **or** camera input, physical QR code, list of QR codes and their locations.
Attend a scheduled orientation event	Location sensor (GPS, Wi-Fi, cellular), internal clock, event schedule with location and time information.
Meet another student	List of contacts (New contacts added using the Bump API (2012) which triggers a connection between two users when a "bump" motion is detected at the same time, using location, accelerometer and time sensors).

Game Elements

This case study chose to primarily investigate the effect of video game achievements when added to non-game mobile applications. The video game achievement has become a popular way to add extra challenges and play time to video games with little expense. Video game achievements are task-reward systems, often separate from the game itself, that generally either reward players with points, unlock bonus in-game material, or simply exist as status symbols. Similar game elements have been explored in previous studies (e.g., Montola, et al., 2009) and are also commonly used in gamification platforms. A number of achievement systems were surveyed from various game networks (Steam, Xbox Live and PlayStation Trophies), video games (Portal, Team Fortress 2 and World of Warcraft) and gamified applications (Foursquare and GiantBomb). This survey revealed a number of elements commonly used in achievement design, such as an achievement title and a clue as to how to unlock the achievement. Also an image and text were often included, revealed when the achievement has been completed. It was also found that achievements can often use informal and humorous language, and can be awarded for a range of different activities from the very mundane to the very difficult. The findings of the review influenced the design of the achievement system for the mobile application. Achievement design also received input and feedback from both orientation staff and existing university students.

Achievements were created by picking a goal, choosing an activity as an input, and building an achievement around it, as shown in Table 5.

These achievements were then expanded upon, with clues, images and unlock text added influenced by the findings of the achievement system review (See Table 6).

Players were introduced to the system through easy, introductory achievements. Immediate feedback was provided via an alert message when

Table 5. Examples of linking utility to achievements via recorded student activities

Goal	Student Activities	Example achievements
Encourage exploration of the campus and services available	Learn the details of a university service	- Enter the campus shuttle bus number - Enter the security phone number
Encourage exploration of the campus and services available	Find a location on campus	- Find the library - Find orientation tent
Encourage participation in the orientation events	Attend a scheduled orientation event	- Attend one event - Attend three events
Encourage participation in the orientation events	Find an object with a barcode	- Collect your student card - Find a library book
Encourage students to meet other students	Meet another student	- Add your first contact - Add your third contact

Table 6. Two examples of achievements and their elements

Details	Achievement 1 Example	Achievement 2 Example
Set	Event Manager (Participate)	Campus and Services (Explore)
Title	Roll call	Hail to the bus driver.
Image	Calendar with one days crossed off	Bus at a bus stop
Clue	Check-in to your first event	We're looking for the three digit route number of the intercampus bus, enter it to unlock this achievement.
Trigger	Events attended = 1	Number entry = 391
Description	All right! You're off to a good start! Attending events is essential to getting the most out of QUT Orientation. The events page will help you keep track of upcoming events and events you've attended. This feature only works if you're at the right place at the right time so don't forget to check in!	Need a lift to the other campus? Never fear, the university provides a free shuttle bus service to assist students and staff travelling between the each campus for the purpose of attending lectures or attending to University business. The service is operated for the University by TransLink. University staff and students are able to travel free but will need to show the driver their university identity card.

an achievement was complete. To make the achievement system challenging, achievements became progressively harder to complete either by requiring more activities (e.g., Attend your *third* event), by providing location hints instead of giving exact location (e.g., this place will fulfill all your sugary desires), or by having cryptic clues (e.g., Title: 025.344 15. Clue: *???* - the title being a catalog code requiring the student to scan a book in the library) (See Figure 3).

The design of the application went through a number of iterations that employed paper and digital prototypes. Expert evaluations and usability tests were conducted, as well as a field

Figure 3. Screenshots of the achievements and leaderboard game elements

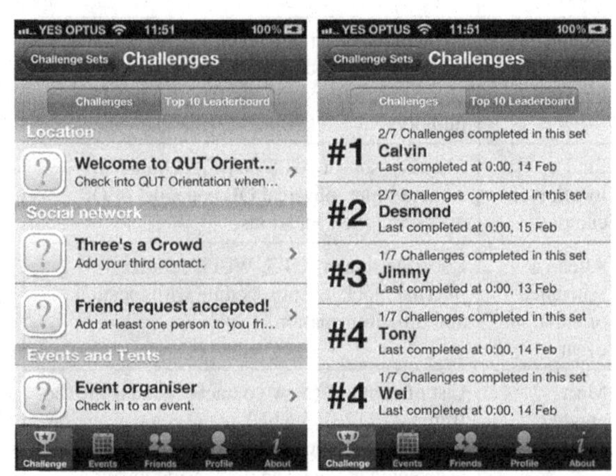

study during orientation before the application was integrated with the university's iPhone application.

Field Study

The prototype was tested in a field trial run during orientation week in 2011. Forty-six first year university students were recruited (male = 31, female = 15), twenty-six of whom trialed the gamified application sometime during orientation week at university, the other twenty trialed an application without game elements. Participant ages ranged from 17 to 45 years old with an average age of 21 years. To try and compare the gamified application to the base application a number of survey questions asked participants to indicate if the application motivated them to complete general activities, such as *the application encouraged me to attend events* and *the application encouraged me to meet new people*. Survey results indicate only small differences between the results of the two applications, with an initial analysis indicating the gamified application was reported as only slightly more motivating at encouraging *exploration*, *visiting new places on campus*, and *meeting new people* and reported as less motivating for encouraging *event attendance* and *learning of services*.

Initial analysis of the gamified application data found that all of the participants completed at least four or more achievements out of the twenty available. The majority (81.8%) completed at least ten or more and all but one participant reported that the achievement system not only added value to their orientation experience, but was also fun to use. Half the participants preferred achievements that involved seeking particular locations by scanning QR codes compared to the other activities. A number of participants reported that for achievements that required numerical input (e.g., finding how many levels a building had, or entering the university bus number) they could simply guess the answer. It was also found that because of the large radius compensating for the location sensor limitations some students cheated and checked-in to multiple different events at the same time. Achievements that were properly enforced, like finding a location with a QR code, were found to be more enjoyable and challenging than those that could be simply guessed. Few participants completed the achievements based around adding friends to the contact list. Unlocking these achievements required participants to 'bump' phones with each other so the system knew that they had not cheated and just added a fake contact. However, this meant that participants couldn't add a friend who wasn't using the application, reducing the usefulness of the application in order to enforce the game elements properly. Participants indicated that more competitive and physical reward elements would motivate them to engage with the game elements more.

Integration with the University Application

Based off feedback from the field study the prototype was redesigned and then integrated with the university's official iPhone application. A number of changes were made to the sensing and game layers of the application based on the feedback from the field trial. This included removing the numerical achievements and replacing them with text challenges that were more challenging. The majority of the event achievements were removed as well, as usability issues and cheating occurred, and the field study results indicated that they did not have much of an effect on encouraging event attendance. A leaderboard was added to encourage competition, and physical rewards were offered, including an iPad 2 and various gift vouchers for those who had completed the most achievements. The application was released and a total of total of forty achievements were added and released during orientation week and over the first three weeks of university. Data gathered during this time included survey responses from thirteen students, and usage data from 121 students. Results suggest that the game elements complement the orientation application and experience, encouraging some students to explore the campus more. Respondents generally found the text answers not as challenging as the QR code challenges. A number of respondents reported that they found seeking locations and scanning QR codes enjoyable. However, issues arose with some participants reporting that QR codes were missing from some locations. A number of issues were raised that should be addressed in future releases including reviewing the game elements and providing further support for the application during its use.

Discussion

The framework provided an easy way to match goals to game elements by identifying input through the combination of the various physical and logical sensors available. This input helped inform the design of the achievement system, relieving the user from the burden of implementing the rules themselves. However, the findings of the study suggest that it is important to find the right sensing technique to enforce the added game elements, as some sensors may not be that reliable and could lead to cheating, as found in the results. By determining multiple ways to sense certain activities, decisions can be made as to which is the most appropriate choice of input for the game elements. Some techniques may not be that enjoyable if they do not enforce the game elements successfully, for example event check-in achievements used a large radius due to accuracy issues with the GPS sensor, which meant cheating occurred. Participants also reported that check-in achievements using the GPS sensor were not as enjoyable as those that used QR codes to verify specific locations. This could be because finding QR codes could be a more exciting activity than simply checking into an event. For the sensing that was not as enjoyable, or not as accurate, alternative sensing options may be considered (e.g., using QR codes to check-in to events instead of GPS), or these achievements could be removed completely. Employing multiple sensors (e.g., scanning a QR code *and* capturing the location of a user using a GPS sensor) could also maximize accuracy and minimize cheating. These findings on choosing appropriate sensors could be included as further guidelines to the sensing layer of the framework.

The choice of game elements also needs to be considered further. Although the study was primarily looking at the effect of achievement systems on the application experience, the question arises whether achievement systems were an appropriate choice or would other game elements have suited better? Based on the results the achievement system was reported as being fun to use and adding value to the orientation experience, however when compared to the application without game elements the application with achievements seemingly had very little influence on the self-reported motivation. It would be worthwhile to evaluate which game elements may suit the underlying goals by drawing further on game design techniques. For example gamification design could borrow further from traditional game design principles and processes, employing evaluation techniques such as playtesting, as well as looking at identifying the player themself - their demographic, player type and skill level. This could be considered further as an addition to the game layer of the framework.

CONCLUSION

Research into gamification provides a number of interesting opportunities, especially when considering how sensing techniques can be used as input for the game elements. Recognizing user contexts using mobile sensors provides a way to record actions and enforce game rules layered on top of non-game activities. This chapter presented an overview of the concept of gamification, related research and gaps, the presentation of a gamification framework to describe and design gamification experiences, and a case study that provided an example of the framework in use. The case study demonstrated how the framework was used to aid in the design of an achievement system added to a mobile application to try and engage new students at university orientation. The framework provided a useful way to design a gamified experience for this case study, identifying goals that could be linked to game elements using the various sensing capabilities available. The game elements and sensing techniques were generally well received, however there were a number of important design and evaluation issues that arose that could be added to the framework to aid further gamification design in the future. These included considering alternative sensing options and considering the way in which game elements were integrated. These results can be fed back into the framework, developing it further so as to study gamification in other case studies in the future.

ACKNOWLEDGMENT

This research was carried out as part of the activities of, and funded by, the Smart Services Cooperative Research Centre (CRC) through the Australian Government's CRC Programme (Department of Innovation, Industry, Science, and Research). The authors wish to thank the QUT Student Engagement Team for their input and support, in particular Peter Gatbonton and Duyen Nguyen.

REFERENCES

Abt, C. C. (1987). *Serious games*. University Press of America.

Adams, E. (2010). *Fundamentals of game design* (2nd ed.). Berkeley, CA: New Riders.

Antin, J., & Churchill, E. (2011). Badges in social media: A social psychological perspective. In *Proceedings of the 29th ACM SIGCHI International Conference on Human-Computer Interaction*. Vancouver, Canada: ACM.

Baldauf, M., Dustdar, S., & Rosenberg, F. (2007). A survey on context-aware systems. *International Journal of Ad Hoc and Ubiquitous Computing*, *2*(4), 263–277. doi:10.1504/IJAHUC.2007.014070.

Bång, M., Svahn, M., & Gustafsson, A. (2009). Persuasive design of a mobile energy conservation game with direct feedback and social cues. In *Proceedings of the 3rd International Conference of the Digital Games Research Association*. West London, UK: ACM Press.

Blythe, M. A., Overbeeke, K., Monk, A. F., & Wright, P. C. (2004). *Funology: From usability to enjoyment*. London: Springer.

Chiu, M.-C., Chang, S.-P., Chang, Y.-C., Chu, H.-H., Chen, C. C.-H., Hsiao, F.-H., & Ko, J.-C. (2009). Playful bottle: A mobile social persuasion system to motivate healthy water intake. In *Proceedings of the 11th ACM International Conference on Ubiquitous Computing* (pp. 185-194). Orlando, FL: ACM.

de Oliveira, R., Cherubini, M., & Oliver, N. (2010). MoviPill: Improving medication compliance for elders using a mobile persuasive social game. In *Proceedings of the 12th ACM International Conference on Ubiquitous Computing*, (pp. 251–260). Copenhagen, Denmark: ACM.

Deterding, S., Dixon, D., Khaled, R., & Nacke, L. (2011). From game design elements to gamefulness: Defining gamification. In *Proceedings of the 15th International Academic MindTrek Conference: Envisioning Future Media Environments*, (pp. 9–15). Tampere, Finland: ACM.

Draper, S. (1999). Analysing fun as a candidate software requirement. *Personal Technologies*, *3*(3), 117–122. doi:10.1007/BF01305336.

Flatla, D. R., Gutwin, C., Nacke, L. E., Bateman, S., & Mandryk, R. L. (2011). Calibration games: Making calibration tasks enjoyable by adding motivating game elements. In *Proceedings of the 24th Annual ACM Symposium on User Interface Software and Technology,* (pp. 403–412). Santa Barbara, CA: ACM.

Fujiki, Y., Kazakos, K., Puri, C., Pavlidis, I., Starren, J., & Levine, J. (2007). NEAT-o-games: Ubiquitous activity-based gaming. In *Proceedings of the 25th ACM SIGCHI International Conference on Human-Computer Interaction*, (pp. 2369–2374). San Jose, CA: ACM.

Gustafsson, A., Katzeff, C., & Bang, M. (2009). Evaluation of a pervasive game for domestic energy engagement among teenagers. *Computers in Entertainment*, *7*(4), 1–19. doi:10.1145/1658866.1658873.

Hunicke, R., LeBlanc, M., & Zubek, R. (2004). MDA: A formal approach to game design and game research. In *Proceedings of the AAAI Workshop on Challenges in Game AI* (pp. 04–04). AAAI.

Indulska, J., & Sutton, P. (2003). Location management in pervasive systems. In *Proceedings of the Australasian Information Security Workshop Conference on ACSW Frontiers 2003* (vol. 21, pp. 143–151). Darlinghurst, Australia: Australian Computer Society, Inc.

Malone, T. W. (1981). Toward a theory of intrinsically motivating instruction. *Cognitive Science*, *5*(4), 333–369. doi:10.1207/s15516709cog0504_2.

Michael, D. R., & Chen, S. (2006). *Serious games: Games that educate, train and inform*. Boston, MA: Thomson Course Technology.

Mieure, M. (2012). *Gamification: A guideline for integrating and aligning digital game elements into a curriculum.* (Masters Thesis). Bowling Green State University, Bowling Green, OH.

Montola, M., Nummenmaa, T., Lucero, A., Boberg, M., & Korhonen, H. (2009). Applying game achievement systems to enhance user experience in a photo sharing service. In *Proceedings of the 13th International MindTrek Conference: Everyday Life in the Ubiquitous Era*, (pp. 94–97). Tampere, Finland: ACM.

Prensky, M. (2001). *Digital game-based learning*. New York: McGraw-Hill.

Rogers, Y., Sharp, H., & Preece, J. (2011). *Interaction design: Beyond human - computer interaction* (3rd ed.). New York, NY: John Wiley & Sons.

Roto, V. (2006). *Web browsing on mobile phones – Characteristics of user experience.* (Doctoral Thesis). Helsinki University of Technology, Helsinki, Finland.

Ryan, R. M., Rigby, C. S., & Przybylski, A. (2006). The motivational pull of video games: A self-determination theory approach. *Motivation and Emotion*, *30*(4), 344–360. doi:10.1007/s11031-006-9051-8.

Schmidt, A., Beigl, M., & Gellerson, H. W. (1999). There is more to context than location. *Computers & Graphics*, *23*, 893–901. doi:10.1016/S0097-8493(99)00120-X.

Vassileva, J. (2012). Motivating participation in social computing applications: A user modeling perspective. *User Modeling and User-Adapted Interaction*, *22*(1), 177–201. doi:10.1007/s11257-011-9109-5.

von Ahn, L., & Dabbish, L. (2004). Labeling images with a computer game. In *Proceedings of the 22nd ACM SIGCHI International Conference on Human-Computer Interaction*. Vienna, Austria: ACM.

ADDITIONAL READING

Bång, M., Svahn, M., & Gustafsson, A. (2009). Persuasive design of a mobile energy conservation game with direct feedback and social cues. In *Proceedings of the 3rd International Conference of the Digital Games Research Association*. ACM.

Caillois, R., & Barash, M. (2001). *Man, play, and games*. Chicago, IL: University of Illinois Press.

Csikszentmihalyi, M., & Csikszentmihalyi, I. S. (1992). *Optimal experience: Psychological studies of flow in consciousness*. Cambridge, UK: Cambridge University Press.

Deci, E. L., & Flaste, R. (1995). *Why we do what we do: The dynamics of personal autonomy*. New York: Putnam's Sons.

Froehlich, J., Dillahunt, T., Klasnja, P., Mankoff, J., Consolvo, S., Harrison, B., & Landay, J. A. (2009). UbiGreen: Investigating a mobile tool for tracking and supporting green transportation habits. In *Proceedings of the 27th International Conference on Human Factors in Computing Systems* (pp. 1043–1052). New York, NY: ACM.

Gustafsson, A., & Bang, M. (2008). Evaluation of a pervasive game for domestic energy engagement among teenagers. In *Proceedings of the 2008 International Conference on Advances in Computer Entertainment Technology* (pp. 232–239). New York, NY: ACM.

Huizinga, J. (1950). *Homo ludens: A study of the play element in culture*. Roy.

Johnson, S. (2006). *Everything bad is good for you*. New York: Penguin Group US.

Kapp, K. M. (2012). *The gamification of learning and instruction: Game-based methods and strategies for training and education*. New York: John Wiley & Sons.

Kazakos, K., Bourlai, T., Fujiki, Y., Levine, J., & Pavlidis, I. (2008). NEAT-o-games: Novel mobile gaming versus modern sedentary lifestyle. In *Proceedings of the 10th International Conference on Human Computer Interaction with Mobile Devices and Services* (pp. 515–518). ACM.

Lane, N. D., Miluzzo, E., Lu, H., Peebles, D., Choudhury, T., & Campbell, A. T. (2010). A survey of mobile phone sensing. *IEEE Communications Magazine*, *48*(9), 140–150. doi:10.1109/MCOM.2010.5560598.

Lindqvist, J., Cranshaw, J., Wiese, J., Hong, J., & Zimmerman, J. (2011). I'm the mayor of my house. In *Proceedings of the 2011 Annual Conference on Human Factors in Computing Systems* (p. 2409). Vancouver, Canada: ACM.

Madeira, R. N., Silva, A., Santos, C., Teixeira, B., Romão, T., Dias, E., & Correia, N. (2011). LEY! Persuasive pervasive gaming on domestic energy consumption-awareness. In *Proceedings of the 8th International Conference on Advances in Computer Entertainment Technology* (pp. 72:1–72:2). New York, NY: ACM.

McGonigal, J. (2011). *Reality is broken: Why games make us better and how they can change the world*. New York: Random House.

Nicholson, S. (2012). *A user-centered theoretical framework for meaningful gamification*. Paper presented at the Games, Learning, Society. Madison, WI.

Ryan, R. M., & Deci, E. L. (2000). Self-determination theory and the facilitation of intrinsic motivation, social development, and well-being. *The American Psychologist*, *55*(1), 68–78. doi:10.1037/0003-066X.55.1.68 PMID:11392867.

Webster, J., Trevino, L. K., & Ryan, L. (1993). The dimensionality and correlates of flow in human-computer interactions. *Computers in Human Behavior*, *9*(4), 411–426. doi:10.1016/0747-5632(93)90032-N.

Woszczynski, A. B., Roth, P. L., & Segars, A. H. (2002). Exploring the theoretical foundations of playfulness in computer interactions. *Computers in Human Behavior*, *18*(4), 369–388. doi:10.1016/S0747-5632(01)00058-9.

ENDNOTES

1. http://foursquare.com
2. http://www.rexbox.co.uk/epicwin
3. http://www.chorewars.com
4. http://badgeville.com
5. http://bigdoor.com
6. http://bunchball.com
7. http://www.ausgamers.com/

Chapter 7
A Context-Aware Smart TV System with Body-Gesture Control and Personalized Recommendation

Wei-Po Lee
National Sun Yat-sen University, Taiwan

Che KaoLi
National Sun Yat-sen University, Taiwan

ABSTRACT

Smart TV enables viewers to conveniently access different multimedia content and interactive services in a single platform. This chapter addresses three important issues to enhance the performance of smart TV. The first is to design a body control system that recognizes and interprets human gestures as machine commands to control TV. The second is to develop a new social tag-based method to recommend most suitable multimedia content to users. Finally, a context-aware platform is implemented that takes into account different environmental situations in order to make the best recommendations.

INTRODUCTION

In recent years, smart TV becomes more and more popular in the market. Different from the traditional TV focusing on media broadcasting, smart TV systems are to deliver diverse multimedia contents from networked devices directly to the end-users, and allow them to access the contents through a friendly interface. It also provides interactive Internet-based services, such as media-on-demand, social networking, on-line gaming, etc. Smart TV is continuously developing to offer more and more functions and services. In this work, we address three important issues closely related

DOI: 10.4018/978-1-4666-4054-2.ch007

to smart TV, including human-machine interfacing, personalized content recommendation, and context awareness, and develop the corresponding mechanisms to further enhance the performance of smart TV system.

Traditionally, people used to adopt device-based control to operate different consumer electronics at home. Afterwards, some researchers have also implemented systems that utilized personal hand-held devices to work as controllers, for example (Wang, Chung, & Yan, 2011; Lee, Wang, 2004). Lately, to provide natural control over the equipment, different ways of human-machine interactions (such as voice-based and gesture-based control) have been proposed to command the equipment without any remote control devices (Wachs, Kolsch, Stern, & Edan, 2011; Dumas, Lalanne, & Oviatt, 2009). Moving towards an even more natural way for interacting with machines, in this work we further design a body control mechanism through which the smart TV can response to how the human users move.

Though the smart TV system can easily present different types of multimedia contents to end-users, however, the large amount of contents leads to the problem of information overload. It is thus important to develop personalization techniques to recommend most suitable contents to users (Kim, Pyo, Park, & Kim, 2011; Angelides, 2003). Many recommendation methods have been proposed, ranging from content-based user modeling to group-based collaboration, and generally speaking the collaboration-based approach is considered more efficient and effective than the content-based user modeling approach. The current trend of organizing and sharing digital contents through user-created metadata (i.e., social tags) indicates that the performance of collaborative recommendation can be further improved by using such metadata to reason about how the users likes specific items. In this work, we will adopt such metadata for multimedia annotation, use the tag information to analyze how the user likes specific items, and exploit such user information to perform collaborative recommendation.

In addition to the recommendation techniques that focus on the multimedia items, context is another important issue to be considered in personalized recommendation. This is especially critical when the mobile smart TVs become popular, and the handheld devices can use a variety of sensory technology to collect and analyze information about their human user. In general, context awareness means the ability of computing systems to acquire and reason about the context information and adapt the corresponding applications accordingly. In other words, it is about capturing a broad range of contextual attributes (such as the user's locations, activities, and their surrounding environments) to better understand what the user is trying to accomplish, and what content suits the user the most in that context (Baldauf, Dustdar, & Rosenberg, 2007; Baltrunas, Ludwig, Peer, & Ricci, 2012; Adomavicius, Sankaranarayanan, Sen, & Tuzhilin, 2005). By integrating context information into the application service, a recommendation system can fulfill the user's needs more efficiently and practically.

Taking the previous issues into account, in this work we develop a smart TV system with several unique features. First, we design a Kinect-based system to recognize human body gestures for TV control. Second, we compare different computational methods for making personalized recommendation on multimedia items. Following the current trend of community-based information sharing, we also propose to exploit social tags to annotate multimedia items for the improvement of recommendation performance. Experimental results show that the tag-based method outperforms other conventional methods. Finally, we implement a context-aware platform and present how to integrate various environmental situations into the proposed method to perform item recommendation accordingly.

BACKGROUND

As mentioned prior, due to the tremendous amount of digital multimedia items a smart TV can broadcast, a recommendation mechanism is thus needed to offer better services. Recommender systems have been advocated in different service domains for many years (Adomavicius & Tuzhilin, 2005; Ricci, Rokach, & Shapira, 2011). In general, the recommendation techniques can be categorized into two types: the content-based and the collaboration-based filtering. The content-based approach is to predict the user's preference on unknown items from his historical records. Therefore, the most important issue is to construct a computational model to perform the prediction. Many machine-learning approaches have been applied to construct user model. Nevertheless, it is worth noting that the content-based approach largely relies on the sufficient examples used for model construction. Meanwhile, this type of approach inevitably recommends items within some specific scopes, and thus loses item diversity (i.e., ignoring items of the unfamiliar classes). Different from the content-based methods, a collaborative approach does not build a personal model for prediction. Instead, it recommends items to the user according to the evaluations (opinions) from other users with similar tastes. In such approach, the measurement of similarity between users is most important, so the system can employ a *k*-nearest neighbor method to find most similar users to perform recommendation. The prediction of an unknown item for a user is thus based on the combination of the ratings of his nearest neighbors. This type of approach has been widely used in different applications, for example (Lee & Wang, 2004; Kwon & Hong, 2011).

In addition to measuring item similarity or user similarity to predict the user preference, we can also analyze further how the user prefers specific items from different feature dimensions (i.e., in more details). Social tags are brief descriptions of items and can be used as features to capture the semantic of the target items, for example (Wu, Zhang, & Yu, 2006; Levy & Sandler, 2009). They are freely supplied by a community of Internet users to aid the access of large collections of media (Macgregor & McCulloch, 2006; Barragans-Martinez, Rey-Lopez, Costa-Montenegro, Mikic-Fonte, Burguillo, & Peleteiro, 2010). Compared to knowledge representation schemes that involve domain experts, the tagging activity shifts the task of classifying domain items from knowledge engineers to the Internet users. It is thus a cost-efficient alternative to the popular and precise knowledge ontology in content annotation (Wu, Zhang, & Yu, 2006; Pudota, Dattolo, Baruzzo, Ferrara, & Tasso, 2010). Also, tagging is neither exclusive nor hierarchical. In some circumstances, it has advantages over hierarchical taxonomies. Therefore, in this work we choose to use social tags to represent item features, and propose a new tag-based collaboration method for multimedia recommendation. Though there are other works using tags for recommendation (e.g., Guan, Wang, Bu, Chen, Yang, Cai, & He, 2010; Hays, Avesani, & Veeramachaneni, 2007), our work is different mainly because we take social tags to reason about a user's preference, not simply to annotate target items.

Context plays an important role in determining the relevance of a service (or function) of an application to a user's need, and any small contextual changes may lead the user to select a different service. Dealing with the context issue involves defining contexts relevant to the application service and identifying the key contexts in which people often use the service. Regarding different mobile applications, context factors can be defined as any information used to characterize the user situation that can influence his decision in requesting a service. There are two types of context factors: personal and environmental ones (7, 20). Personal context is the personal state or condition of the user himself (such as his emotional and physical states), whereas environmental context means the full set of a user's external circumstances (such as the location, distraction, and crowding that indicate the geographical settings). Currently, the former is relatively hard to measure, while the latter can be automatically detected and has been applied to several application domains, for example (Kurkovsky & Harihar, 2006; Kim, Kim, & Ryu, 2009).

THE PROPOSED CONTEXT-AWARE SMART TV SYSTEM

To provide smart TV services, we develop a system framework with client-server architecture. The client part is responsible for gathering and processing user-side information in real time, and the server part mainly manages user profiles and performs computation for personalization. Figure 1 illustrates the system framework. As other smart TV systems, the proposed framework provides several popular multimedia functions, such as digital content broadcasting, social networking, Web browsing and searching, e-book reading, and so on. In addition, it has three important features: recognizing and interpreting a user's body gestures for TV control, collecting social opinions to perform collaborative recommendation, and integrating environment information to make context-aware recommendation. The details are described in the subsections next.

Body-Gesture Control

Gesture recognition is a natural way to support communication between human beings and machines. Through the process of appropriate interpretation, the user's gestures are made known to the system. Our system includes such a mechanism for the user to interact with and control a TV. Gestures can be static or dynamic. In the latter case, gestures are produced continuously and each of them is affected by the preceded ones. To infer the aspects of gesture, the system needs to sense the user's movement and collect his position data. This is often achieved by using the traditional sensing devices attached to the user (such as magnetic field trackers), or by using cameras with computer vision techniques. The latter offers a more natural way of human-machine interaction. Microsoft's Kinect is a popular vision-based device and able to perform real-time pose recognition, therefore we implement a Kinect-based system to recognize human gestures for TV control. Figure 2 illustrates the flow.

To recognize the user's gestures, the system needs to temporally segment the gestures to extract a time-varying sequence of parameters that describe the relevant body parts. The recognition of user gestures from these parameters then becomes a pattern recognition task that involves transforming the input into the appropriate representation

Figure 1. The proposed system framework for smart TV

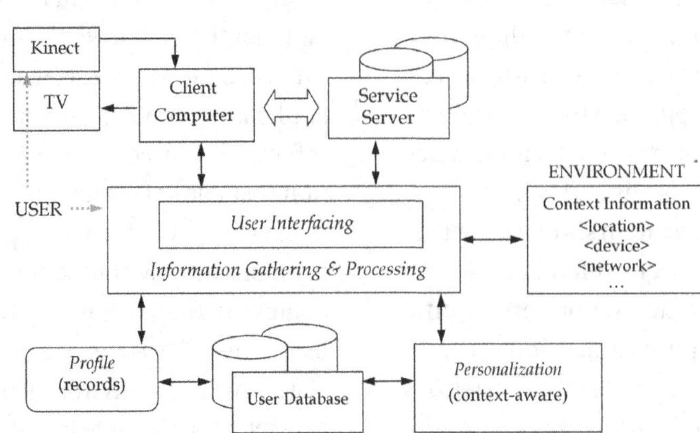

Figure 2. The flow of how the user gestures are interpreted

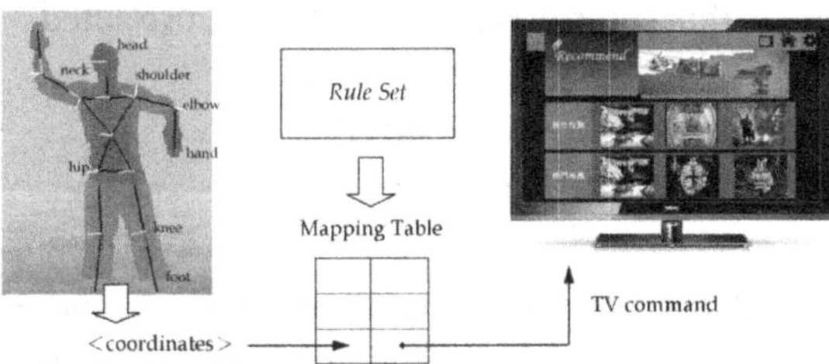

and classifying it from a database of predefined gestures. As mentioned before, we exploit Kinect to achieve such mappings. Kinect segments a depth image into a dense probabilistic body part labeling; it localizes spatial modes of each part distribution and thus generates confidence-weighted proposals for the 3D locations of each skeletal joint (Shotton, Fitzgibbon, Cook, Sharp, Finocchio, Moore, Kipman, & Blake, 2011). In this work, we utilize Kinect SDK to extract body-part data of the human demonstrator (i.e., the coordinates of the head, neck, shoulder, elbow, hand, hip, knee and foot indicated in Figure 2). Then we pre-program a rule set to interpret the data as user gestures (by calculating the relative positions of the body-parts), and encode these gestures into appropriate TV commands. To avoid inconsistent commands as the consequence of multiple users' competing with each other to control the TV, the system can only serve one person (the first human figure recognized) at the present stage. Table 1 lists the major mappings used in our work. It can be expanded by creating and including more human gestures in the same way. In this table, the "arrows" describe how the user moves his arms and the "circles" indicate that the user stretches his arms. The reference data points of the body parts are also shown.

Personalization Recommendation

To perform personalized content recommendation, the first step is to create a personal profile to be a common reference point. Therefore, it is critical to collect profile data explicitly or implicitly and keep it up-to-date following a user's changing needs and contexts. The next step is to use the obtained information to infer the user's preference. Here, we take the collaborative method to achieve personalized recommendation. Two types of collaboration-based methods are developed and assessed, as described in the following subsections.

Table 1. Gestures and the corresponding TV function

command/ data points	gesture	commend/ data points	gesture
start program: right/left hand right/left elbow head		*volume control:* (decrease) right/left hand right/left elbow hipcenter	
change channel: (previous) right/left hand right/left shoulder		*scroll web page:* right/left hand right/left shoulder	
change channel: (next): right/left hand right/left shoulder		*home:* right/left hand right/left elbow right/left shoulder hipcenter	
volume control: (increase) right/left hand right/left elbow hipcenter		*exit:* right/left hand right/left shoulder hipcenter	

Collaborative Recommendation

Collaborative recommendation (or collaborative filtering, CF) performs predictions for a specific user based on the evaluations (ratings) done by other users with similar tastes. For a user u_a, users with most similar preferences or interests are selected as a neighbor set $Neig(u_a)$, and their combined opinion on a certain item m_{recom} is used to predict whether u_a will like this item. That is, the rating of preference of a specific item m_{recom} is defined as:

$$R_{pre}(u_a, m_{recom}) = \bar{R}_{pre}(u_a) + z_1 \times \sum_{u_n \in Neig(u_a)} Sim(u_a, u_n) \cdot (R_{pre}(u_n, m_{recom}) - \bar{R}_{pre}(u_n)) \quad (1)$$

In the previous equation, $R_{pre}(u, m)$ represents the preference of user u on item m; $\bar{R}_{pre}(u)$ is the average preference rating of user u about all items he has rated; $Sim(u_a, u_n)$ is the similarity between two users u_a and u_n; and z_1 is the normalized factor and can be calculated as:

$$z_1 = 1 \Big/ \sum_{u_n \in Neig(u_a)} \left|Sim(u_a, u_n)\right| \quad (2)$$

There are many methods to calculate the similarity mentioned before, and a common one is the Pearson correlation coefficient. Here, we take this method to measure the similarity between two users u_a and u_n as shown in Box 1.

In Equation (3), $Com(u_a, u_n)$ is the set of items that both users u_a and u_n have already rated. This coefficient is between 1 (the preferences of both users are exactly the same) and -1 (their preferences are opposite each other); and a value of zero means that their preferences are not correlated.

Tag-Based Collaborative Recommendation

Though the previously mentioned CF method is efficient for content recommendation, it does not consider the reasons behind the user's comments in detail. Other methods that can provide such detailed analysis are thus in need. Folksonomy (i.e., community-based method) is a very popular technique nowadays to annotate items (or contents). To exploit the current trend in adding metadata to the shared contents, we develop a new approach that incorporates user-specific tags with CF to conduct item recommendation.

To enable users to share tags and keep the annotation consistent, in this work we adopt the method of suggestive-tagging that collects and provides a table of popular tags (as shown in Table 2) for multimedia annotation. The user can select tags from the table and give a value from 1 (lowest) to 5 on each tag to explicitly express his evaluation.

Table 2. The suggested tags for annotation

story	climax	originality
profundity	dialogue	pace
thinking	role	portraying
popularization	entertainment	characteristic
touchingness	satire	humor
horror	cast	acting
visual	action	stunt
music	atmosphere	director

Box 1.

$$Sim_{CF} = \frac{\sum_{m_l \in Com(u_a,u_n)} (R_{pre}(u_a, m_l) - \bar{R}_{pre}(u_a)) \times (R_{pre}(u_n, m_l) - \bar{R}_{pre}(u_n))}{\sqrt{\sum_{m_l \in Com(u_a,u_n)} (R_{pre}(u_a, m_l) - \bar{R}_{pre}(u_a))^2} \sqrt{\sum_{m_l \in Com(u_a,u_n)} (R_{pre}(u_n, m_l) - \bar{R}_{pre}(u_n))^2}} \quad (3)$$

The current tag table is created manually with reference to online comments for movies. It can be further expanded or evolved automatically from users with certain community-based strategies.

Similar to the collaborative recommendation method described before, our Tag-CF method also measures the similarity between users and finds a neighbor set $Neig(u_a)$ for a specific user u_a to predict his preference on a certain item m_{recom}. The same equation for measuring rating preference in CF is used here to calculate $R_{pre}(u, m)$. But it is notable that the Tag-CF method uses tags to look for similar users rather than item preference as in the traditional CF method.

To consider both factors of user preference and tag evaluation together, we combine them to develop a new method for calculation of similarity between two users and use it to work with the collaborative recommendation. The user similarity is now described as:

$$Sim_{TagCF} = [(z_2 \times \sum_{m_l \in Com_{LL}(u_a,u_n)} TC(u_a,u_n,m_l) + z_3 \times \sum_{m_l \in Com_{DD}(u_a,u_n)} TC(u_a,u_n,m_l)) / 2] - z_4 \times \sum_{m_l \in Com_{LD}(u_a,u_n)} TC(u_a,u_n,m_l)$$

$$\text{if } Com_{LL}(u_a,u_n) \neq \varphi \text{ and } Com_{DD}(u_a,u_n) \neq \varphi \tag{4}$$

$$z_2 \times \sum_{m_l \in Com_{LL}(u_a,u_n)} TC(u_a,u_n,m_l) - z_4 \times \sum_{m_l \in Com_{LD}(u_a,u_n)} TC(u_a,u_n,m_l), \text{ if } Com_{DD}(u_a, u_n) \neq \varphi \tag{5}$$

$$0, \text{ if } Com_{LL}(u_a, u_n) \neq \varphi \tag{6}$$

In the previous equations, $Com_{LL}(u_a, u_n)$ represents the set of items that u_a and u_n have evaluated and the two users both like; $Com_{DD}(u_a, u_n)$ includes the items that both users do not like; and $Com_{LD}(u_a, u_n)$ includes the items that one of the two users like. The conditions indicate that the measurement is based on the situation that the two users have rated the same items. In addition, the normalized factors z_2, z_3, and z_4 are represented respectively as the following:

$$z_2 = 1 \Big/ \sum_{m_l \in Com_{LL}(u_a,u_n)} \left|TC(u_a,u_n,m_l)\right| \tag{7}$$

$$z_3 = 1 \Big/ \sum_{m_l \in Com_{DD}(u_a,u_n)} \left|TC(u_a,u_n,m_l)\right| \tag{8}$$

$$z_4 = 1 \Big/ \sum_{m_l \in Com_{LD}(u_a,u_n)} \left|TC(u_a,u_n,m_l)\right| \tag{9}$$

As can be observed, these equations accumulate tag evaluations of those items that both users have the same preference ratings, and then decrease the effect caused from the items they have different preferences. In these equations, $TC(u_a, u_n, m_l)$ measures the similarity between u_a's and u_n's tag evaluations about m_l. The calculation is based on the Tanimoto coefficient ([24]), that is an extended Jaccard coefficient often used to calculate the similarity of vectors with asymmetric binary attributes. In our work, if the tags are only used (evaluated) by one user, a default value of 3 will be automatically assigned to another user in comparison. However, if the tags are not used by any of the users, they will simply be ignored. In this way, the $TC(u_a, u_n, m_l)$ is defined as in Box 2. Here, $R_{tag}(u, m, t_g)$ means the evaluation result of user u on tag t_g that is used to annotate item m_l.

Context Awareness

Since the mobile Internet has become increasingly popular, therefore, a variety of environmental contexts that do not happen in the stationary Internet should be taken into account now in the deployment of context recommendation. This

Box 2.

$$TC(u_a, u_n, m_l) = \frac{\sum_g R_{tag}(u_a, m_l, t_g) \cdot R_{tag}(u_n, m_l, t_g)}{\sum_g R_{tag}(u_a, m_l, t_g)^2 + \sum_g R_{tag}(u_n, m_l, t_g)^2 - \sum_g R_{tag}(u_a, m_l, t_g) \cdot R_{tag}(u_n, m_l, t_g)} \quad (10)$$

section describes the contexts considered in our system, including user location, audience, mobile device, and network condition as discussed later. They have been used to develop a set of rules to re-rank the recommendation list derived from the user preference.

The first type of contexts is location information. A location-aware system functions according to the certain geographical location it is situated in. One way to achieve location awareness is to use the location of a mobile user as parameter for service provision, for example, using handset-based positioning solution (such as global positioning system, GPS) or network-based positioning solution (such as GSM cellular system). Here, we do not use the detailed location parameters; rather we categorize the location context into the public and the private types. We then use this information to restrict the playing of multimedia contents. According to the MPAA (Motion Picture Association of America) rating systems, we categorize multimedia products into five levels: G (general audiences), PG (parental guidance suggested), PG-13 (parental strongly cautioned), R (restricted), and NC-17 (only people 17 and older are admitted), and map them with appropriate location types. This is to consider the impression of the people around the user who is accessing the multimedia with a mobile device.

The second type of context, social context, considers the crowd (or audience) around the user. It means that sometimes a user may invite other people to access multimedia content (e.g., to watch a movie) together and the system needs to consider whether the content suitable to all of them. Similar to the situation in the previous location context, the recommendation list needs to be adjusted and some contents need to be filtered out from it. In this work, we use the MPAA rating criterion to remove the contents not suitable for many viewers, and re-rank the rest items in the list according to their popularity.

The third type of context is related to the infrastructure delivering the service. Today, it is very popular to use powerful mobile devices, such as mobile phones, slim notebooks, or portable TVs, to access multimedia content in the mobile environment. Devices used by mobile users are diverse and heterogeneous. They have different screen size, memory, media support, connection speed, and perhaps the most important capability, computational power, to deal with the multimedia content. For the high-resolution media content, a more powerful computational mechanism is needed to prevent the delay of video and audio in media playing. Battery power is another important issue to note because accessing multimedia through mobile devices is a very power-consuming application. Under such circumstances, mobile devices, due to hardware limitations, cannot support some contents. Therefore, a context-aware system must take the device context into consideration in making recommendation.

In addition to the hardware devices, the communication condition is also a critical factor that decides the quality of service directly. High-resolution online contents are generally preferred, but they require the support of high network bandwidth. Regarding this, multimedia products can therefore be produced in different modes to fit in the available network bandwidth. For example,

YouTube started to support high-resolution films since 2008 when the network condition was largely improved. It allows the user to choose appropriate resolution mode for the multimedia he is playing, depending on the network and device conditions. Similarly, our system also considers the communication condition to recommend most suitable content to users.

EXPERIMENTS AND RESULTS

System Implementation

The proposed smart TV system has been implemented and verified. In addition to the multimedia content broadcasting system, our framework integrates some popular interactive software, currently including Google search engine, YouTube, Facebook, PDF reader, etc. At the first stage of verification, we have used the Kinect SDK to construct a mechanism on a PC (with a CPU of Intel Pentium Dual 2.0 GHz and 2GB RAM) for TV control. As described in section 3.1, this mechanism recognizes and interprets the user's body gestures, and turns them into control commands. The appropriate software is then driven and the user can operate it through his body gestures. Figure 3 presents two examples, in which the user requested a keyboard on the screen and typed a keyword to perform Google search (left), and chose a TV program from the recommendation list (right).

In addition to the previously mentioned implementation, we then expanded our system to conduct the personalized content recommendation in a context-aware environment. Figure 4 illustrates the enhanced client-server architecture. In this system, the tag-based collaborative method has been used to develop the recommendation module, as it has been shown to give the best performance.

To identify the networking condition, the system on the server sends certain packets to the client and waits for its responses. The system can then estimate the bandwidth available for the user. Meanwhile, the system also retrieves information recorded in the HTTP header received from the client, to classify the operating system embedded in the most popular client devices (currently including Windows NT, Mac OS, Linux and Solaris for non-mobile devices; and Windows CE, iPhone Mac OS, Palm OS, Pocket PC, EPOC, and Linux Operation for the mobile devices), and to recognize the type of the user device accordingly. Based on the prior information of network condition and user device, our system is able to suggest the most suitable type of resolution to the user.

As mentioned, the GPS or GSM cellular system can be used (with the electronic map) to provide detailed location context. However, because the handheld devices with embedded positioning system are not widely used in our daily life (they are mostly used for tracking purpose), for practical reason, in our current implementation we have not integrated the positioning system-based location information to our system. Instead, we ask the

Figure 3. Examples of using body gestures to control TV

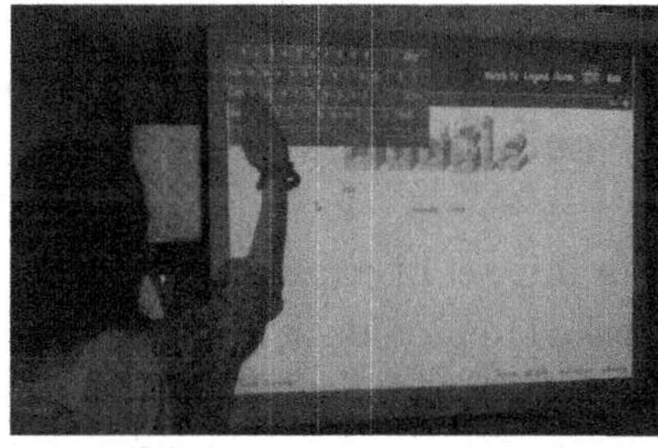

Figure 4. System implementation of our context-aware personalized recommendation

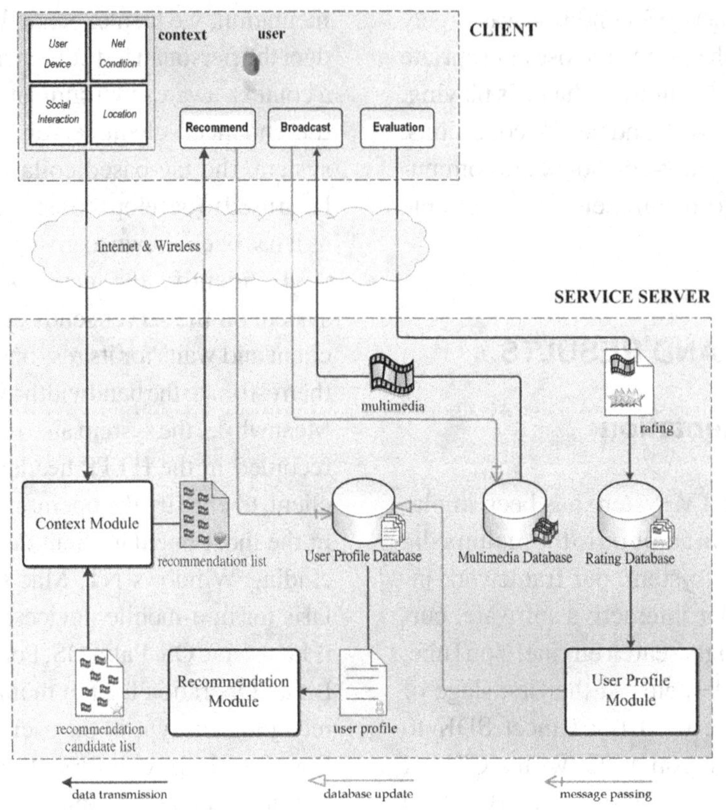

user to provide his environmental context to the system by making some selections from predefined choices on the interface (as shown later).

Figure 5 shows two screenshots of the system presented to the user. As is presented, the upper-left icons provide predefined options of social and location contexts. The social context here tries to capture the social condition around the user. The user can inform the system whether he is alone (single user) or with other people (multiple users) through these options. Similarly, the user can inform the system whether he is in a public or private domain. The recommended movies are listed later the icons according to the contexts the user specifies. In Figure 5, the interface windows show the recommendation results for two different situations (with language translation): the first is the user choosing both "multiple users" and "public place"; and the second, the user choosing "single user" and "private place". The frame for showing the movie is allocated in the middle area of the interface window. The information of the user device, currently including the kind of operating system and the network bandwidth, is detected and provided on the right hand side, and the resolution for viewing the movie is then suggested accordingly. Figure 6 shows the recommendation examples for considering device and network conditions. In addition, the tags used to annotate the movie are also provided below the movie frame. The user can give his feedback through the evaluation of these tags.

Performance Evaluation on Recommendation

In this set of experiments, each participant was asked to provide at least 40 movie items (chosen from a default list with 320 items, or specified by user manually) he has experienced and evaluated

Figure 5. The presented results with different context situations

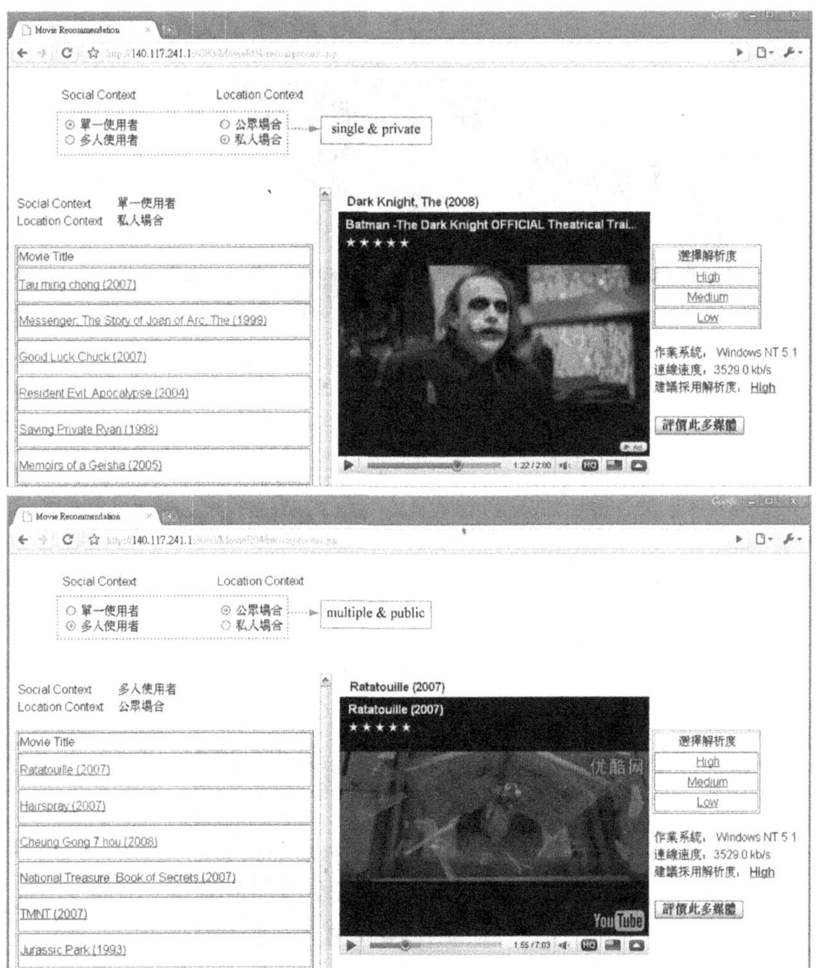

before. To avoid the data imbalanced problem and to produce more objective evaluations for different methods, we asked the user to give roughly the same number of example items for each class (like or dislike). For each item, the user needed to further express his degree of preference from the common five-scale values measurement (the degree of preference decreases from 5 to 1). This value will be used to predict user preference in the collaboration-based approach as indicated. In addition to the overall preference on the items, in the data collection process, the users were also asked to arbitrarily pick some tags from the suggested list and then provide his preferences on these tags. This is to investigate how the user commented a specific item from different feature dimensions. In the collected dataset, the user gave 12 tag-evaluations for each movie item on average.

Two types of recommendation methods (content-based and the collaboration-based) have been used. The first series of experiments is to evaluate the performance of the proposed tag-based method for the situation in which a content-based strategy is adopted for recommendation. In the experiment, the traditional keyword-based method and the tag-based method were used to represent the multimedia content respectively. For the keyword-based method, the publicly available online database Internet Movie Database (IMDb) was used, and the keywords for each movie were

Figure 6. The device information is shown and the suggestion is provided

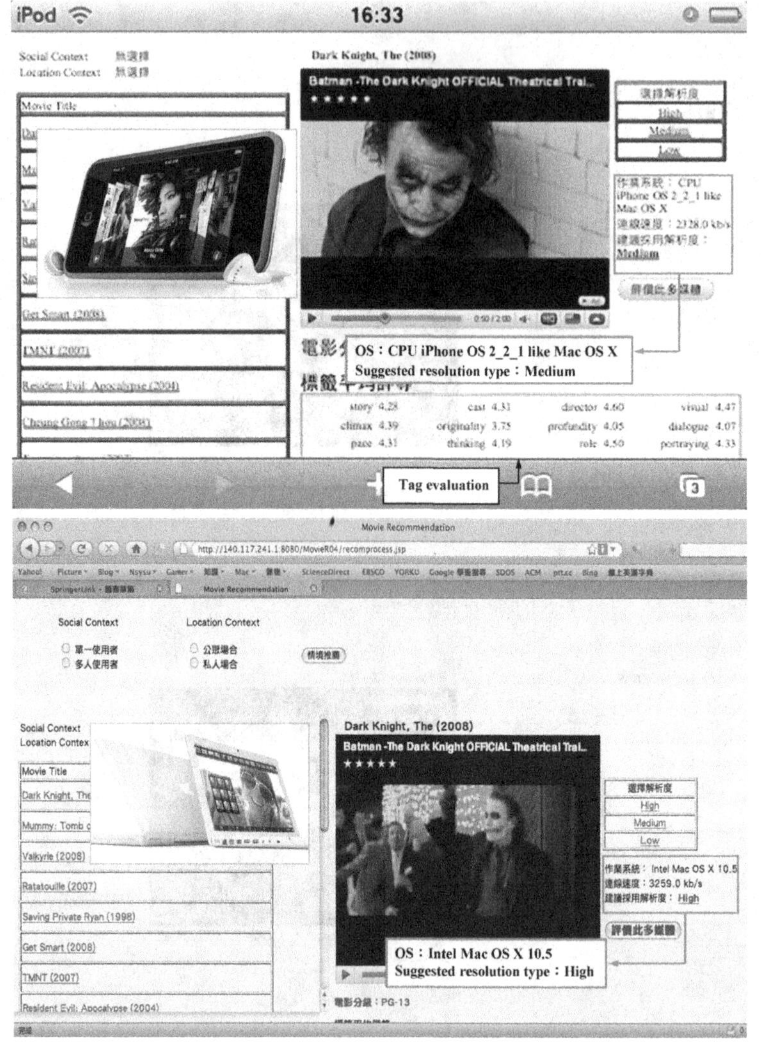

extracted to represent the movie accordingly. On the other hand, the tags listed in Table 2 were provided to users and they can arbitrarily use the tags to annotate the movies. These tags were then collected and used to build the user model for preference prediction.

To find out which of the previous representations (i.e., keywords and tags) can deliver better performance, three computationally efficient content-based methods widely used in prediction, including decision tree, Support Vector Machine (SVM), and Naïve Bayes (NB) classifier, have been used to work with the two representations. To obtain a more objective assessment, the 10-fold cross validation evaluation method was employed. Figure 7 presents the results in which the details of three popular performance measurements in classification: accuracy, precision and recall are provided. As can be seen, in all measurements the tag-based method outperforms the keyword-based method when the previous three machine learning techniques were used for user modeling. It shows that the proposed representation can more pertinently capture user characteristics in movie recommendation.

Figure 7. Comparison of different content-based methods

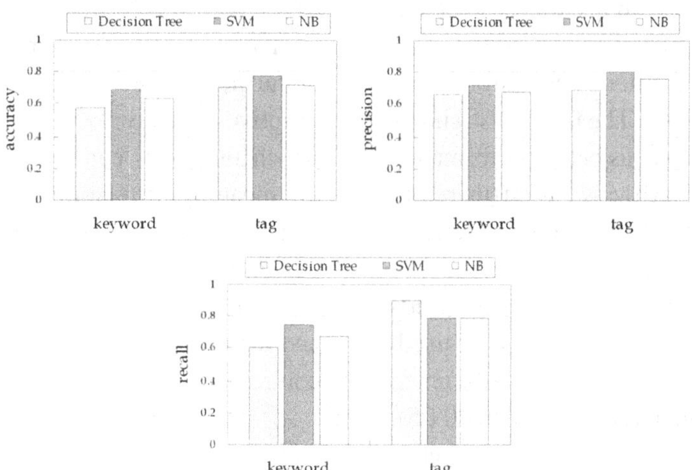

In addition to the content-based strategy, the second series of experiments is based on user-collaboration. This series was conducted to examine the efficiency of tag-based method when it works with the collaborative filtering strategy. In this series of experiments, we firstly employed the methods described in section 3.2 to measure the similarity between users. If this measuring result exceeds a certain threshold, their preferences were considered similar. This way, the nearest neighbors of a certain user can be determined. To examine the effect of information sharing by the traditional collaborative filtering method and the proposed tag-based collaboration method, we arranged three sets of experiments with different user-similarity thresholds (0.7, 0.6 and 0.5-they were the ones with best results in our preliminary tests).

Figure 8 presents the test results in which three criteria are used for performance measurement as in the experiments by content-based methods. As can be seen, the collaborative filtering methods perform better than the content-based ones. We can also observe from this figure, the tag-based

Figure 8. Comparison of the tag-based CF and the traditional CF method

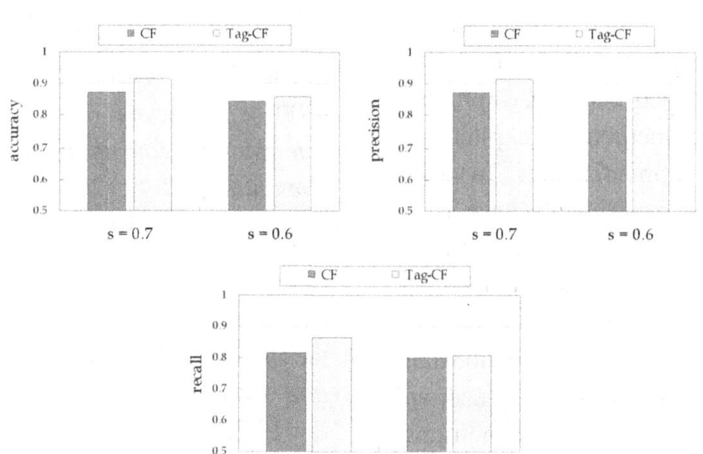

method can offer better recommendations than the traditional CF method in accuracy, precision and recall for the situation of threshold 0.7. And the results of the two methods show significant difference (t-test, α <0.05). The figure also shows that the tag-based method has better performance when a similarity threshold 0.6 was used (therefore more user opinions were taken into consideration), and the results obtained from the two methods are significantly different. But when lower thresholds (0.5 or less, meaning lower similarity) were used, the two methods show no significant difference after a statistical examination (t-test, α<0.05). These results indicate that social tags are better media to capture user characteristics and they can be used to measure user similarity more precisely. Therefore, the method based on tag measurement can deliver better performance in preference prediction.

CONCLUSION

In this work, we develop a smart TV system with three important features. The first is a body gesture-based control system, implemented to enrich the way of operating smart TVs. This control mechanism recognizes human body gestures and turns them into machine commands. We also emphasize the need of developing personalized recommendation service to help end-users to access most suitable digital contents. Considering the current trend of organizing and sharing digital content through user-created metadata, we propose a new recommendation method that exploits social tags to annotate multimedia items. To verify the proposed method, experiments have been conducted to compare it with different methods. As the social tags can capture user preferences from different semantic dimensions, it can thus give better performance on recommendation than others. In addition to the user's preference, we also consider different environmental conditions, including user location, audience, device, and network connection, and expand our work to be a context-aware system.

The work presented here shows some prospects of further research. We are now developing an adaptive strategy to maintain the tag table, in which the users can freely give their tags and the meaningful relations among the tags can be extracted and established automatically and periodically. Moreover, we plan to build an enhanced recognition system to distinguish multiple users and determine the priority for these users in TV control.

REFERENCES

Adomavicius, G., Sankaranarayanan, R., Sen, S., & Tuzhilin, A. (2005). Incorporating contextual information in recommender systems using a multidimensional approach. *ACM Transactions on Information Systems*, *23*(1), 103–145. doi:10.1145/1055709.1055714.

Adomavicius, G., & Tuzhilin, A. (2005). Toward the next generation of recommender systems: A survey of the state-of-the-art and possible extensions. *IEEE Transactions on Knowledge and Data Engineering*, *17*(6), 734–749. doi:10.1109/TKDE.2005.99.

Angelides, M. C. (2003). Multimedia content modeling and personalization. *IEEE MultiMedia*, *10*(4), 12–15. doi:10.1109/MMUL.2003.1237546.

Baldauf, M., Dustdar, S., & Rosenberg, F. (2007). A survey on context-aware systems. *International Journal of Ad Hoc and Ubiquitous Computing*, *2*(4), 263–277. doi:10.1504/IJAHUC.2007.014070.

Baltrunas, L., Ludwig, B., Peer, S., & Ricci, F. (2012). Context relevance assessment and exploitation in mobile recommender systems. *Personal and Ubiquitous Computing*, *16*(5), 507–526. doi:10.1007/s00779-011-0417-x.

Barragáns-Martinez, A. B., Rey-López, M., Costa-Montenegro, E., Mikic-Fonte, F. A., Burguillo, J. C., & Peleteiro, A. (2010). Exploiting social tagging in a web 2.0 recommender system. *IEEE Internet Computing*, *14*(6), 23–30. doi:10.1109/MIC.2010.104.

Dumas, B., Lalanne, D., & Oviatt, S. (2009). Multimodal interfaces: a survey of principles, models and frameworks. In Lalanne, D., & Kohlas, J. (Eds.), *Human machine interaction (LNCS)* (*Vol. 5440*, pp. 3–26). Berlin: Springer. doi:10.1007/978-3-642-00437-7_1.

Guan, Z., Wang, C., Bu, J., Chen, C., Yang, K., Cai, D., & He, X. (2010). Document recommendation in social tagging services. In *Proceedings of the Nineteenth International Conference on World Wide Web* (pp. 391-400). IEEE.

Hays, C., Avesani, P., & Veeramachaneni, S. (2007). An analysis of the use of tags in a blog recommender system. In *Proceedings of International Joint Conference on Artificial Intelligence* (pp. 2772-2777). IEEE.

Kim, E., Pyo, S., Park, E., & Kim, M. (2011). An automatic recommendation scheme of TV program contents for (IP)TV personalization. *IEEE Transactions on Broadcasting*, *57*(3), 674–684. doi:10.1109/TBC.2011.2161409.

Kim, H. K., Kim, J. K., & Ryu, Y. U. (2009). Personalized recommendation over a customer network for ubiquitous shopping. *IEEE Transactions on Service Computing*, *2*(2), 140–151. doi:10.1109/TSC.2009.7.

Kurkovsky, S., & Harihar, K. (2006). Using ubiquitous computing in interactive mobile marketing. *Personal and Ubiquitous Computing*, *10*(4), 227–240. doi:10.1007/s00779-005-0044-5.

Kwon, H. J., & Hong, K. S. (2011). Personalized smart TV program recommender based on collaborative filtering and a novel similarity method. *IEEE Transactions on Consumer Electronics*, *57*(3), 1416–1423. doi:10.1109/TCE.2011.6018902.

Lee, I., Kim, J., & Kim, J. (2005). Use contexts for the mobile Internet: a longitudinal study monitoring actual use of mobile Internet services. *International Journal of Human-Computer Interaction*, *18*(3), 269–292. doi:10.1207/s15327590ijhc1803_2.

Lee, W. P., & Wang, J. H. (2004). A user-centered remote control system for personalized multimedia channel selection. *IEEE Transactions on Consumer Electronics*, *50*(4), 1009–1015. doi:10.1109/TCE.2004.1362492.

Levy, M., & Sandler, M. (2009). Music information retrieval using social tags and audio. *IEEE Transactions on Multimedia*, *11*(3), 383–395. doi:10.1109/TMM.2009.2012913.

Macgregor, G., & McCulloch, E. (2006). Collaborative tagging as a knowledge organisation and resource discovery tool. *Library Review*, *55*(5), 291–300. doi:10.1108/00242530610667558.

Pudota, N., Dattolo, A., Baruzzo, A., Ferrara, F., & Tasso, C. (2010). Automatic keyphrase extraction and ontology mining for content-based tag recommendation. *International Journal of Intelligent Systems*, *25*(12), 1158–1186. doi:10.1002/int.20448.

Ricci, F., Rokach, L., & Shapira, B. (2011). Introduction to recommender systems handbook. In Ricci, F., Rokach, L., Shapira, B., & Kantor, P. (Eds.), *Recommender systems handbook* (pp. 1–35). Berlin: Springer. doi:10.1007/978-0-387-85820-3_1.

Shotton, J., Fitzgibbon, A., Cook, M., Sharp, T., Finocchio, M., & Moore, R. Blake, A. (2011). Real-time human pose recognition in parts from single depth images. In *Proceedings of IEEE International Conference on Computer Vision and Pattern Recognition*. IEEE.

Wachs, J. P., Kolsch, M., Stern, H., & Edan, Y. (2011). Vision-based hand-gesture applications. *Communications of the ACM*, *54*(2), 60–71. doi:10.1145/1897816.1897838 PMID:21984822.

Wang, S. C., Chung, T. C., & Yan, K. Q. (2011). A new territory of multi-user variable remote control for interactive TV. *Multimedia Tools and Applications*, *51*(3), 1013–1034. doi:10.1007/s11042-009-0435-0.

Willett, P., Barnard, J. M., & Downs, G. M. (1998). Chemical similarity searching. *Journal of Chemical Information and Modeling*, *38*(6), 983–996. doi:10.1021/ci9800211.

Wu, X., Zhang, L., & Yu, Y. (2006). Exploring social annotations for the semantic web. In *Proceedings of the Fifteenth International Conference on World Wide Web* (pp. 417-426). IEEE.

KEY TERMS AND DEFINITIONS

Collaborative Recommendation: It is the process of making recommendation for information or patterns using techniques involving collaboration among multiple agents, viewpoints, data sources, etc.

Context Awareness: It is about capturing a broad range of contextual attributes (such as a user's positions, activities, and his surrounding environment) to better understand what the user is trying to accomplish. Context may be applied with mobile computing with any moving entities, especially with bearers of smart communicators.

Gesture Recognition: Gesture recognition pertains to recognizing meaningful expressions of motion by a human, involving the hands, arms, face, head, and/or body.

Mobile Multimedia: It refers to the multimedia information exchange over wireless networks or wireless Internet. The advances of mobile computing devices and mobile networks, has made it possible to increase the range and complexity of mobile multimedia applications and services provided to end-users.

Personalization: Personalization technology enables the dynamic insertion, customization or suggestion of content in any format that is relevant to the individual user, based on the explicit preference details given by the users or the implicit user behavior recorded by a system.

Smart TV: It is either a television set with integrated Internet capabilities, or a set-top box for television that offers more advanced computing ability and connectivity than a contemporary basic television set.

Social Tagging: It is a term to describe the marking, saving and archiving of certain resource through the use of metadata for annotation. So Internet users can understand the content of the resource without first needing to download it for themselves.

Chapter 8
Intelligent Mobile Learning Systems for Learners with Style

Tracey J. Mehigan
University College Cork, Ireland

Ian Pitt
University College Cork, Ireland

ABSTRACT

The advent of modern wireless technologies and devices has seen a shift towards the design and development of mobile educational systems. The uptake of mobile technology will enable learners to reap the benefits of anytime, anyplace, ubiquitous learning.

The development of personalized user models, designed to meet specific learner needs through analysis of individual learning styles, can lead to improved learning outcomes for students. However within electronic and mobile learning (mLearning) systems, the development of didactic profiles for this purpose is currently achieved through the pre-completion of questionnaires by system users. This is time consuming, however, and where user specific information is excluded, standardized learning content is presented to the user, potentially leading to decreased motivation.

Recent research has indicated that it is possible to identify certain aspects of a user's learning style automatically, based on the way they interact with a system. Using various measures, researchers have been able to determine user's scores on the Global/Sequential and Visual/Verbal dimensions of the Felder-Silverman Learning Style Model (FSLSM). Spada et al. (2008) showed that that it is possible to gather learning style data through analysis of mouse-movement patterns. However, while such methods are useful in traditional eLearning environments, they cannot be employed in mobile environments. Other interactive technologies, such as eye tracking and accelerometers, offer a potential means to gather data on interaction and so facilitate the development of automatically adaptive mLearning environments.

DOI: 10.4018/978-1-4666-4054-2.ch008

This chapter discusses the development of intelligent personalized user models for mLearning. Previous research findings are reviewed, indicating that it is possible to identify aspects of a user's learning style though biometric technologies. A user interface model is presented, designed to intelligently detect the learning-style of individual's using a mobile learning environments and adapt learning content accordingly. The application of the model to a mLearning system is described.

INTRODUCTION

Meeting the needs of specific learners can lead to advantages over traditional "one-style-fits-all" teaching practices. In most cases, adaptive learning systems are based on the identification of learning-styles. The main purpose of such adaptive learning systems is the adjustment of the learning process to suit the individual learner. Individuals are presented with appropriate learning objects according to their fit within the scales of a learner-style model. Such systems can increase learners' motivation and thus improve learning outcomes.

Adaptive systems have mostly relied on the completion of learner-style and personality questionnaires to profile individual learners. However, this method of assessment can be time consuming, and thus off-putting for students, resulting in reduced motivation to learn. Research has indicated that it is possible to detect a user's learning-style through their behavior patterns when interacting with a learning system. This research has mainly focused on the use of such methods within eLearning environments. However, most existing approaches to the automatic detection of learning style (e.g., Spada et al., 2008) cannot be applied in mobile environments because of the user interaction techniques involved, for example, analysis of mouse movement patterns. It is therefore necessary to develop a method of detection suitable for use with mobile environments.

Background

Research has indicated that it is possible to identify a learner's learning-style within an eLearning system. In most cases, the research has involved the categorization of learning-style in accordance with the Felder-Silverman Learning Style Model (FSLSM). In this model, "learners are characterized by values on the four dimensions. These dimensions are based on major dimensions in the field of learning-styles and can be viewed independently from each other" (Graf, 2007). Most practical work has focused on the Global/Sequential dimension of the model.

Mouse movement patterns were examined by Spada et al., (2008) as a means of gathering data for the detection of learning-styles. This research found a high degree of correlation between the way in which an individual uses a mouse and their learning-style, as determined using the Felder-Solomon Index of Learning-Styles (FSILS) questionnaire. The researchers were thus able to predict scores on the Global/Sequential dimension of the FSLSM with a high level of confidence, based on measurements of mouse acceleration. This approach was used to determine the Global/Sequential dimension of an individual's learning-style solely through the way in which they used a mouse whilst interacting with an educational Website. Changes in mouse coordinate as the user moved the input device were recorded. Students also completed the FSILS Questionnaire, which was incorporated into the Website. The results indicate a strong correlation between maximum vertical speed of mouse movement (y axis) and Global/Sequential dimension score. The correlation coefficient was found to be $r=-0.8$, indicating that students with a lower maximum vertical speed tend to be more Sequential, while students with a higher maximum vertical speed tend to be more Global.

Other researchers have considered user behavior patterns in LMS (Learning Management Systems), including scrolling and time spent on pages, as a means of gathering user interaction data (Graf et al., 2008). Bayesian Networks (Garcia et al., 2005) and Feed Forward Neural Networks (Villivarde et al., 2006) have also been considered in the research.

Popescu (2008) and Cha et al., (2006) have focused on the Visual/Verbal dimension of the FSLSM. Both set out to identify visual and verbal learners by comparing the time spent on visual and verbal learning objects, and use the results to provide personalized content in eLearning environments.

These methods have shown positive outcomes for eLearning environments, in particular the work of Spada et al., (2008). However, in most cases the methods employed in traditional eLearning environments cannot be employed in mobile environments, for example, because mouse-based interaction is excluded. Other biometric interaction-based technologies, including eye tracking and accelerometer technologies, can be used to overcome interaction limitations presented by mobile devices, facilitating the creation of adaptive mobile learning environments. These technologies have been successfully employed within learning environments to determine user's scores on the Global/Sequential dimension (Mehigan et al., 2009; Mehigan & Pitt, 2010; Mehigan & Pitt, 2012) and the Visual/Verbal dimension (Mehigan et al., 2009; Mehigan & Pitt, 2011; Mehigan & Pitt, 2012) of the FSLSM. Therefore, these methods make it possible to identify learning-style automatically in mobile environments.

Mobile Learning

mLearning can be defined as "eLearning in your pocket" (Quinn, 2000, para. 1.) and as "any service or facility that supplies a learner with general electronic information and educational content that aids in the acquisition of knowledge regardless of location and time" (Lehner et al., 2002, pp. 24). However, there is no widely accepted definition, and those that exist are under debate.

mLearning can be categorized as situated and/or ubiquitous. Situated learning occurs as a function of activity, context and culture (Lave, 1991). In most cases, situated learning in mobile systems attempts to provide context-relative information based on the learner's particular location. GPS (Global Positioning Systems) and sensor-based technologies such as RFID (Radio Frequency IDentification) can be used to enable the learner to interact with their environment during the learning process. For example, the EXPLORE system (Costabile et al., 2008) supports middle school students on field trips to an Italian archaeological park, making the learning experience more effective and efficient. Other examples of this type of system provide context relative information for university-based orientation (Attwell & Saville-Smith 2003) and for groups in museum environments (Ting et al., 2008).

In some cases however, mLearning is focused on learner-centric tutoring systems. This is evident in the "Adaptive Geometry Game" (Ketamo, 2002) which provides a simple ITS (Intelligent Tutoring System) for kindergarten children to help them learn geometric shapes through polygon recognition using handheld computers. Commercial examples of this category also exist. For example, Dr. Kawashima's "Brain Training" for Nintendo DS could be placed within this category. "Brain Training" purports to "train" the individual brain to reduce its "age" based on the measurement of the learner's ability and skill. This system has seen a high uptake by the "older adult" population across the globe (Tokyo, 2006). Millar & Robertson (2010) indicate that using the system can lead "to improvements in both accuracy and speed of computation, and self-esteem" (p. 251).

Collaborative systems can also be represented in mLearning. These systems allow groups of students or students and teachers to combine their efforts whilst working towards a particular outcome or goal. For example, the WiTEC system (Liu et al., 2003) extends systems such as eClass, thus creating a wireless technology-enhanced classroom that "empowers the teacher and students to apply technologies to a variety of traditional and innovative learning and teaching activities seamlessly" (p. 380). By facilitating one-to-one and one-to-many computing, mobile devices enable flexibility in that they allow students and teachers to engage in highly collaborative activities (Soloway et al., 2001). Rochelle (2003) points to research that indicate both teachers and students respond well to the use of collaborative systems.

Other innovative applications for mobile-based education exist on a commercial and non-commercial level. These include probeware, knowledge aggregation, and participatory simulations (Klopfer and Squire, 2008).

mLearning vs. eLearning

Currently, there is a strong impetus toward mLearning existing as an extension of eLearning, thus placing mLearning "somewhere on eLearning's spectrum of portability" (Traxler, 2005, pp. 1-6). mLearning is also seen as a facilitator to dLearning (distance learning).

eLearning is still a developing field in its own right, with theories still under development as researchers attempt to diversify it from traditional learning. eLearning itself has seen an explosion in the development of applications while the development of theories has been slow. mLearning appears at this early stage to be taking the same track.

However, there are specific differences between eLearning and mLearning. At a pedagogical level the main differences relate to mobility. The mobile device lends portability, removing the location confines associated with classroom/terminal-based eLearning and thus decreases the limitations of learning location. mLearning encourages conversation-based learning, leading to spontaneous and flexible communication between students, and/or students and teachers. This offers significant benefits over eLearning which can be passive and scheduled. mLearning will characteristically aim at specific kinds of knowledge depending on location, situation, device and learner. Therefore, "mLearning is much more than simply eLearning through mobile devices. Mobile eLearning will become increasingly different from conventional eLearning and will create a new learning environment, an environment where learners can access content, teachers, and other learners anywhere and anytime" (Mikic et al., 2007, p. 2).

Learning-Styles and the Felder-Silverman Model

Felder and Silverman (1988) define learning-styles as the characteristic strengths and preferences in the ways individuals take in and process information. In general, learning-styles reflect how students acquire, retrieve and retain information.

There are many learning-style and personality models used for the development of adaptive eLearning systems. These include Kolb's model (Kolb, 1999), the Myer-Briggs Type Indicator Model (Pittenger, 1993), the Cognitive Style Analysis Model (Graf et al., 2009), and the FSLSM (Felder-Silverman Learner Style Model) (Felder & Silverman, 1988).

The FSLSM, originally intended for use with engineering students, has become a popular learning-style model across many disciplines in recent years. FSLSM has become increasingly popular for student analysis in the area of eLearning systems.

The FSLSM is based on four distinct learning-style dimensions, Active/Reflective, Sensitive/Intuitive, Global/Sequential, and Visual/Verbal, each of which encompasses distinct characteristics at each extreme of the dimension. For example:

- **Active Learners:** Learn best by doing something with information while *Reflective* learners prefer to think about information quietly first.

- **Sensing Learners:** Tend to like learning facts while *Intuitive* learners prefer to discover possibilities and relationships.
- **Sequential Learners:** Require information to be presented in small incremental steps of complexity. *Global* learners usually achieve a learning outcome through large leaps and bounds.
- **Visual Learners:** Remember best what they see, while *Verbal* learners gain more from text and auditory (verbal) explanations.

The main focus of the research outlined in this chapter is the Visual/Verbal dimensions of the FSLSM.

The Felder-Solomon Index of Learning Styles

The Felder Solomon Index of Learning Styles (FSILS) questionnaire is the main instrument used to measure learning-style in relation to the FSLSM. Initially published in 1991, an updated online version became available for use as a non-commercial instrument in 1997 (Felder & Spurlin, 2005). On completion of the online questionnaire, a results screen is displayed to the user (Figure 1). The results are outlined to the user as follows:

- A learner placed between a score of 1 and 3 on a dimension of the results screen is deemed balanced on that dimension.
- A person placed between 5 and 7 on one side of a dimension, will be deemed to have a moderate leaning in favor of the relevant side of that dimension scale.
- A learner placed between 9 and 11 on a dimension scale will be deemed to have a high leaning toward the relevant side of the dimension scale.

For example, the FSILS questionnaire results illustrated in Figure 1 indicate that the learner is balanced on the Active/Reflective and the Global/Sequential dimensions with scores of 1 and 3 respectively. The learner is indicated as moderately Sensitive with a score of 7.

The learner also shows a moderate leaning to the Verbal side of the Visual/Verbal dimension, again with a score of 7.

Biometric-Based Devices

Derived from the Greek words "bios" and "metric", life measurement is the direct translation of the word biometrics. There are two main categories within the field of biometrics, physical biometrics and behavioral biometrics. The focus of the research outlined is behavioral biometrics. To-date, behavioral biometrics has mainly been used

Figure 1. Individual FSILS questionnaire results

```
ACT                                  X                                REF
      11   9   7   5   3   1   1   3   5   7   9   11
                              <-- -->

SEN               X                                                   INT
      11   9   7   5   3   1   1   3   5   7   9   11
                              <-- -->

VIS                                              X                    VRB
      11   9   7   5   3   1   1   3   5   7   9   11
                              <-- -->

SEQ                         X                                         GLO
      11   9   7   5   3   1   1   3   5   7   9   11
                              <-- -->
```

for verification purposes and has been generally concerned with measuring the characteristics an individual acquires naturally over a time. This can be based on the measurement of patterns, for example, typing rhythm, gait etc. Technologies such as the accelerometer and the eye tracker can potentially provide a means of tracking behavior in this regard.

Accelerometers

An accelerometer sensor device measures the acceleration and speed of motion of an object across the x, y and z axes. This is achieved by measuring non-gravitational accelerations caused by movement and vibration of the accelerometer device. When the device is moved, an electrical output is produced proportionately to the rate of acceleration. An accelerometer can also measure vibration, shock, rotation and tilting, causes of acceleration.

In recent years, we have seen the emergence of the accelerometer in mobile technology. Laptops include motion shock sensors, while the iPhone and the iPad include an accelerometer to enable "tilt technology" to change screen orientation. The Nintendo Wii also incorporates an accelerometer device in its WiiMote for fun, interactive, wireless gaming. More recently we see accelerometers shipping as standard in high end Symbian 60/80 series mobile devices such as the Nokia N97. Since the emergence of accelerometers within mobile phone devices, research has been conducted into the potential use of such technologies to enhance game play interaction ("GamePlay Talk", n.d., para. 5), text entry and browsing (Sung-Yung et al., 2007), scrolling, zooming and scaling (Eslambolchilar et al., 2004). A non-intrusive device, the accelerometer can be harnessed to provide an efficient data gathering system, offering great potential to track user learner-styles on both the Global/Sequential and the Visual/Verbal dimensions of the FSLSM, in mLearning environments.

Eye Tracking Technology

Eye tracking technology is widely used in many disciplines, from special-needs education to commerce (Chandon et al., 2007). Eye tracking's origins lie in gaze motion research, conducted by Javel in 1879, which showed that reading text involves fixations and saccades, rather than a smooth motion of the eye in relation to the text. When reading, if a user pauses over an information area of interest a fixation occurs; a rapid movement between points of fixation represents a saccade (Salvucci & Goldburg, 2000, pp. 71-78). The emergence of eye tracking relates directly to research conducted by Javel. "Eye tracking works by reflecting invisible infra-red light onto an eye, recording the reflection of the pattern with a sensor system" ("Tobii", n.d., para. 1). The research outlined in this chapter was conducted using the Tobii eye tracking system.

Eye Tracking with Mobile Devices

To date, most eye tracking systems for use with mobile devices have relied on the use of standing devices and head mounted cameras attached, for example, to a baseball cap or spectacles. The FaceLab eye tracker and ASL's MobileEye are examples of devices that are operated in this manner. Desktop trackers can also be used with mobile devices, for example the FaceLab eye trackers have been used in cooperation with the Apple iPhone to track a user's interaction with simple mobile device applications, while Tobii eye trackers have also been used to track user interaction with Mobile devices.

Eye tracking systems are becoming smaller, less cumbersome and thus feasible for use with mobile devices, offering a new method of mobile device interaction. For example, Tobii Glasses ("Tobii", n.d.) offer a more suitable and streamlined approach to eye tracking with mobile devices as they are worn in a similar fashion to standard eyeglasses. Recent work conducted by (Miluzzo et al., 2010) facilitates the use of forward facing mobile device cameras for eye tracking purposes. As mobile technology advances, it is likely that built-in mobile eye tracker technology could be included. Other alternatives are currently emerging. Nokia, for example, are exploring Gaze Tracking Eye Ware Technology to reduce the need for intrusive head mounted technologies. Such developments indicate that there is potential to use eye tracking in mLearning environments. Eye trackers could be used to identify learning-styles through analysis of gaze and visualization patterns relating to avatar movement and environmental observation.

ADAPTIVE MLEARNING SYSTEMS

Adaptive systems for mLearning environments have to date focused mainly on the ability to adjust delivered or displayed content based on the specific device requirements. That is, there has been an emphasis on the dissemination of information, be it contextual or otherwise, dependent on device specification. This has led to technology becoming an important consideration in the design of mLearning systems. For example, de Oliveira and Amazonas (2008) propose a learning environment architecture to select learning objects for mobile devices. Their approach considered several aspects of adaptation based on the characteristics of the device itself.

In recent years, there has been a switch in focus within mLearning research toward the didactic profiling of learners. This is evident in research into MLMS (Mobile Learning Management Systems) environments conducted by Kinshuk et al. (2009), which extends the work of Graf and Kinshuk (2008). Kinshuk examines the potential of automatic adaptation with a focus on user learner-style. There is also an example of didactic profiling in the work of Meawad and Stubbs (2006) who use Bayesian networks for profiling based on learning-style models.

A number of studies have been conducted by the authors to assess the potential of biometric technologies for the detection of user learning-styles. The studies have used both accelerometer and eye tracking devices, and have examined their potential for identifying both Visual/Verbal and Global/Sequential learners in accordance with the FSLSM. However, in this chapter we focus on the Visual/Verbal dimension.

Detection of Visual/Verbal Learners via Accelerometer Devices

Based on work conducted by Popescu (2008) and Cha et al. (2006), the authors have assessed the potential of accelerometer devices to detect such learners in eLearning environments. Popescue (2008) and Cha et al. (2006) both assess Visual and Verbal learners through user behavior based on time spent on visual and verbal learning objects for the provision of personalized content in eLearning environments.

A study was conducted to assess the potential of the accelerometer for the automatic detection of learning-styles based on a learner's interaction with an eLearning application. An application was developed to incorporate a Web-based learning screen and a data acquisition model designed to log accelerometer coordinates and time duration for each user to an external file. The study was conducted to test the following hypotheses:

- **Visual Learners:** (As determined in terms of the FSLSM) will exhibit longer total time duration on visual learning content (images/graphics) than Verbal learners.
- **Verbal Learners:** (As determined in terms of the FSLSM) will exhibit longer total time duration on textual learning content than their Visual counterparts.

Positive results were obtained. A strong correlation coefficient of r= 0.78519 was achieved in respect of participants total time duration on images and their score of the Visual/Verbal dimension of the FSILS. A strong inverse correlation of r=-0.81204 was achieved in respect of the second condition, time duration on textual content using the accelerometer vs. participant score on the FSILS. Further information on the methodology used in the study can be found in Mehigan and Pitt (2011).

Detection of Visual/Verbal Learners via Eye trackers

A study was also conducted to assess the use of eye tracking technology in the detection of Visual/ Verbal learners. This work was again based on work conducted by Popescu (2008) and Cha et al. (2006).

The study was designed to assess the potential of the eye tracker for the automatic detection of learning-styles, based on a student's interaction with an eLearning application. An application was developed to incorporate a Web-based learning screen and a data acquisition model designed to log user visualization patterns using the Tobii system. The study was conducted to test the following hypotheses:

- Visual learners, as defined by the FSLSM, exhibit longer total time (fixation) duration on visual learning content (images/graphics) than their Verbal counterparts.
- Verbal learners (as defined by the FSLSM) will exhibit longer total time (fixation) duration on textual learning content than their Visual counterparts.

The study findings showed a strong correlation coefficient of r = 0.723 in respect of subjects' fixation duration on visual content and participants' score on the FSILS questionnaire. A strong inverse correlation of r = -0.77504 was established where participants' overall duration on textual content was measured against their score on the FSILS questionnaire. Further information on the methodology used in the study can be found in Mehigan et al. (2011).

THE MAPLE MODEL: FACILITATING INTELLIGENT SYSTEMS

A user interface model was developed based the findings from the studies described before. The model was designed for application to mLearning environments to facilitate the intelligent and automatic detection of user learning-style based on a user's biometric interaction data. The model is suitable for use with any mobile platform and any device category including IOS, Android OS and Web-based applications. While intended for use in mobile systems, the model could equally be applied to eLearning environments.

The model structure is comprised of a number of stages as indicated in the model diagram (Figure 2).

MAPLE's Structure

The model is comprised of three main components, the adaptation engine, and the learning environment and user interaction facilities. Each component has associated rules, which are applied at each stage of the model's implementation as described next.

Figure 2. The MAPLE model

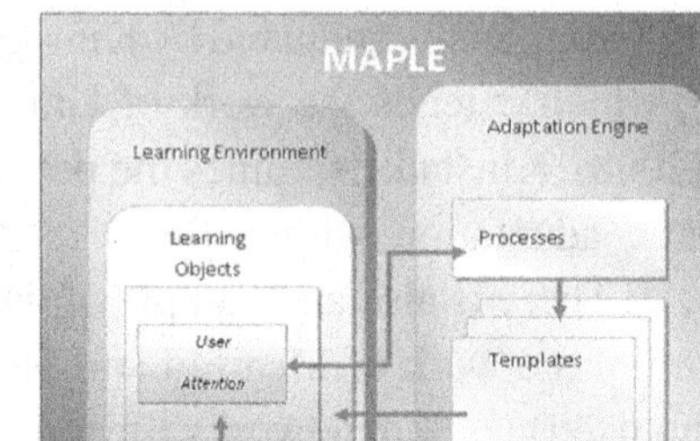

The Adaptation Engine

The adaptation engine represents the backend and intelligent aspect of the model. It holds learning object templates designed to match the required display to learner's requirements based on the adaptation engine's assessment of their learning-style. These processes are based on the data obtained from the user's interaction with a system through the learning environment associated with it.

Stored Templates

The stored templates comprise learning objects specific to the learning topic and course requirements. The templates can include visual learning objects suitable for display to visual learners, and verbal learning objects, suitable for display to verbal learners. For example, the visual content can comprise pictures, graphs, flowcharts, timelines, video, and demonstrations. The verbal learning objects might include text and auditory (verbal) explanations. This content meets the requirements for content presentation in terms of FSLSM teaching-styles, as outlined in Felder & Silverman (1988). Display content can be combined to include a balance of visual and verbal content where a student is deemed (by a system employing the MAPLE model) to have no preference for either category.

Processes

The processes gather and process data obtained from the "Learning Environment", for example, the attention paid to content by users, as demonstrated through their interaction with the learning objects displayed to them. Based on the research outlined before, the total user interaction time on particular learning objects (Visual/Verbal learners) can be used by the processes to detect the learner style of the user. While the focus of this chapter is on the Visual/Verbal dimension, the model can also be used to detect Global/Sequential learners, based on a user's maximum vertical (y axis) speed when viewing displayed content.

This is in accordance with the results attained in the studies outlined previously. The outcome of the processes determines the template to be displayed to the user as part of the overall learning environment.

The Learning Environment

The learning environment comprises the user interaction aspects and represents the frontend of any system based on the MAPLE model. The learning environment displays learning-object-content to users. The display has two phases, the initial assessment display, and a subsequent display of adapted content to suit the needs of the specific learner based on the outcome of the adaptation engine's processes.

User Interaction and User Attention

User interaction is facilitated via biometric technologies. These can include, but are not restricted to, the accelerometer and/or eye tracking technologies. User attention reflects the overall attention and focus of a learner on a specific learning object within the displayed content. A prototype mLearning environment has been developed to incorporate the MAPLE model on this basis. The prototype detects user learning-style on the Visual/Verbal dimension of the FSLSM based on learners' accelerometer interaction.

APPLYING MAPLE TO A PROTOTYPE MLEARNING ENVIRONMENT

A prototype application was developed based on the MAPLE model. The learning system application comprised elements of a course in statistics, with the topic area based on the Normal Distribution curve. Sample user interface screens are illustrated in Figure 4. The application makes use of mobile devices' built in accelerometer sensor to facilitate the main user interaction function. This particular application is designed to detect a user's learning-style on the Visual/Verbal dimension of the FSLSM based on their accelerometer interaction. On assessment of the user's learning-style, the system adapts the object-based learning content displayed to the user. The prototype mLearning system is designed to run on the Apple iPhone and iPod. However, the MAPLE model can be applied to any mobile platform and/or mLearning environment.

Creating the Prototype Learning Environment

A view-based iPhone learning system application was developed in Objective C using XCode and Interface Builder via the IOS4 SDK.

When the learning environment is initialized the system presents a splash screen to welcome the user (Figure 3). To commence interaction the user is required to touch the "Next" button. The button changes the view (or screen display) to the initial user interaction screen of the learning environment. Within the learning environment users are presented with a set of learning objects held within an initial learning-style assessment screen (Figure 4). The screen is divided into two parts representing a balance of graphical and textual learning content. Each screen section is contained within a boundary box. A cursor image is also visible on the screen, and this too is contained within a boundary box.

Figure 3. Prototype mLearning application welcome splash screen

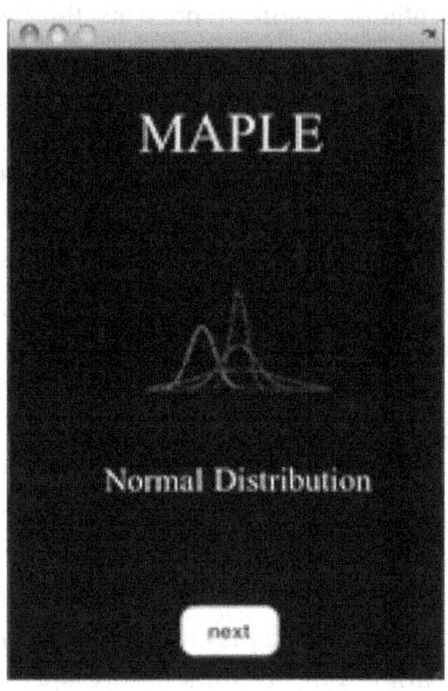

Figure 4. Prototype user assessment screen

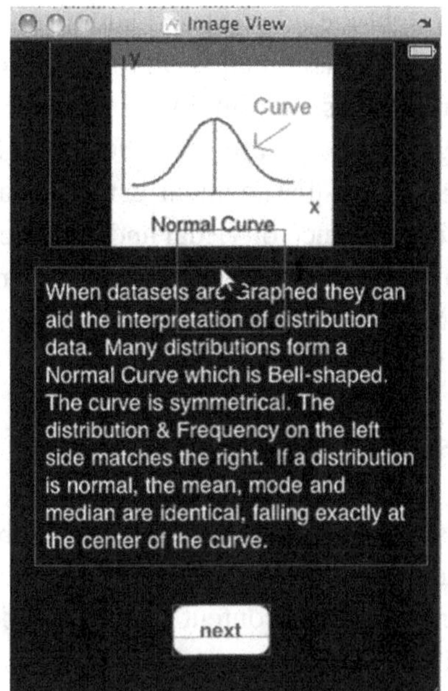

User Interaction and Adaptation Processes for Prototype Implementation

The boundary boxes for each learning object area and the cursor are each created programmatically. The cursor image boundary box is defined by its parameters, which are set with accelerometer-relative variables. This enables the re-calculation of the boundary-box coordinates depending on the current location of the cursor on screen. The location of this image (and thus its boundary) is controlled and animated via the user's accelerometer interaction.

The learning objects are individually defined by a boundary box, one to represent textual content and a second to represent visual content. However, as learning object content is stationary, the coordinates of the boundary box are fixed to specific locations on the screen. The accelerometer interaction is used to animate the cursor image to its new position when the device is active. This facilitates the user interaction component of the MAPLE system model. As the user interacts with the system the cursor image is moved on screen to the specific learning object on which they have placed their attention. A function is applied via a condition statement to monitor when the cursor image intersects the boundary of either the textual or the graphical content.

Each boundary box is assigned an individual timer variable that is given an initial value of 0. When an intersection occurs between a specific learning object boundary box (text or visual) and the cursor boundary box, the timer is incremented. When the cursor image moves away from the intersection area, the process ends and the value of the timer variable is stored until a further intersection process occurs for example, the user invokes the action of the "Next" button. This allows the system to determine the user's total overall focused attention duration for each category of learning object. The level of user attention on each learning object is measured in milliseconds.

Facilitating the Adaptation of Content

Once the learner completes their interaction with the assessment screen (by clicking the "Next" button in the case of the prototype system), measurement data is processed by the adaption engine. The learning object timer values are compared and subsequently the system decides the learning-style of the user. For the purposes of developing the prototype, focus was placed on the Visual/Verbal dimension extremes, that is, whether a learner is highly visual or highly verbal on that dimension. However, the development of a full-scale learning environment could take into consideration the other dimension score measures, for example if a learner has a moderate preference for one style or is balanced on a dimension. In the case of the prototype, if the visual learning object's timer value is higher than that of the verbal learning object's timer value, the system deems the user to be a visual learner, and vice versa. Learning screen templates for each category of learner are held within the adaption engine. Based on process outcomes, a new screen is issued to the learning environment by the system adaptation engine based on the system's assessment of the user's learning-style. The screen is displayed to the user.

The user's learning-style is logged to the console running from XCode when the application is run via a connection to the development environment. The learner screen templates are shown in Figure 5.

Inclusion of Extra Features

Once the user invokes the "Next" button on completion of the template screen, the user is then issued with a multiple choice test screen (held as a template, Figure 6).

The user is presented with possible answers to a question. A segmented control widget is used to log the user's answer to the question. For the purpose of the prototype, the answer selected by

Figure 5. Prototype template screens: a) verbal learner screen, b) visual learner screen

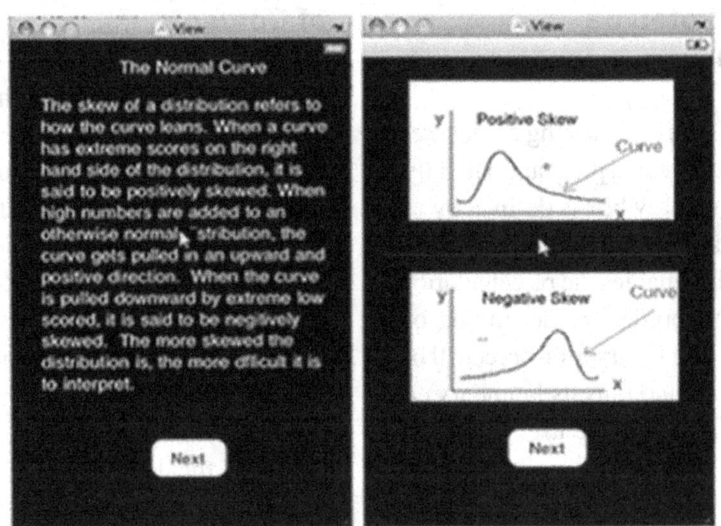

Figure 6. Prototype template screens, multiple choice test screen

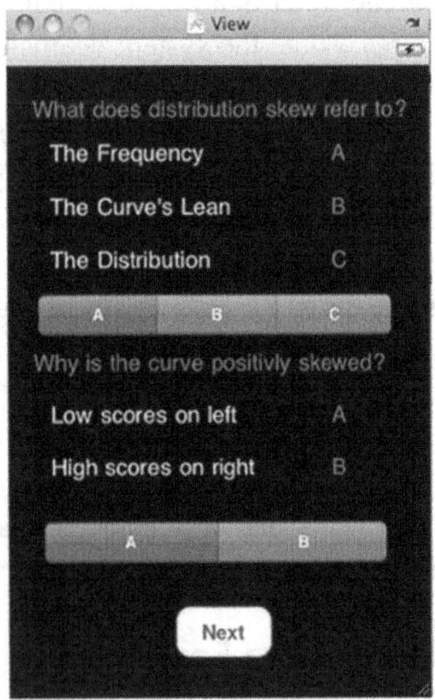

the user is logged by the system. Such extra information could be used to extend the didactic profile of the user.

While touch-based buttons are used for the purpose of navigation in the prototype system, this could be facilitated through the use of biometric commands, such as accelerometer-based gestures, voice control and/or gaze control in the development of future systems.

In the development of the prototype system the main focus was on the Visual/Verbal dimension of the FSLSM. However, MAPLE could also be applied to the Global/Sequential dimension of FSLSM, or both dimensions. Consideration could also be given to the mode of interaction; for example, MAPLE could be applied for use with eye tracking technology. MAPLE could also potentially be employed in eLearning environments for data gathering based on mouse movement patterns.

FUTURE RESEARCH DIRECTION

Learning-styles have been examined with a focus on the FSLSM, and its associated measurement instrument, the FSILS. The research outlined in this chapter has focused on two dimensions of the FSLSM, the Visual/Verbal dimension and the Global/Sequential dimension. To date, the MAPLE model has been developed for use with these dimensions. Future research will examine the potential of incorporating the remaining two dimensions of the FSLSM, that is, the Sensitive/

Intuitive dimension and the Active/Reflective dimension.

Incorporating Other Biometric-Based Sensor Devices

There are many methods for gathering biometric information through the interaction of system users, including physiological measures such as heart rate, skin conductance and speech (Mostow et al., 2011). However, these methods can be intrusive for users. NeuroSky offer a headset that uses single-channel Electroencephalogram (EEG) technology to measure voltage signals that occur between an electrode that rests on the forehead and electrodes in contact with the ear. This technology has been used to exploit EEG input for use in a reading tutor to reliably discriminate between students' reading patterns. Mostow et al., (2011) indicate that, based on their finding, they can detect transient changes in cognitive task demands or specific attributes. They report that the statistically reliable relationship between reading difficulty and EEG data illustrates the device's potential for the detection of mental states relevant to tutoring, including engagement, comprehension and learning. It is also reported that such information could help generate information based on the interplay between emotion, cognition, and learning. Work conducted by Campell et al. (2010) on the NeuroPhone system combines a similar EEG device, the "Emotive EPOC" headset, with the iPhone to establish a "Brain-Mobile" phone interface that includes a thought-controlled address book dialing application. They outline the potential of this type of system to measure understanding for foreign language teaching purposes, and the measurement of emotional state. Consequently, there is an opportunity to assess the potential of physiological-based biometric devices for example, the "NeuroSky" headset, for learning-style analysis in mLearning environments.

The incorporation of non-intrusive biometric devices in this regard could potentially increase motivation for users, resulting in an enhanced learning experience and improved learning outcomes.

Facilitating Game-Based Learning

While the prototype application of the MAPLE model represents its implementation in a learner-centric Intelligent Tutoring System (ITS), the model could also be applied to different mLearning categories for example, game-based mLearning. As 3D-enabled mobile devices emerge (such as the LG Optimus 3D), there is a potential added value for game-based mLearning applications, extending and enhancing the immersive aspect of games. This is complemented by the emergence of technology specifications such as the recent introduction of Bluetooth 4.0, an ultra-low power specification promising data transmission with lower battery consumption. Such advances in technology offer the potential to use biometric technology with mobile devices in order to gather user data related to avatar movement and environmental observation. This would facilitate the generation of didactic profiles of learners based on their individual learning-style.

As technology advances, research based on Around Device Interaction (ADI) (Kratz et al., 2009; Ketabdan et al., 2010) has emerged in the literature. The research examines the potential of ADI as an efficient interaction technique for mobile devices. This method of interaction could facilitate an extension to the interaction space of mobile devices beyond their physical boundary. Current research is focused on forms of interaction in which the device screen is not in line of user's sight, and ADI techniques can be used in combination with other interaction methods, for example touch interaction. This type of interaction technique could be significant for game-based learning and, like the accelerometer, could facilitate data gathering for learning-style analysis purposes.

CONCLUSION

This chapter has reviewed the current state of adaptive learning systems. Learning-styles have been examined with a focus on the FSLSM, and its associated measurement instrument, the FSILS. Future research will be directed toward the inclusion of the remaining dimensions of the FSLSM, these being the Active/Reflective and the Sensitive/Intuitive dimensions. As the FSLSM contains elements of other models, there is also potential to incorporate aspects of other learning-style and personality models.

The term biometrics has been defined. Biometric-based devices have been explained and examples of such devices outlined. Results from studies conducted to detect user learning-styles via biometric devices have been presented. Based on the findings of the research discussed, a user interface model called MAPLE was designed to facilitate the development of adaptive mLearning systems through user interaction. The model's structure has been described. The development of a prototype application incorporating the MAPLE model has been highlighted. The application is focused on the statistical topic "the normal distribution curve". Consideration has been given, as part of future research considerations, into the potential extension of MAPLE to facilitate further dimensions of the FSLSM, different categories of mLearning environments, for example game-based mLearning, and the inclusion of other biometric devices.

While the prototype application of the MAPLE model represents its implementation in a learner-centric Intelligent Tutoring System (ITS), the model could also be applied to different mLearning categories for example, situated and game-based mLearning.

ACKNOWLEDGMENT

Funding for this research was provided by the Irish Research Council for Science, Engineering, and Technology funded by the National Development Plan in collaboration with the Digital Hub.

REFERENCES

Atwell, J. (2005). *Mobile technologies and learning: A technology update and m-learning project summary* (Tech. Rep. No. 1.). London, UK: Technology Enhanced Learning Research Center, Learning and Skills Development Agency.

Campbell, A., Choudhury, T., Hu, S., Lu, H., Mukerjee, M., Rabbi, M., & Raizada, R. (2010). *Neurophone: Brain-mobile phone interface using a wireless EEG headset*. Paper presented at MobiHeld 2010 ACM. New York, NY.

Cha, H. J., Kim, Y. S., Park, S. H., & Yoon, T. B. (2006). Learning styles diagnosis based on user interface behaviours for the customisation of learning interfaces in an intelligent tutoring system. *Lecture Notes in Computer Science*, *4053*, 513–524. doi:10.1007/11774303_51.

Chandon, P., Hutchenson, J. W., Ardlow, E. T., & Young, S. H. (2007). *Measuring value point of purchase marketing with commercial eye tracking data* (Research Paper No.2007/22/MKT/AC-GRD). Paris, France: INSEAD Business School Global Research and Development Alliance.

Costabile, M. F., De Angeli, A., Lanzilotti, R., Ardito, C., Buono, P., & Pederson, T. (2008). *Explore! Possibilities and challenges of mobile learning*. Paper presented at the Twenty-Sixth Annual SIGCHI Conference on Human Factors in Computing Systems. Florence, Italy.

DeOliveira, A., & Amazonas, A. (2008). *Proposal of an adaptive mlearning architecture on the characteristics of mobile devices, to learning styles, to the performance and knowledge of students.* Paper presented at the International Technology, Education, and Development Conference (INTED). Valencia, Spain.

Felder, R. M., & Silverman, L. K. (1988). Learning and teaching styles in engineering education. *English Education*, *78*(7), 674–681.

Felder, R. M., & Solomon, B. A. (n.d.). *Index of learner styles*. Retrieved from http://www.engr.ncsu.edu/learningstyles/ilsWeb.html

Felder, R. M., & Spurlin, J. (2005). Applications, reliability, and validity of the index of learning styles. *International Journal of Engineering*, *21*(1), 103–112.

GamePlay Talk. (n.d.). Retrieved from http://www.gesturetekmobile.com/pressreleases/press_mar312008.php

Graf, S., & Kinshuk. (2006). *An approach for detecting learning styles in learning management systems*. Paper presented at the Sixth International Conference on Advanced Learning Technologies. Kerkrade, The Netherlands.

Graf, S. (2007). *Adaptivity in learning management systems focusing on learning styles*. (Unpublished Doctoral Dissertation). Vienna University of Technology, Vienna, Austria.

Graf, S., & Liu, T. C., Kinshuk, Chen, N.S., & Yang, S.J.H. (2009). Learning styles and cognitive traits – Their relationship and its benefits in web-based educational systems. *Computers in Human Behavior*, *25*(1), 1280–1289. doi:10.1016/j.chb.2009.06.005.

Just, M. A., & Carpenter, P. A. (1980). A theory of reading: From eye fixations to comprehension. *Psychological Review*, *87*(1), 329–355. doi:10.1037/0033-295X.87.4.329 PMID:7413885.

Ketabdar, H., Roshandel, M., & Yuksel, K. A. (2010). Towards using embedded magnetic field sensor for around mobile device 3D interaction. Paper presented Mobile HCI 2010. New York, NY.

Ketomo, H. (2002). mLearning for kindergarten's mathematics teaching. In *Proceedings of the IEEE International Workshop on Wireless and Mobile Technologies in Education*, (vol. 1, pp. 167-168). IEEE.

Klopfer, E., & Squire, K. (2008). Environmental detectives—The development of an augmented reality platform for environmental simulations. *Educational Technology Research and Development*, *56*(2), 203–228. doi:10.1007/s11423-007-9037-6.

Kolb, D. A. (1999). *The Kolb learning style inventory: Version 3*. Boston: Hay Group.

Kratz, S. G., & Rohs, M. (2009). Hoverow: Expanding the design space of around-device interaction. Paper presented Mobile HCI 2009. New York, NY.

Lave, J. (1991). *Situated learning: Legitimate peripheral participation*. New York: Cambridge U.P. doi:10.1017/CBO9780511815355.

Lehner, F., Nösekabel, H., & Lehmann, H. (2003). Wireless e–learning and communication environment: WELCOME at the University of Regensburg. *e-Service Journal, 2*(3), 23-41.

Liu, T., Wang, H., Liang, J., Chan, T., Ko, H., & Yang, J. (2003). Wireless and mobile technologies to enhance teaching and learning. *Journal of Computer Assisted Learning*, *19*, 1–14. doi:10.1046/j.0266-4909.2003.00038.x.

Meawed, F. E., & Stubbs, G. (2005). *An initial framework for implementing and evaluating probabilistic adaptivity in mobile learning*. Paper presented at the Fifth IEEE International Conference on Advanced Learning Technologies (ICALT'05). Kaohsiung, Taiwan.

Mehigan, T., Barry, M., Kehoe, A., & Pitt, I. (2011). *Using eye tracking technology to identify visual and verbal learners*. Paper presented at IEEE International Conference on Multimedia & Expo. Barcelona, Spain.

Mehigan, T., Barry, M., & Pitt, I. (2009). *Individual learner styles inference using eye tracking technology*. Paper presented at I-HCI 09, 3rd Conference of the Irish HCI Community. Dublin, Ireland.

Mehigan, T., & Pitt, I. (2011). *Learning on the move: Harnessing accelerometer devices to detect learner styles for mobile learning*. Paper presented at I-HCI 11, 5th conference of the Irish Human Computer Interaction Community. Cork, Ireland.

Mehigan, T. J., & Pitt, I. (2010). Individual learner style inference for the development of adaptive mobile learner systems. In Guy, R. (Ed.), *Mobile Learning - Pilot Projects and Initiatives* (pp. 167–183). Santa Rosa, CA: Informing Science Press.

Mikic, F., Anido, L., Valero, E., & Picos, J. (2007). *Accessibility and mobile learning standardisation, introducing some ideas about device profile (DP)*. Paper presented at the Second International Conference on Systems, ICOS'07. Sainte-Luce, Martinique.

Miller, D. J., & Robertson, D. P. (2010). Using a games console in the primary classroom: Effects of 'brain training' programme on computation and self-esteem. *British Journal of Educational Technology*, *41*(2), 242–255. doi:10.1111/j.1467-8535.2008.00918.x.

Mostow, J., Chang, K., & Nelson, J. (2011). Toward exploiting EEG input in a reading tutor. Paper presented AIED 2011. New York, NY.

Mulizzo, A., Campell, T., & Wang, E. (2010). *EyePhone: Activating mobile phones with your eyes*. Paper presented at MobiHeld 2010. New Delhi, India.

Pittenger, D. J. (1993). The utility of the Myers- Briggs type indicator. *Review of Educational Research*, *63*(4), 467–488. doi:10.3102/00346543063004467.

Popescu, E. (2008). An artificial intelligence course used to investigate students' learning style. *Lecture Notes in Computer Science*, *5145*, 122–131. doi:10.1007/978-3-540-85033-5_13.

Quinn, C. (2000). *mLearning: Mobile, wireless, in-your-pocket learning*. Retrieved from www.Linezine.com

Rochelle, J. (2003). Unlocking the learning potential of wireless mobile devices. *Journal of Computer Assisted Learning*, *19*, 260–273.

Salvucci, D. D., & Goldburg, J. H. (2000). Identifying fixations and saccades in eye tracking protocols. In *Proceedings of the Eye Tracking Research and Applications Symposium* (pp. 71–78). New York: ACM Press.

Spada, S., Sánchez-Montañés, M., Paredes, P., & Carro, R. M. (2008). Towards inferring sequential-global dimension of learning styles from mouse movement patterns. *Lecture Notes in Computer Science*, *5149*, 337–340. doi:10.1007/978-3-540-70987-9_48.

Tewissen, F., Lingnau, A., Hoppe, U., Mannhaupt, G., & Nischk, D. (2001). *Collaborative writing in a computer-intergrated classroom for early learning*. Hoboken, NJ: Lawrence Erlbaum Associates.

Ting, S. L. J., Kwok, S. K., Lee, W. B., Tsang, A. H. C., & Lee, Y. H. (March). *Dynamic mobile RFID-based knowledge hunting system in ubiquitous learning environment*. Paper presented at INTED'08. Valencia, Spain.

Tobii. (n.d.). Retrieved from www.tobii.com

Tokyo. (2006, March). Oldies get a gaming goodies. *The Sunday Morning Herald*.

Traxler, J. (2005). MLearning its here but what is it? *Interactions: University of Warwick Centre for Academic Practice*, *25*(9), 1–6.

Trompler, C., Muhlhauser, M., & Wegner, W. (2002). *Open client lecture interaction: An approach to wireless learners-in-the-loop*. Paper presented at the International Conference on New Educational Environment. Lucerne, Switzerland.

Villaverde, J. E., Godoy, D., & Amandi, A. (2006). Learning styles' recognition in e-learning environments with feed-forward neural networks. *Journal of Computer Assisted Learning*, (22): 197–206. doi:10.1111/j.1365-2729.2006.00169.x.

ADDITIONAL READING

Abowd, G., Dey, A., Brown, P., Davies, N., Smith, M., & Steggles, P. (1999). Towards a better understanding of context and context-awareness. *Journal of Handheld and Ubiquitous Computing*, *1707*(1), 304–307. doi:10.1007/3-540-48157-5_29.

Boucheix, J., & Lowe, R. K. (2010). An eye tracking comparison of external pointing cues and internal continuous cues in learning with complex animations. *Learning and Instruction ET & Pointing Cues*, *20*(2), 123–135. doi:10.1016/j.learninstruc.2009.02.015.

Busato, V. V., Prins, F. J., Elshout, J. J., & Hamakera, C. (1999). The relation between learning styles: The big five personality traits and achievement motivation in higher education. *Personality and Individual Differences*, *26*(1), 129–140. doi:10.1016/S0191-8869(98)00112-3.

Canham, M., & Hegarty, M. (2010). Effects of knowledge and display design on comprehension of complex graphics. *Learning and Instruction ET & Pointing Cues*, *20*(2), 155–166. doi:10.1016/j.learninstruc.2009.02.014.

Chang, C., & Sheu, J. (2002). *Design and implementation of adhoc classroom and eschoolbag systems for ubiquitous learning*. Paper presented at the IEEE International Workshop on Wireless and Mobile Technologies in Education (WMTE). New York, NY.

Coffield, F., Moseley, D., Hall, E., & Ecclestone, K. (2004). *Learning styles and pedagogy in post-16 learning: A systematic and critical review*. Retrieved from http://www.hull.ac.uk/php/edskas/learning%20styles.pdf

De Koning, B., Tabbers, H., Remy, M., & Riker, F. (2010). Attention guidance in learning from a complex animation: Seeing is understanding? *Journal of Learning and Instruction*, *20*(2), 111–122. doi:10.1016/j.learninstruc.2009.02.010.

Dede, C. (2013). *Immersive interfaces for engagement and learning*. Science Magazine.

Drewes, H., De Luca, A., & Schmidt, A. (2007). *Eye-gaze interaction for mobile phones*. Paper presented at the 4th International Conference on Mobile Technology, Applications and Systems. Singapore.

Dunn, R., & Griggs, S. A. (2004). *Synthesis of the Dunn and Dunn learning-style model research: Who, what, when, where, and so what?* New York: St. John's University, Center for the Study of Learning and Teaching Styles.

Eldridge, M., & Grinter, R. (2001). *Studying text messaging in teenagers*. Paper presented to the Conference on Human Factors in Computing Systems (CHI 2001). Seattle, WA.

Eslambolchilar, P., & Murray-Smith, R. (2004). *Tilt-based automatic zooming and scaling in mobile devices - A state-space implementation*. Paper presented at Mobile HCI 2004. Strathclyde, UK.

García, P., Amandi, A., Schiaffino, S., & Campo, M. (2005). *Using Bayesian networks to detect students' learning styles in a web-based education system*. Paper presented at the 7th Symposium on Artificial Intelligence. Rosario, Argentina.

Gee, J. P. (2003). What video games have to teach us about learning and literacy. *ACM Computers in Education*, *1*(1), 1–4.

Goth, C., Frohberg, D., & Schwabe, G. (2006). *The focus problem in mobile learning*. Paper presented at the Fourth International Workshop on Wireless Mobile and Ubiquitous Technology in Education. Athens, Greece.

Gulliver, S. R. (2004). Stars in their eyes: What eye tracking reveals about multimedia perceptual quality. *IEEE Transactions on Systems, Man, and Cybernetics. Part A, Systems and Humans*, *34*(4), 472–482. doi:10.1109/TSMCA.2004.826309.

Gutl, C., Pivec, M., Trummer, C., Garcia-Barrios, V. M., Modritscha, F., Pripfl, J., & Umgeher, M. (2005). AdeLE (adaptive e-learning with eye tracking), theoretical background, system architecture and application scenarios. *European Journal of Open. Distance and E-Learning*, *2*(1), 1–6.

Haider, H., & Frensch, P. A. (1999). Eye movement during skill acquisition: More evidence for the information-reduction hypothesis. *Journal of Experimental Psychology. Learning, Memory, and Cognition*, *25*(1), 172–190. doi:10.1037/0278-7393.25.1.172.

Hooper, R. (2008). Educational technology: A long look back. *British Journal of Educational Technology*, *39*(2), 234–236. doi:10.1111/j.1467-8535.2008.00813.x.

Hoppe, U., Lingnau, A., Machado, I., Paiva, A., Prada, R., & Tewissen, F. (2000). *Supporting collaborative activities in computer integrated classrooms - The nimes approach*. Paper presented at the 6th International Workshop on Groupware (CRIWG). Madeira Island, Portugal.

Howard, R. A., Carver, C. A., & Lane, W. D. (1996). *Felder's learning styles, Blooms taxonomy, and the Kolb learning cycle: Tying it all together in a CS2 course*. Paper presented at the Twenty-Seventh SIGCSE Technical Symposium on Computer Science Education. Philadelphia, PA.

Ivory, M. Y., & Hearst, M. A. (2001). The state of the art in automating usability evaluation of user interfaces. *ACM Computing Surveys*, *33*(4), 470–516. doi:10.1145/503112.503114.

Jacob, R. J. K., & Karn, K. S. (2003). Eye tracking in human-computer interaction and usability research: Ready to deliver the promises. In Hyona, J., Radach, R., & Deubel, H. (Eds.), *The mind's eye: Cognitive and applied aspects of eye movements* (pp. 573–605). London, UK: Elsevier.

Jarodzka, H., Scheiter, K., Gerjets, P., & Van Gog, T. (2010). In the eyes of the beholder: How experts and novices interpret dynamic stimuli. *Journal of Learning and Instruction*, *20*(2), 145–146.

Junho, S. (2013). Roadmap for e-commerce standardization in Korea. *International Journal of IT Standards and Standardization Research*, *3*(2).

Meehan, P., & Moloney, K. (2010). *Basic principles of operation and application of the accelerometer*. Retrieved from http://www.lit.ie/departments/seit-docs/Accelerometer-Eng-Leaving-Cert-Topic-2010.pdf

Meyer, S. A. (1959). *A program for elementary arithmetic, present and future*. New York, NY: Wiley.

Muhlhauser, M. (2004). *eLearning after four decades: What about sustainability?* Paper presented at the World Conference on Educational Multimedia. Lugano, Switzerland.

Pahler, H., McDaniel, M., & Rohrer, D. (2009). Learner styles: Concepts and evidence. *Psychological Science in the Public Interest*, *9*(3), 105–119.

Porto, M. (2008). *Implementing eye-based user aware e-learning*. Paper presented at CHI'08. Florence, Italy.

Prensky, M. (2001). Digital natives, digital immigrants. *Horizon*, *9*(5), 1–6. doi:10.1108/10748120110424816.

Sung-Jung, C., Murray-Smith, R., Chang Kyu, C., Young Hoon, S., Kuanghy, L., & Yeun Bao, K. (2007). *Dynamics of tilt based browsing*. Paper presented at CHI'07. Seattle, WA.

Williamson, S. D., Squire, K. R., Halverson, R., & Gee, J. P. (2005). Video games and the future of learning. *PHI Delta Kappa*, *87*(2), 105–111.

Zhai, S. (2003). What's in the eyes for attentive input. *Communications of the ACM*, *46*(3), 34–39. doi:10.1145/636772.636795.

KEY TERMS AND DEFINITIONS

Accelerometer: An accelerometer is a biometric sensor device that measures the acceleration of movement and vibration on the x, y, and z-axes.

Adaptive Systems: A system that adapts to meet the needs of specific users based on an existing or automatically generated user model.

Eye tracking: The charting, by a system, of a user's eye-movements as they visualize/interact with a screen.

Human Computer Interaction (HCI): HCI represents the relationship that exists between a human and their interaction with a computer/ machine.

Learning-Styles: Learning-styles represent a student preference for how they receive information and their preference for the subsequent processing of that information.

mLearning: Learning system that incorporate mobile devices/technologies to facilitate ubiquitous learning.

User-Model: A representation of a system user's individual characteristics for use by that system.

Section 3
Accessible Technology

This section provides some emerging mobile-multimedia applications for accessible computing.

Chapter 9
Monitoring User's Emotions Using Brain Computer Interfaces

Katie Crowley
University College Cork, Ireland

Ian Pitt
University College Cork, Ireland

ABSTRACT

This chapter discusses the use of commercial Brain Computer Interfaces to monitor the emotions and interactions of a subject as they use a system. Tracking how a user interacts with a system, and the emotion-based responses that are invoked as they interact with the system, yield very valuable datasets for the development of intelligent, adaptive systems. The proliferation of mobile devices as an emerging platform offers scope for the development of the relationship between Brain Computer Interfaces and mobile technology, towards ubiquitous, minimally invasive, mobile systems.

INTRODUCTION

Recent growth in the eLearning sector has resulted in the rising popularity of interactive and adaptive systems as learning tools. Mobile learning, interaction design and serious game design are all areas that are expanding as the mobile device market grows. Learning and assessment go hand in hand, and the importance of monitoring the learning progress of a user is paramount to any pedagogical tool. In fact, monitoring the interaction of a user with any system is a very important task. Gathering information on how the user interacts with a system/device can provide invaluable data to the system, the designer, the instructor and even the user themselves. With growth of ubiquitous computing, how a user feels is also very valuable feedback data for intelligent and adaptive systems.

DOI: 10.4018/978-1-4666-4054-2.ch009

The growth of mobile devices, along with advances in minimally invasive wearable sensors, offers promising scope for the development of intelligent, adaptive, sensor-based mobile devices. What applications are there for such devices? This chapter discusses Brain Computer Interface systems and some of their applications when used with mobile devices. We discuss the state of the art in mobile Brain Computer Interfaces and their role in tracking user response and interaction with a system, as well as highlighting the benefits, drawbacks and scope of using BCI mobile devices to monitor user's emotions and interactions. Studies will be reviewed that reflect how BCIs can be used to measure secondary data (e.g. emotional response data) relevant to assessment, learning and emotional response.

BACKGROUND

User Interaction

As a subject uses a system, a large amount of data can be gathered including assessment and interaction data. This data can be categorized into primary and secondary data. Primary data is any essential information that must be collected by the assessment process. This would include answers or responses given by the user to the system, for example, multiple-choice questionnaires. A system is usually designed with the purpose of gathering such primary data. Secondary data is data that is often indirectly captured within a scenario or during an interaction, and that helps advance the system or assessor's understanding of the user's action.

There are various reasons why it relevant to monitor how a user interacts with a system. Monitoring how a subject uses a system makes the system interactive and potentially adaptive, and this feedback is important for intelligent human-computer interaction. The most effective models for interactive systems work with a three phase approach: (1) Evaluate (2) Compare (3) Adapt. The evaluation phase can consist of assessing primary or secondary data. Monitoring secondary data such as how a user interacts with a mobile device can be a minimally invasive method of developing an intelligent, adaptive, interactive system.

An array of literature exists in the human-computer interaction domain that highlights the relevance and importance of monitoring and tracking user interaction and experience (Goecks, 2000; Cockburn, 2001; Jacob, 2003; Thüring, 2007; Coyle, 2008). Dedicated disciplines have been established that are concerned with understanding how the user interacts with a system in order to improve system design (interaction design), user experience (usability), augment user feedback (human factors) and enhance learning, to name but a few. Maglio et al (2000) believe that the future of user interaction with pervasive devices lies in *attentive user interfaces* (AUI), systems that attend to user actions by monitoring the user through sensing mechanisms such as gaze, speech, hand gestures, or by multimodal devices that track keystrokes, mouse input or even accelerometer actions.

Attentive systems can monitor user behavior, model user goals and interests, anticipate user needs, provide users with information and interact with users. A system that monitors user interactions can, for example, assess the ability or learning style of a subject and adapt the difficulty (or other parameter of the system) accordingly (Mehigan, 2012). A system that monitors interactions can profile a user in terms of their response to various aspects of the system; including emotion-based responses such as attention and meditation, which can reflect the applied effort of the subject (Crowley & Sliney, 2011).

Brain Computer Interfaces

A Brain Computer Interface is a communication system in which messages or commands that an individual sends to the external world do not pass through the brain's normal output pathways of peripheral nerves and muscles (Crowley, 2010; Molina, 2009; Wolpaw, 2002).

The human brain is made up of billions of nerve cells (neurons), which emit electrical impulses and hemodynamic measurements when interacting. The electrical impulses form a measurable voltage on the scalp, which can be detected by Electroencephalogram (EEG) devices (both invasive and non-invasive methods). Hemodynamic measurements include functional Magnetic Resonance Imaging (fMRI), positron emission imaging (PET), and functional near-infrared brain monitoring (fMIRS). Brain Computer Interfaces (BCIs) that use non-invasive and minimally invasive EEG methods are the preferred systems due to their cost effectiveness, ease of acquisition and high time resolution. BCI systems acquire EEG signals from the brain and translate them into digital commands that can be recognised and processed on a computer or computers using advanced algorithms (Wolpaw, 2002). The result of such commands can allow a user to modulate his/her brain activity to accomplish an intended action. BCIs depend on the interaction of two adaptive controllers: the user and the BCI system. The user must generate the intent; encoded as a brain signal and the system must then translate these signals into commands that can accomplish the user's intention.

BCIs were once confined to the research laboratory as cumbersome and expensive EEG systems that required an expert operator. In recent years, a number of commercial, low-cost EEG type devices have come to the market. The medical use of EEG dates back to the last century; however, it is only since the 1970s that EEG signals have been examined to explore Brain Computer Interfaces (BCIs). Medical grade EEG analysis examines a number of electrophysiological parameters including *sensorimotor activity, P300, Steady State Visual Evoked Potentials (SSVEPs),* and *slow cortical potentials.* These are beyond the scope of this chapter, which focuses on commercial, low-cost BCI devices that can be used with mobile devices. For further information on these signals see the additional reading section (Bashashati, 2007; Molina, 2009; Wolpaw, 2002; etc.).

The latest commercial, low-cost devices are portable, minimally invasive BCIs that have many applications for human-computer interaction. Current BCI devices typically consist of fewer sensors than the research or medical grade equivalents. The cheapest commercial BCIs consist of one or two sensors that are fitted to a lightweight headset that can be comfortably worn by a user for a length of time. Essentially, BCIs measure the brain-wave activity of a subject, and some of this activity reflects emotion-based responses of the user. The emotional response of a user as they interact with a system is potentially very valuable secondary data. Recent work has demonstrated the effectiveness of Brain Computer Interfaces as minimally invasive technologies to monitor emotion-based responses from a user during a task or system interaction (Crowley & Sliney, 2011; Crowley & Sliney, 2010).

An EEG signal essentially is a voltage that is measured on the surface of the scalp, arising from neural activity e.g. mental state, cognitive activity etc. Fluctuations in the EEG signal occur within defined frequency bands that have been associated with brain states such as attention, engagement, frustration, meditation and so on. Changes in the signal within these frequencies bands can be measured by EEG devices, which reflect changes in neural activity.

The latest BCIs in the form of low-cost EEG headsets, have gained recent popularity in the gaming industry. Two companies, Emotiv and NeuroSky, have released commercial BCIs, which read neural activity to control consumer devices that can be used for gaming. With the advancement of mobile devices and sensors, such commercial EEG devices are also currently being developed for use with mobile phone applications. This combination presents huge scope for the development of intelligent mobile systems that are adaptive, minimally invasive and portable. BCI technology is now pervading a variety of domains such as entertainment, assistive technologies, eLearning, security and HCI.

NeuroSky technologies have developed a minimally invasive, dry, biosensor to read neural activity representing states of attention and meditation (relaxation). NeuroSky's first headset, *MindSet,* was a low-cost, easy to use EEG developed for non-clinical human-computer interaction. The MindSet consisted of three electrodes that touch the skin at three different locations: beneath the ears and the forehead. A newer version of the MindSet, the *MindWave*, consists of a single dry sensor position at the forehead, known as FP1, to capture pre-frontal cortex activity in the front of the brain, where higher thinking occurs. Emotions, mental states, concentration, etc. are all dominant in this area. The MindWave captures neural signals from 0 to 100Hz and provides information on a user's Delta, Theta, Alpha, Beta and Gamma brainwave band power levels. The signals captured are used as inputs to NeuroSky's proprietary algorithms that generate *eSense* values; custom measures of *attention* and *meditation* (NeuroSky, 2009). The algorithm returns one number per second on a scale from 0 to 100, representing the level of attention or meditation of the subject (Rebolledo-Mendez, 2009) (See Figure 1).

Emotiv produce the *EPOC* headset, which uses 14 data collecting electrodes and 2 reference electrodes. The electrodes are placed on the head in roughly the international 10-29 system which is an EEG standard system. The EPOC headset, like the MindWave, transmits encrypted data wirelessly to a Bluetooth device. Each headset operates with its own proprietary chip. The EPOC headset detects various facial expressions, as well as levels of engagement, frustration, meditation and excitement. The Emotiv headset also has a built-in gyroscope that detects the change of orientation of the subject's head.

Figure 1. NeuroSky MindWave headset (left) and Emotiv EPOC headset (right). © 2012 NeuroSky and © 2012 Emotiv. Used with permission.

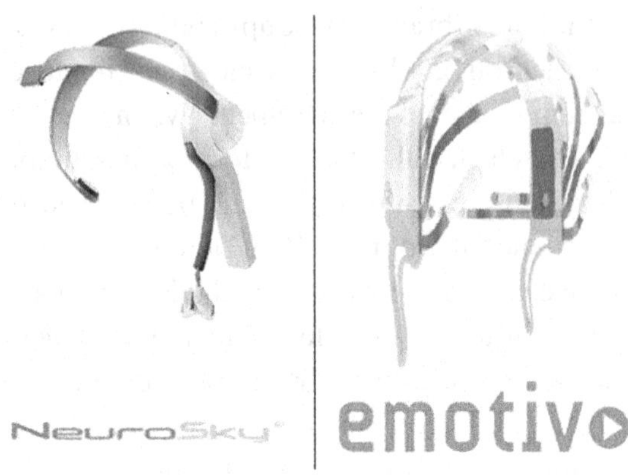

We have chosen to use NeuroSky's headset, the MindWave for our experiments as it offers the most user-friendly option for novice users, and therefore we feel is more commercially viable with mobile devices and other HCI applications. While the MindWave offers the option of reading the raw, unfiltered EEG waves, we have used the custom eSense meters in our studies to examine how accurately they can measure the emotions of attention and meditation.

MAIN FOCUS

Physiological Data and Biometrics

The term *biometric* derives from the Greek words "bio", meaning biological life, and "metric" meaning measure or measurement. A biometric identifier is a distinctive, measurable characteristic used to label and describe an individual. There are two categories of biometrics: physiological and behavioral. Behavioral biometrics measure characteristics an individual acquires naturally over time or action patterns, for example, the dynamics of writing one's signature or typing on a keyboard. Biometrics associated with emotional response are primarily physiological. These could include iris patterns, voice, facial expressions and even neural signals.

When a subject is aroused or experiences an emotional response a number of physiological elements may be affected. An expressive reaction may occur, where a subject may smile or frown, accompanying the emotional response with facial, vocal or postural changes. Physiological reactions may also occur, such as changes in heart rate, skin conductance, blood pressure and pupillary response. These physiological reactions, along with brain waves fluctuations are some of the factors influenced by the autonomic nervous system (ANS) that is activated by emotions.

For many years, scientists and psychologists have used physiological sensors to measure and trace some of the most common changes that occur during emotional response. The polygraph system (used for lie detection) is a well-known example of physiological measurement. It traces blood pressure, pulse, respiration, and skin conductivity, usually for truth verification purposes. The subject is seated and a number of electrodes and devices are connected to him/her. Of these individual physiological measures, the systems that are most used are pulse and skin conductivity. These are measured using pulse rate monitors and galvanic skin response systems respectively.

In experimental laboratory settings, pulse rate is usually measured in real-time using a sensor that is often attached to the subject's finger. Galvanic Skin Response (GSR) measures skin conductivity also from the fingers. This is done by attaching an electrode to the finger of the non-dominant hand of a subject. A current is passed through the skin and it's resistance to passage is measured. Any change in this resistance is recorded. The change is caused by the degree to which a person's sweat glands are activated. If a person is relaxed, their skin has low electrical conductance and high resistance. The more active the sweat glands, the less resistance yielded. GSR devices measure the level of arousal or emotional activation of a subject.

In our studies to date we have used a number of biometric devices to gather data about the emotional response of a user during a task. Experiments have been designed using a number of different stressors intended to elicit emotional responses from subjects. Stressors have included psychological tests, affective images, affective sounds and so on. Biometric devices that we have used include GSR, Eye Trackers and commercial BCIs. Eye Trackers have been used to measure and trace the pupillary response (or pupil dilation) of a subject as a measure of stress (Backs, 1992).

The main issues with GSR devices are their invasive and restrictive natures. Substantial literature in psychophysiology supports the use of these devices to measure emotional response for analysis and tracking purposes. As these devices measure physiological response in real-time it is therefore possible that they could act as physiological controllers for systems and devices. However, the requirement that the subject sit as still as possible, in a dedicated laboratory setting makes these systems unsuitable for mobile devices. The Eye Tracking system that we use in our laboratory is the *Tobii* system, which uses a dedicated eye-tracking monitor. Consequently, this also requires the use of a dedicated laboratory setting and movement restrictions for the user. Tobii have released *Tobii* Goggles that are head mounted and do not require a specialized monitor, consequently they are less restrictive, however they are still currently financially prohibitive for the mass market. A more appropriate solution for widespread mobile development is to use commercial BCIs to measure and trace emotional response.

Biometrics and Mobile Devices

While applied psychologists and psychphysiologists have used biometric suites such as the polygraph for a number of years, it is only in recent years with the advancement of technology that newer biometric technologies have begun to be explored. With the decreasing cost and size of technology, these newer, more innovative biometric measures can now be used with many mobile devices. Biometrics such as retinal, finger and palm print scanners, eye-trackers, voiceprints, and hand-writing pattern analysis are all gaining popularity as technology advances. Many biometrics are now being used as security measures and for personal identification purposes. As the demand for biometric technology increases, so does its potential for integration with mobile devices. Biometric measures such as accelerometers and eye-trackers are now available for use with mobile devices. Such integrations offer great scope for future development of security and commercial technologies, medical devices, learning and many leisure tools such as gaming.

HCI researchers and developers know that the more information and data you can gather about your user, and how they interact with the system the better the system will be. A security device that can gather multimodal biometric data on a user will be more robust than a simple model password-protected system for example. A pedagogical tool or game that can use biometrics to determine the engagement level of a subject, or assess if the difficultly level of the game is inappropriate for the user, will ultimately be a more adaptive and intelligent system (Mandryk, 2006). Systems that can automatically adjust their settings depending on feedback from a biometric measure are key to truly intelligent and adaptive computing. Integrating such adaptivity with mobile devices has ever increasing scope as technology advances. Mehigan (2012) has shown that biometric technologies such as accelerometer and eye-tracking devices, when integrated with mobile devices geared toward mLearning, can return informative user information including learning style and didactic profiling. A further layer can be added to such systems by including affective computing (computing that is concerned with emotions [Picard, 1995]). An adaptive, intelligent system that can automatically adjust to a user's preferences based on emotion-based biometrics has almost limitless potential. Affective BCI systems can adjust their settings to keep the user motivated, involved and engaged in their interaction with the system, as well as providing emotion-based feedback on factors such as attention, satisfaction or frustration. Intelligent, affective BCI systems that acquire the past and current emotional state of the user could potentially predict user intentions and minimize errors as well as required effort and interaction with the system.

Applications of BCIs

BCIs are physiological, biometric interfaces that allow communication directly between the human brain and an external device (Brower, 2005). Such devices can take many forms, from desktop applications, gaming hardware, mobile phones or assistive devices. The wireless, minimally invasive nature of current low-cost commercial BCIs offers huge advantages to any user requiring hands-free control over devices. Such systems present great scope to people with disabilities, particularly motor disabilities, offering interfaces to control assistive devices such as wheelchairs, or even prosthetics.

In addition to their role as assisting/control devices, BCIs are now proposed for a wider range of applications. Research on BCI technology is active across a number of disciplines including biotechnology, computer science, cognitive science, biomedical engineering and neuroscience. NeuroSky, the manufacturers of one of the BCIs mentioned earlier, have a commercial focus on the gaming and leisure markets, while actively assisting research on BCI applications for healthcare, cognitive science and computer science.

Some specific examples of modern BCI applications include:

- Cyberkinetics Neurotechnology Systems has developed the *Brain Gate Neural Interface System*, a medical device designed to help patients with spinal cord injuries to control external devices.

- The Honda Motor Corporation and ATR Computational NeuroScience Laboratories have researched using brain signals to control a robot's movements.
- Scientists at Helsinki's University of Technology are using a subject's neural signals to control virtual keyboards by simply thinking about hand movement.
- Canadian researchers use BCI systems as the basis of biometric identity-authentication.
- Neural Signals groups have released a BCI-based speech-restoration project that uses invasive (implanted) EEG sensors.

BCIs with Mobile Devices

Brain Computer Interfaces, as we have described, essentially create a direct communication pathway between the brain and an external device. To date, gaming has been the most widely used application for BCI devices. However, as the power of modern mobile devices grows, the scope to extend BCI technology into the field of mobile communication also expands.

Recently, with advances in integrated circuit technology, mobile devices combined with DSP (Schneiderman, 2010) and built-in Bluetooth functionality has become very popular in the consumer market. The ubiquity, mobility and processing power of mobile phones make them a potentially valuable tool for creating portable BCIs for real-world environments.

Brain-mobile phone interfaces use BCIs to detect neural signals as a subject interacts with their mobile device. The BCI signals can then be wirelessly transmitted from the headset to a mobile device to be processed, interpreted, and used to drive mobile applications.

Based on these requirements, researchers at Dartmouth College have developed a prototype application, *NeuroPhone*, which automatically dials contacts in a mobile phone's address book based solely on neural cues from the user (Campbell, 2010). These cues are sent via a minimally invasive wireless EEG headset from the user's brain to the mobile phone. During trials, the dialing accuracy of the brain-mobile phone interface measured over 70 percent when the subject was sitting and about 60 percent when the subject as standing. While there are still a number of challenges to be overcome with brain-mobile phone interfaces, these preliminary results are promising.

While NeuroPhone is a research development product, commercial BCI vendors are also moving into the mobile device market. NeuroPhone uses Emotiv technology, and at the date of publication of this article, it has not released its own mobile device controller. Other companies however, are taking full advantage of this growing commercial market. *XWave,* released in 2010, is the result of a partnership between *PLX Devices Inc.*, a cutting-edge consumer electronics company, and *NeuroSky*, a developer of commercial BCI headsets.

The XWave is a wired BCI interface for Apple iOS devices. It uses NeuroSky's dry sensor and connects to the iPhone via the regular jack. PLX's reasoning for the wired device is that the Bluetooth connection on mobile devices can be temperamental. Since the release of XWave, NeuroSky have released its own mobile version of their MindWave headset, *MindWave Mobile.* It is marketed as "the world's first comprehensive brainwave reading device for iOS and Android platforms", however, just like the MindWave, and its predecessor the MindSet, it also works on PC and Mac desktop platforms.

BCI developers are utilizing the mobile aspect of these commercial devices to design BCI-based systems that work with existing mobile technology and devices. This level of portability offers great scope for developing adaptive, interactive systems that can be tailored to the user's emotional response.

Monitoring Emotions with BCIs

For years, psychologists have promoted the importance of emotional skills, an area that has often been overlooked often by scientists. In the past, emotions were regarded by scientists as being non-scientific, and carried less weight than rational thought and logical thinking. However in recent years, as cross-disciplinary research increases, and particularly in the area of human-computer interaction, scientific research into human emotional response is growing.

Rosalind Picard, who is based in the MIT Media Lab, coined the term *affective computing* to describe "computing that relates to, arises from, or influences emotions" (Picard, 1995). The principles behind affective computing and the analysis of emotion-based responses is that understanding how a user feels when using a system leads to more intelligent, more adaptive and ultimately more useful systems.

In recent years, scientists along with psychologists have shown the importance of emotions in many areas from learning to usability. Affective computing is concerned with how these relationships can be maintained and exploited in conjunction with technology. Recognizing affective feedback is important for intelligent human-computer interaction. A tutoring application that can recognize emotion, for example, will be a far more effective pedagogical system. A system that recognizes the emotional response of a student may adapt content or pace to make the learning process more engaging and effective for the user.

It is not just learning that can benefit from affective feedback. Everything from gaming to online applications can benefit from the knowledge of how a user responds emotionally when interacting with the system. The relevance and importance of tracking user interaction is well documented. Adding emotional response as a feedback measure makes interactive systems even more powerful. Knowing how a user responds emotionally when interacting with a system, or any product, enhances our understanding of the level of enjoyment, engagement, frustration and so on that the user may be experiencing.

Emotions are psychophysiological responses associated with a variety of expressed subjective feelings and observable behaviors, usually accompanied by changes in autonomic body state. Many factors can influence human emotions. Mood, temperament, personality, habituation, personal circumstance, disposition, motivation, and even hormonal changes can all influence our emotional response. Emotional responses can also be elicited with various triggers or stimuli that can act on a variety of emotions such as fear, happiness, stress, anger, sadness, frustration and so on.

Psychologists and cognitive scientists have typically used subjective methods to measure emotions, such as self-reports, checklists and questionnaires. Applied psychologists and psychphysiologists use objective methods of instrumentation such as the polygraph, and its individual measures (GSR, pulse rate, etc.) to monitor emotional response in real-time. Objective methods examine the physiological manifestations of emotional response from both the autonomic (GSR, pulse rate, temperature, respiration, facial expression, voice) and central nervous systems (EEG/BCI).

Emotional theory is a huge area of psychology and cognitive science, and despite extensive research there is presently no universally accepted model to categorize emotions. Some theories define finite sets of emotions, while other propose that emotions are better measured as differing in degree on one or another dimension. The specifics of emotional theory are beyond the scope of this chapter, however some additional reading references can be found after the reference section.

One model that is regularly used is the model of *affect* by Russel (1980). Affect is also the term used by Picard (1995) in relation to computing that "relates to, arises from or influences emotions". Russel's model uses two orthogonal dimensions to model emotions: *valence* (pleasantness or hedonic value) and *arousal* (physiological activation relating to the affective state). This "arousal", or physical response evoked by the emotional reaction is the measure examined by systems such as GSR.

As previously outlined, GSR devices are not very suitable for use with mobile devices, and are quite restrictive on the subject. The minimally invasive, portable nature of commercial BCI headsets offers greater scope for emotion detection and measurement. Much research has been carried out using medical grade EEG to detect emotions (Horlings, 2008; Murugappan, 2008; Savran, 2006). As commercial BCI headsets become more popular, research on using these devices for emotion detection is increasing. While categorizing specific emotional responses is challenging due to the lack of clear definitions of individual emotions, it is possible to identify some responses that can be attributed to "standard" emotion labels.

We discussed earlier in the chapter how fluctuations in different EEG bands are attributed to different neural activity. Some of these bands relate to emotion-based responses, and concentrating on these frequencies, we can capture emotional response data. The *delta* band (0-4Hz) traces deep sleep/unconsciousness; *theta* (4-8Hz) traces drowsiness/deep relaxation; *alpha* (8-13 Hz) traces relaxation/meditation; *beta* (13-30Hz) traces focus/attention/concentration; *gamma* (30-100Hz) relates to sensory processing and short-term memory and finally *mu* (8-13Hz) shows rest state of motor neurons.

There are a number of EEG correlates of emotion and specific mental activities that produce distinctive patterns in the signal. These actions, and their corresponding EEG correlates are termed as electrophysiological sources of control (Bashashati, 2007). As previously mentioned, detailed discussion of these sources is beyond the scope of this chapter, however some key EEG correlates of emotions include:

- **Frontal EEG Asymmetry:** Where front activity is characterized in terms of decreased magnitude of the alpha band (8-13Hz) has been consistently found to be associated with emotional states (Coan, 2004).
- **Event Related Desynchronisation/ Synchronisation:** Relatively greater right hemisphere synchronization for negative stimuli and greater left hemisphere synchronization for positive stimuli.
- **Event Related Potentials (ERP):** Affective stimuli show an effect on ERPs under emotional response (Olofsson, 2008). Both valence and arousal have been noted to produce detectable changes in the waveform, particularly around P300 for emotional arousal.

These EEG correlates of emotions are primarily used by signal processing experts using medical grade EEG devices. The commercial units currently being produced are aiming to present some bespoke analysis methods for more general users that will appeal to a wider audience. Commercial BCI headsets (Emotiv and NeuroSky) can trace EEG band activities. NeuroSky, aiming more for the commercial market, have developed their own *eSense* scale to make it easier for users to interact with the headset. The eSense scale filters the beta and alpha EEG bands and outputs values for *attention* and *meditation* respectively. These scales operate from 0 to 100, representing the level of attention or meditation of the subject (NeuroSky, 2009).

Studies to date have shown that minimally invasive BCIs can track the attention and meditation (which we interpret/read as a measure of stress) levels of a subject during a task (Crowley & Sliney, 2010), as well as highlighting the connection between attention and meditation and how they impact on the applied effort of the subject during a task (Crowley & Sliney, 2011). The initial aim of these studies was to examine the suitability of the NeuroSky headset as a minimally invasive method of monitoring the emotion-based responses of attention and meditation. Two psychological tests, The Stroop Colour-Word Interference Test and The Towers of Hanoi were conducted to assess the headset's capability to track the attention and meditation of the subject during the task.

In these studies, we examined the use of bio-signal analysis in measuring the emotional response of a subject, with an emphasis on stress.

The detection and quantification of stress can be crucial in applications that are stress sensitive, such as devices controlled by speech, precision tactile control and BCIs. The detection of stress can also be beneficial in other areas such as computer-assisted learning and assessment. Knowing if a student is stressed or not paying attention while completing a learning task can be very beneficial data for a system.

We administered two psychological tests to the subjects during these studies: The Towers of Hanoi and The Stroop Colour-Word Interference Test. Subjects were recruited from students of the College of Science, Engineering, and Food Science in University College Cork. 20 users participated in the study (20 completed the Stroop and 17 undertook The Towers of Hanoi), with each test averaging 20 minutes. The BCI used to record the meditation and attention levels was NeuroSky's MindSet. A backup headset was used to check for consistent calibration.

The Towers of Hanoi, is a mathematical game or puzzle frequently used in psychological research on problem solving (Janssen, 2010). It consists of three rods, and a number of disks of different sizes, which can slide onto any rod. The puzzle starts with the disks in a neat stack in ascending order of size on one rod, the smallest at the top, making a conical shape. The objective of the puzzle is to move the entire stack to another rod, while obeying the following rules: only one disk may be moved at a time; each move consists of taking the upper disk from one of the rods and sliding it onto another rod (on top of the disks that may already be present on that rod); no disk may be placed on top of a smaller disk. The first time a subject completes the Towers of Hanoi they usually find it quite difficult and often cannot complete the task. We administered this task three times in succession in order to see if there was a) an improvement in the ability of the subject to complete the task and b) a reduction in the stress response of the subject as the task became more manageable.

The second test, The Stroop Colour-Word Interference Test (Stroop, 1935) is a well-known psychological test of selective attention, cognitive flexibility and processing speed. The test exploits the fact that for most readers, the reading of a word has become an automatism. The (electronic version of the) test consists of words displayed on a screen, where a word stimulus and colour stimulus are presented simultaneously (i.e. the name of a colour displayed in the font colour of another colour; the word "red" displayed in a blue font). The assessment requires the user to identify the colour stimulus and not the word stimulus (i.e. name the colour that is displayed and not the word that is written). The cognitive mechanism involved in this task is directed attention. The subject must manage their attention by inhibiting one response in order to say or do something else.

Both of these psychological tests aimed to elicit a stress response from the subject. NeuroSky's MindWave was used to monitor the attention and meditation (measure of stress) of the subject as they completed each task. The attention and meditation levels for each subject were recorded for both tests, and each iteration of the Towers of Hanoi. Baseline measures of meditation were also monitored during routine stress-free tasks such as completing usability forms. It was noted that during the stress-free tasks, the meditation signal did not drop below the 40 mark (scale measured from 0 – 100). This threshold value was then used to categorise user response in terms of meditation. Depending on the percentage of time the user's levels of meditation dropped below the threshold value of 40 a subject could be categorized as "poor attention/stressed", "normal attention/calm" or "high attention/calm" (Crowley & Sliney, 2010). The study demonstrated clear differences in the level of stress between the first and third iterations of the Towers of Hanoi for the majority of subjects. In the first iteration, subjects were typically categorized as stressed or very stressed, whereas by the third iteration, with practice and learning, they were typically demonstrating a calm response. This is the pattern we anticipated to observe for the Towers of Hanoi. Two users had prior experience of the puzzle and understood the test and a solution. For these subjects we expected that the stress level would be low even on the first attempt – both methods of categorization verified this prediction classifying these users as calm in each of the three attempts.

An adapted electronic Stroop Test was used, based on the work of Rothkrantz (2004), where the difficulty of the task linearly increased over time. 20 subjects participated in he Stroop Test that ran for five minutes. The difficulty of the task increased as the interval of time between the appearances of the colours was shortened. Every minute, the interval of time between each word was reduced by half a second, decreasing from two and a half seconds intervals during minute one, to intervals of half a second in minute five. During the five minutes of the Stroop Test, subjects were required to wear the NeuroSky headset and speak into a microphone to capture the speech signal. The attention and meditation levels of the subject were recorded using the NeuroSky headset for the duration of the test. The tester noted the accuracy of each response (if the colour displayed matched the response given) from the subject for post-test analysis.

In our study we could not consistently or predictably induce stress for each subject during minute five of the Stroop test. For some subjects, the results reflected those obtained by Rothkrantz, i.e. an increase in stress was observed during minute five of the task. Other subjects maintained a consistent level of stress throughout the task, while a further group decreased their stress level during the task. By monitoring the attention and meditation of the subject during the task using the headset, and by observing the behavior of the subject, it was evident that for a large number of subjects the attention level decreased as the difficulty of the task increased. These levels of attention and stress also relate to the applied effort of the subject during the task (Crowley & Sliney, 2011). Close examination of the two signals highlighted a correlation between attention, stress and effort, which in turn offered an explanation for the inconsistent stress responses across the test subjects. Awareness of the relationship between attention, stress and applied effort is very valuable data for interaction design, particularly for learning-based systems in terms of user engagement and motivation. Capturing emotion-based BCI data such as the level of engagement and/or frustration of the learner is paramount to the development of a robust pedagogical system.

Identifying a threshold value for eSense meters, such as the stress threshold value of 40 during the Towers of Hanoi task also offers future potential (Crowley & Sliney, 2011). There is scope for using emotion-based bio-signals as such controllers/ triggers, as previous work has outlined (Sliney 2011). Setting a threshold value for attention or meditation, for example, could act as an indicator or trigger to adapt an aspect of a system to suit the individual user. For example, a user may be struggling to keep up in a learning module may experience a drop in attention levels, or an increase in stress levels. An intelligent, adaptive system using mobile devices could track this fluctuation and adapt the difficulty level of the program accordingly. There are many uses for such an adaptive system, eLearning and Serious Games to name but two, and these areas have increasing mobile potential.

Our studies have examined a number of biometric devices for detecting stress responses in subjects including GSR, Eye Trackers, Pulse Rate Monitors and BCIs. We have assessed the suitability of commercial BCI devices such as the NeuroSky headset to monitor the emotional response of the user and we have found it to be effective for non-clinical use. During these studies we observed the relationship and correlation between the eSense meter emotions of attention and meditation. The more attentive a subject is, the more likely they are to experience stress as their level of engagement in the task is increased. For subjects whose attention level has dropped, higher meditation (less stress) values have been observed, showing that a decrease in applied pressure is less likely to elicit a stress response from a subject. These observations are not only relevant to personality theory and user interaction theory but they are also valuable insights for developers of interactive, adaptive and pedagogical systems to note.

Adding the benefits of tracing user interaction using mobile devices to monitor emotion-based bio-signals would result in truly adaptive and intelligent systems that include realism, immersion and often simulation. Unlike Eye Trackers, GSR or Pulse Rate Monitors, BCIs are minimally invasive, relatively comfortable for a user to wear for a prolonged period, low-cost and most importantly for mobile devices; they are portable and usually wireless.

Issues/Problems

Perhaps the most pressing issue with current BCI technology is the "background noise" detected by the BCI sensors. Scientific EEG devices have been finely tuned to account for minute changes in neural signals. As costs are cut, and devices made smaller and less invasive, quality and accuracy have also deteriorated. The accuracy of detection of neural cues with commercial BCIs can be diminished when users are simultaneously conducting numerous tasks, which may be as simple as walking, or even excessive head movement. More advanced filters are required to discriminate against neural "background noise".

As with many research-grade technologies that have been downgraded for commercial use, quality has suffered for cost. While current BCI technology is not yet cost efficient for the mass market, the devices are certainly becoming a more achievable purchase. Of the main competitors in the BCI market, there is a difference in price and accuracy. The Emotiv headset is more expensive, has more sensors, and offers more analysis points. However, the NeuroSky device is more commercially viable for the mass market, with two sensors and a lower price tag. The challenge for NeuroSky will be to maintain this lower price while increasing the accuracy of their device and software. Recent model versions have shown significant increases in robustness and reliability.

A further issue that raises concerns for future development of affective sensors is the matter of privacy and ethics. If a system uses an emotion-detecting biosensor how much information will be provided to the user of the system about the data that is being gathered? Will the system to be restricted to gathering only certain emotions that it declares to the user in an ethical contract before they interact with the system, or will there be systems that gather clandestine emotional data on a user. A question arises about who will have access to the recognition results and how these results will be stored. What use is made of the recognized emotions is another area of concern.

If a subject uses a BCI for gaming purposes, for example, is the user made aware of the emotion sensing capabilities of the hardware? If not, is this an invasion of privacy? Can all software vendors be trusted to only gather the biometric data that they outline in a user agreement, and not to gather any undisclosed data? It would seem almost certain that some ethical contract outlining privacy will be required for affective systems.

Reynolds et al (2004) theorize that "interaction technologies represent an implicit or provide an explicit contract between the designer and user" and that the designer makes "moral and ethical decisions in the development of an interaction technology". They developed ethical contracts from the theory of *contractualism*, which grounds moral decisions on mutual agreement. Their findings indicate that users report significantly more respect for privacy in systems with an ethical contract when compared to a control. Developers of systems that monitor and/or gather data on user's emotions could find that users may fear a privacy breach with such devices. The work of Reynolds et al (2004) indicates that including ethical contracts may alleviate these concerns.

FUTURE RESEARCH DIRECTIONS

While commercial BCIs have come a long way from the cumbersome, laboratory-confined EEG devices there is still much room for improvement. In terms of hardware, the devices should become smaller, faster and cheaper. The software should become more accessible to external developers, and improvements in noise-reducing filters will also lead to more robust systems. The future aim for BCI development should be to make these novel technologies more practical for everyday use for a mass audience.

When commercial BCIs advance enough to appeal to the mass market the development potential of truly adaptive, intelligent systems will be realized. A system that can automatically adjust their settings depending on feedback from a biometric measure and that can integrate such adaptivity with mobile devices has ever increasing scope as technology advances. We have seen how eye-tracking, BCIs, GSR and other studies using accelerometers (Mehigan, 2012) work well with mobile devices as feedback triggers for adaptive systems (Crowley & Sliney 2011, Sliney 2011). A further layer can be added to such systems by including an affective analysis component. An adaptive, intelligent system that can automatically adjust to a user's preferences based on emotion-based biometrics is the future goal of our research.

The importance and relevance of monitoring user interactions has been recognized for many years in the HCI community. User interaction data has been considered secondary data that can be gathered by a system as a subject uses it. Our recent studies have highlighted the importance of emotion-based secondary data for adaptive systems. While non-mobile devices may not benefit from monitoring user interaction data such as accelerometer interaction, the importance of emotion-based or affective secondary data is relevant to almost every HCI-based system. "Even if the primary assessment purpose does not include emotional response, as secondary data it can be considered of value" (Sliney, 2009).

Future development plans include a *Multi-modal Biometric Feedback System* (MBFS) (See Figure 2) that gives a user real-time feedback data on their emotional response as they use a mobile (or non-mobile) system could work with many auxiliary biometric devices. Our work to date has used BCIs, GSR, and Eye Tracking as monitoring tools for emotion-based responses (studies examining stress response). The potential to also include accelerometers that have been shown to work well with mobile devices offers a multimodal approach to affective biometric assessment and feedback.

External devices would gather biometric data from the user, transmit this data to the system, which in turn would analyze the biometric signals and in almost real-time return feedback data to the user about their emotional response to a given action/stimulus. The adaptive nature of this system would then have sufficient data to adjust

Figure 2. Multimodal biometric feedback system encompassing real-time UI and affective feedback

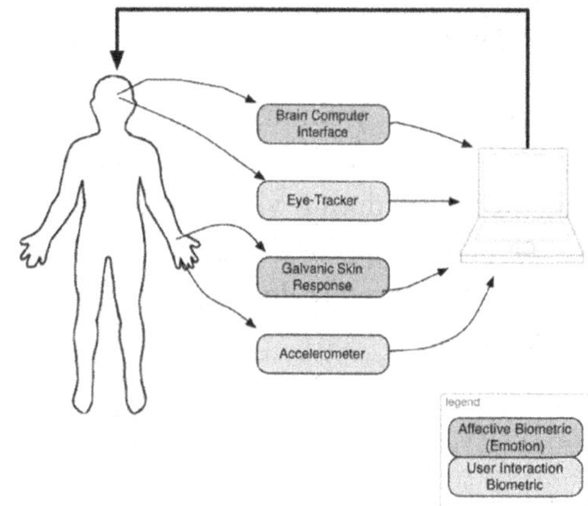

according to the needs of the user. Consideration would also be given to the ethical implications of such a system, and guidelines and approaches such as those set out Reynolds (2004) by would be implemented to include ethical user agreements.

CONCLUSION

Studies have demonstrated that biometric devices can offer a non- or minimally intrusive method of collecting user data for adaptive systems. It has been shown that biometric technologies such as accelerometer and eye tracking can be successfully integrated with mobile devices to gather user specific data for purposes such as profiling (Mehigan, 2012) and feedback (Crowley & Sliney, 2010, 2011). Commercial and research brain-mobile phone interfaces have demonstrated the compatibility of BCIs with mobile devices. BCIs have also been shown to be suitable to monitor and trace user's emotion-based responses during tasks (Crowley & Sliney, 2010, 2011). The potential to combine these layers and to develop an adaptive, intelligent, affective system that works with a mobile device, and automatically adjusts to a user's preference and profile offers huge scope for the future of mobile computing, assistive technologies, eLearning and mLearning, biomedical, speech-based and security systems.

REFERENCES

Backs, R. W., & Walrath, L. C. (1992). Eye movement and pupillary response indices of mental workload during visual search of symbolic displays. *Applied Ergonomics, 23*(4), 243–254. doi:http://dx.doi.org/10.1016/0003-6870(92)90152-L

Bashashati, A., Fatourechi, M., Ward, R. K., & Birch, G. E. (2007). A survey of signal processing algorithms in brain–computer interfaces based on electrical brain signals. *Journal of Neural Engineering*, *4*(2), R32–R57. doi:10.1088/1741-2560/4/2/R03 PMID:17409474.

Brower, V. (2005). When mind meets machine. *EMBO reports, 6*(2), 108–110. doi:http://dx.doi.org/10.1038/sj.embor.7400344

Campbell, A. T., Choudhury, T., Hu, S., Lu, H., Mukerjee, M. K., Rabbi, M., et al. (2010). NeuroPhone: Brain-mobile phone interface using a wireless EEG headset. In *Proceedings of the Second ACM SIGCOMM Workshop on Networking, Systems, and Applications on Mobile Handhelds* (pp. 3–8). New York, NY: ACM. doi:10.1145/1851322.1851326

Coan, J. A., & Allen, J. J. B. (2004). Frontal EEG asymmetry as a moderator and mediator of emotion. *Biological Psychology*, *67*(1-2), 7–49. doi:10.1016/j.biopsycho.2004.03.002 PMID:15130524.

Cockburn, A., & Mckenzie, B. (2001). What do web users do? An empirical analysis of web use. *International Journal of Human-Computer Studies*, *54*(6), 903–922. Retrieved from http://www.sciencedirect.com/science/article/pii/S1071581901904598 doi:10.1006/ijhc.2001.0459.

Coyle, M., Freyne, J., Brusilovsky, P., & Smyth, B. (2008). *Social information access for the rest of us: An exploration of social.* Springer. doi:10.1007/978-3-540-70987-9_12.

Crowley, K., Sliney, A., Pitt, I., & Murphy, D. (2010). Evaluating a brain-computer interface to categorise human emotional response. In Proceedings of Advanced Learning Technologies (ICALT), (pp. 276–278). IEEE. doi: doi:10.1109/ICALT.2010.81.

Crowley, K., Sliney, A., Pitt, I., & Murphy, D. (2011). Capturing and using emotion-based BCI signals in experiments: How subject's effort can influence results. In *British Computer Society Human Computer Interaction* (pp. 1–6). Swinton, UK: British Computer Society. doi: http://dl.acm.org/citation.cfm?id=2305316.2305341

Goecks, J., & Shavlik, J. (2000). Learning users' interests by unobtrusively observing their normal behavior. In *Proceedings of the 5th International Conference on Intelligent User Interfaces (IUI '00)*. ACM. DOI=10.1145/325737.325806 http://doi.acm.org/10.1145/325737.325806

Horlings, R., Datcu, D., & Rothkrantz, L. J. M. (2008). Emotion recognition using brain activity. In *Proceedings of the 9th International Conference on Computer Systems and Technologies and Workshop for PhD Students in Computing* (pp. 6:II.1–6:1). New York, NY: ACM. doi:10.1145/1500879.1500888

Jacob, R. J. K., & Karn, K. S. (2003). Eye tracking in human – computer interaction and usability research : Ready to deliver the promises. In Hyona, J., Radach, R., & Deubel, H. (Eds.), *The Mind's Eye: Cognitive and Applied Aspects of Eye Movement Research* (pp. 573–603). Academic Press.

Janssen, G., & De Mey, H. (2010). Celeration of executive functioning while solving the tower of Hanoi: Two single case studies using protocol analysis. *International Journal of Psychology & Psychological Therapy*, *10*(1), 19–40.

Maglio, P., Matlock, T., Campbell, C., Zhai, S., & Smith, B. (2000). *Gaze and speech in attentive user interfaces*. Springer. doi:10.1007/3-540-40063-X_1.

Mandryk, R. L., Inkpen, K. M., & Calvert, T. W. (2006). Using psychophysiological techniques to measure user experience with entertainment technologies. *Behaviour & Information Technology*, *25*(2), 141–158. doi:10.1080/01449290500331156.

Mehigan, T. J., & Pitt, I. (2012). Detecting learning style through biometric technology for mobile GBL. *International Journal of Game-Based Learning*, *2*(2), 55–74. doi:10.4018/ijgbl.2012040104.

Molina, G. G., Nijholt, A., & Twente, U. (2009). Emotional brain-computer interfaces. In *Proceedings of Affective Computing and Intelligent Interaction and Workshops* (pp. 1–9). IEEE.

Murugappan, M., Rizon, M., Nagarajan, R., Yaacob, S., Hazry, D., & Zunaidi, I. (2008). *Time-frequency analysis of EEG signals for human emotion detection*. Springer. doi:10.1007/978-3-540-69139-6_68.

NeuroSky. (2009). *White paper: NeuroSky's eSense meters and detection of mental state September 2009*. Retrieved from http://company.neurosky.com/files/neurosky_esense_whitepaper.pdf

Olofsson, J. K., Nordin, S., Sequeira, H., & Polich, J. (2008). Affective picture processing: An integrative review of ERP findings. *Biological Psychology*, *77*(3), 247–265. doi:10.1016/j.biopsycho.2007.11.006 PMID:18164800.

Picard, R. (1995). *Affective computing*. Cambridge, MA: MIT Press.

Rebolledo-Mendez, G., Dunwell, I., Martínez-Mirón, E. A., Vargas-Cerdán, M., de Freitas, S., Liarokapis, F., & García-Gaona, A. (2009). Assessing NeuroSky' s usability to detect attention levels in an assessment exercise. *Assessment, 5610*, 1–10. doi: doi:10.1007/978-3-642-02574-7_17.

Reynolds, C., & Picard, R. (2004). Affective sensors, privacy, and ethical contracts. In *Proceedings of the 2004 Conference on Human Factors and Computing Systems - CHI '04*. ACM. doi:10.1145/985921.985999

Rothkrantz, L. J. M., Wiggers, P., van Wees, J. W. A., van Vark, R., & Vark, R. J. V. (2004). *Voice stress analysis*. Springer.

Russell, J. A. (1980). A circumplex model of affect. *Journal of Personality and Social Psychology, 39*(6), 1161–1178. doi:10.1037/h0077714.

Savran, A., Ciftci, K., Chanel, G., Mota, J. C., & Viet, L. H. Sankut, B., Rombaut, M. (2006). Emotion detection in the loop from brain signals and facial images. In *Proceedings of eNTERFACE 2006*. Retrieved from http://www.enterface.net/results/

Schneiderman, R. (2010). DSPs evolving in consumer electronics applications. *IEEE Signal Processing Magazine, 27*(3), 6–10. doi:10.1109/MSP.2010.936031.

Sliney, A., & Murphy, D. (2011). Using serious games for assessment. In M. Ma, A. Oikonomou, & L. C. Jain (Eds.), Serious Games and Edutainment Applications (pp. 225–243). Springer. Retrieved fromhttp://dx.doi.org/ doi:10.1007/978-1-4471-2161-9_12.

Sliney, A., Murphy, D., & O'Mullane, J. (2009). Secondary assessment data within serious games. In O. Petrovic & A. Brand (Eds.), Serious Games on the Move (pp. 225–233). Springer. Retrieved fromhttp://dx.doi.org/ doi:10.1007/978-3-211-09418-1_15.

Stroop, J. R. (1935). Studies of interference in serial verbal reactions. *Journal of Experimental Psychology, 28*(1), 643–662. doi:10.1037/h0054651.

Thüring, M., & Mahlke, S. (2007). Usability, aesthetics and emotions in human–technology interaction. *International Journal of Psychology, 42*(4), 253–264. doi:10.1080/00207590701396674.

Wolpaw, J. R., Birbaumer, N., McFarland, D. J., Pfurtscheller, G., & Vaughan, T. M. (2002). Brain-computer interfaces for communication and control. *Clinical Neurophysiology, 113*(6), 767–791. doi:http://dx.doi.org/10.1016/S1388-2457(02)00057-3

ADDITIONAL READING

Graimann, B., Allison, B., & Pfurtscheller, G. (2011). *Brain-computer interfaces: Revolutionizing human-computer interaction*. Berlin: Springer.

Healey, J., & Picard, R. (1998). Digital processing of affective signals. In *Proceedings of the Acoustics, Speech and Signal Processing, 1998* (pp. 3749–3752). IEEE. doi: 10.1109/ICASSP.1998.679699

Holtzman, J. M., & Goodman, D. J. (1994). *Wireless and mobile communications*. Dordrecht, The Netherlands: Kluwer Academic Publishers. doi:10.1007/978-1-4615-2716-9.

Jones, M., & Marsden, G. (2006). *Mobile interaction design*. New York: John Wiley & Sons.

Krepkiy, R. (2008). *Brain-computer interface: Design and implementation of an online BCI system for the control in gaming applications and virtual limbs*. Retrieved from http://d-nb.info/991305191/04

Ledoux, J. E. (2000). Emotional circuits in the brain. *Annual Review of Neuroscience*, (23): 155–184. doi:10.1146/annurev.neuro.23.1.155 PMID:10845062.

Molina, G. G., Nijholt, A., & Twente, U. (2009). Emotional brain-computer interfaces. In *Proceedings of Affective Computing and Intelligent Interaction and Workshops* (pp. 1–9). ACII.

Ortiz, S. Jr. (2007, January). Brain-computer interfaces: Where human and machine meet. *IEEE Spectrum, 17-21*. doi: doi:10.1109/MC.2007.11.

Picard, R. W. (2003). Affective computing: challenges. *International Journal of Human-Computer Studies*, *59*(1-2), 55–64. doi:10.1016/S1071-5819(03)00052-1.

Picard, R. W., Member, S., Vyzas, E., & Healey, J. (2001). Toward machine emotional intelligence: Analysis of affective physiological state. *IEEE Transactions on Pattern Analysis and Machine Intelligence*, *23*(10), 1175–1191. doi:10.1109/34.954607.

Tan, D. S., & Nijholt, A. (2010). *Brain-computer interfaces: Applying our minds to human-computer interaction*. Berlin: Springer.

Wasinger, R. (2007). *Multimodal interaction with mobile devices: Fusing a broad spectrum of modality combinations*. Boca Raton, FL: IOS Press.

Wolpaw, J., & Wolpaw, E. W. (2012). *Brain-computer interfaces: Principles and practice*. Oxford, UK: Oxford University Press. doi:10.1093/acprof:oso/9780195388855.001.0001.

Yasui, Y. (2009). A brainwave signal measurement and data processing technique for daily life applications. *Journal of Physiological Anthropology*, *28*(3), 145–150. doi:10.2114/jpa2.28.145 PMID:19483376.

Yiend, J. (2010). The effects of emotion on attention: A review of attentional processing of emotional information. *Cognition and Emotion*, *24*(1), 3–47. doi:10.1080/02699930903205698.

KEY TERMS AND DEFINITIONS

Adaptive System: An adaptive system is a flexible system that improves its performance or effectiveness by monitoring and adjusting its own configuration and operations in response to feedback from user interaction or its own environment.

Affective Computing: Affective computing is Human Computer Interaction in which a device is concerned with the emotions of the user. It is the study and development of systems and devices that can recognise, interpret, process, and simulate human emotions.

Biometrics: The term biometric derives from the Greek words "bio", meaning biological life, and "metric" meaning measure or measurement. It refers to the automated recognition, profiling or analysis of individuals based on their biological and behavioural traits.

Brain Computer Interface: (Also known as *Brain Controller Interface)* is an interface that enables direct communication pathways between the human brain and an external device for analysis, control, and diagnostic purposes.

Electroencephalography (EEG): EEG is a neurological procedure that records electrical potentials (brain waves) across the scalp.

Emotional Response: A reaction of a subject to a particular stimulus or situation that results in feeling or feelings, accompanied by physiologic changes that may or may not be outwardly manifested but that motivate or precipitate some action or behavioral response.

Galvanic Skin Response: A change in the electrical resistance of the skin caused by activation of the sympathetic nervous system (the emotional response(s)) of a subject.

Chapter 10
WayFinder:
A Navigational Interface for Visitors and Blind Students on Campus

Tracey J. Mehigan
University College Cork, Ireland

Ian Pitt
University College Cork, Ireland

ABSTRACT

Navigating a university campus can be difficult for visitors and incoming students/staff, particularly those who are blind or vision-impaired. Universities around the world, including University College Cork (UCC), generally rely on physical sign-posting and map-based information (available as a download from the university Website) to direct visitors, staff, and students. These methods are not appropriate for those with vision-impairments. Mobility training is provided by UCC's Disability Support Service (DSS) to enable blind/vision-impaired staff and students to safely and independently navigate campus, but the training is route-specific, time-consuming, and expensive. A navigation solution that facilitates all campus users, both sighted and vision-impaired, could be provided via mobile and wireless technologies. Research has been conducted to assess the needs of those navigating campus, to evaluate relevant technologies, and to assess the state-of-the-art in regard to the provision of navigational information. The results suggest that no off-the-shelf solution exists that fully meets the requirements of UCC. Existing systems fall short in various respects, in particular in the accuracy and reliability of the localization information and the nature of the feedback provided to the user. This chapter summarizes the results obtained from the review. A system is described which has been designed, in the light of the review findings, to enable visitors, staff, and students (both sighted and vision-impaired/blind) to safely and independently navigate the campus using a smartphone. This system has potential for use at other universities and institutions. The development and initial testing of the user interface layer of the system is described.

DOI: 10.4018/978-1-4666-4054-2.ch010

INTRODUCTION

Navigating around a university campus can be difficult for visitors and incoming staff and students, and is a particular challenge for those with vision impairments. A large number of visitors and new students traverse the UCC campus each year (See Table 1).

Universities around the world rely on physical signs and map-based information to direct visitors, staff and students around campus. Maps are generally displayed in billboard fashion on campus, and may also be available for download from a university's Website. This represents the main method of guiding people around campus. Visual signage is not necessarily the best way of helping people who are unfamiliar with routes, and maps in particular are not appropriate for those with vision impairments.

University College Cork (UCC), like other universities, currently relies on signage and map-based information for campus navigation. However, the campus at UCC presents particular problems as it has an undulating topography and no linear structure to its layout. This is compounded by the continued spread of campus, with many of the newer buildings are located outside the main campus boundary. There are also problems within buildings, where navigation between rooms and service locations can be difficult.

Table 1. Visitors and new and disabled students traversing campus at UCC 2010/2011

Campus User		Total
Average annual visitors:	Visitors taking tours through the Visitor Centre	15,000+
	Glucksman Gallery visitors	70,000+
Visiting students		1040
International students		2400
Adult & Continuing Education students		1988
Blind students:	Totally blind students	10
	Vision impaired	28

Mobility training is currently provided by UCC's Disability Support Service (DSS) to those staff and students who need it. It is designed to enable blind and vision-impaired staff and students to safely and independently navigate campus. However, the training is time-consuming for all parties involved, is expensive to provide, and is not available to visitors. The training is also route-specific, which means that staff and students may need re-training when changes are made to course timetables.

Mobile and wireless technologies offer a potential solution to the problems faced by those navigating UCC's campus. Such a solution could facilitate all users, both sighted and blind/vision-impaired, whether existing/regular users or new/one-time users. Using smart-phone technology would allow people to use their own mobile devices for navigation purposes on campus. Information could be provided in an appropriate manner for each individual user: visual, auditory, haptic (or combinations of these) based on the needs of the individual user. Device features, for example the compass and accelerometer, could be combined to accurately identify the user's location on campus. However, there are challenges in implementing such a system at UCC, including the spread of campus across an increasing number of sites, the non-linear layout, and the undulating topography. This is compounded by the problems faced when navigating inside buildings.

This chapter reviews relevant wireless technologies to assess their potential for use in a campus navigation system. The state-of-the-art in relation to navigation systems for vision-impaired users is also reviewed. Results from a study of the needs of blind and vision-impaired campus users are presented. On the basis of the reviews and user-needs assessment, the design and development of a prototype navigation-based system for such users is described, along with studies to assess its usability.

BACKGROUND

There have been a number of attempts to develop efficient and cost-effective navigational systems for blind and vision-impaired people. Several commercial systems are available, such as the Trekker Breeze GPS System and the Mobile Geo, and a number of experimental systems have been developed to facilitate navigation in particular environments, such as exhibitions, museums, and universities. Many different technologies have been used, including Radio Frequency Identification (RFID), Bluetooth, Dead–Reckoning systems using accelerometer devices, and Global Positioning Systems (GPS). Many of these technologies are intended solely or primarily for outdoor navigation, but some can be used indoors, e.g., Nokia's high-accuracy positioning system delivered via Bluetooth 4.0 (Kalliola, 2011). Some systems are specifically aimed at helping blind/vision-impaired pedestrians to navigate safely and independently, while others are designed for a wider range of users.

The majority of systems use Text-to-Speech (TTS) for output. As well as indicating position, they typically provide information on service areas, such as restaurants, and other points of interest along the route. Visual information is provided for sighted users.

Most mobile navigation systems offer improvements over traditional sign-and map-based approaches. However, significant problems still exist. Relying solely or primarily on speech feedback presents problems for many users. It can lead to sensory overload and conflict with other demands on sensory channels, for example, the use of environmental cues for positioning purposes. Currently, there is no system available that provides terrain-based information to the user. Difficulties may arise when using such systems in conjunction with long canes and other mobility aids. Most systems are cumbersome and can also lead to the isolation of user from the campus community. In addition, the accuracy of the positioning information is often inadequate, and there is no system that provides seamless navigation between indoor and outdoor environments. This would require a transfer between different technologies. Mobile device users also encounter problems resulting from high power drain and consequent short battery life when using such a navigation system.

EVALUATING POTENTIAL TECHNOLOGIES FOR NAVIGATION SYSTEM DEVELOPMENT

There are a number of technologies suitable for the collection of localization and positioning information for navigation purposes. Some of these technologies have been employed in existing systems.

Radio Frequency Identification (RFID) Tags

RFID systems comprise non-contact and non-line-of-sight technology. RFID uses a tag (transponder) composed of an electronic chip and an aerial. The tag is electronically programmed with specific and unique data/information. The tags are the backbone of the technology and come in a variety of shapes, sizes and read ranges. Tags can rely on batteries and energy potentially gathered from other sources to support the transmission of information to the reader for distances up to 100 meters. Beside tags, RFID systems consist of a number of components including handheld or stationary RFID tag readers, data input units, and system software (Mau et al., 2008).

There are a number of problems associated with the use of RFID technologies. There is no standard, off-the-shelf solution available; therefore, any solution for navigation purposes would have to be custom-built from various suppliers (Gerst et al., 2005). Consequently there is a difficulty in assessing the advantages and disadvantages of different RFID solutions. Due to different frequency ranges and data transfer modes, readers can become confused if too many signals are received simultaneously. There can be an element of interference with other radios held within mobile devices, for example, Bluetooth. There is a lack of readers compatible with mobile/smart devices. There are limitations to the reading range associated with tags/readers. There can be a failure to detect tags when located close to water/metal (Floerkemeier and Lampe, 2004). The cost of hardware is relatively high and non-transparent and standardization issues also exist with RFID technologies.

Near Field Communication (NFC) for RFID Systems

NFC, while not restricted to RFID technology, can be facilitated through RFID technology enabling communication between devices within proximity, usually up to 20cms, for data exchange. Communication is achieved through magnetic field induction. It is a short-range, standards-based wireless connectivity technology. NFC operates at 13.56MHz (high frequency) band. NFC provides users with access to digital content by touching or bringing devices into proximity.

While very accurate for data transfer, this technology is not a suitable on its own for campus navigation system mainly due to the proximity restriction of 20cm. Some researchers have found ways around the proximity limitation for certain applications. For example D'Atri et al., (2007) created a navigation system for blind people who used tag-readers built into the tip of a white cane or the user's shoe, while the tags were placed under the carpet. However, the need for special hardware greatly restricts the value of such a system, limiting its use to suitably equipped vision-impaired users while excluding visitors, etc. On this basis, a system based on these technologies is not a suitable option for a navigation system at UCC's campus.

GPS (Global Positioning System)

GPS is widely employed in outdoor navigation systems. Several systems using GPS have been developed as orientation aids for vision-impaired students, including the HumanWare systems ("Humanware", n.d.) and the Ariadne GPS ("Ariadne GPS", n.d.) systems. GPS uses geostationary satellite signals for the provision of outdoor navigation and mapping systems. The resolution of GPS localization is limited to a few meters with relative delays when using receivers giving low real-time responsiveness.

Conventional GPS can suffer from non-continuous signal receipt and multipath effects (where signals bounce off buildings or are weakened by passing under tree cover, etc.), reducing its accuracy in determining receiver position. This is a particular problem in urban environments, since navigating among buildings can result in loss of signal or reflected signals. Most vision-impaired and blind people tend to walk close to buildings (which serve to define the path), increasing the risk of navigation errors. In some cases such issues can be overcome through the use of Pseudolite-based systems (Dixon and Morrisson, 2008), however this is not always a viable option as hardware costs can be high. Inaccuracy regarding target location can also be an issue. Not all areas are well mapped, and there is no mapping for UCC's campus.

GPS-based systems are becoming more accurate, giving resolution down to a few meters or less, and are now built into most handsets. However, the issues associated with the technology, in particular its accuracy level, mean that it is not suitable for an inclusive campus navigation system at UCC. While it has the benefit of requiring no additional infrastructure to support it, GPS also suffers from the drawback that it cannot be used for indoor navigation.

Ultra-Low-Power Bluetooth 4.0

Bluetooth (IEEE 802.15.1) technology is currently the leading and only proven short-range wireless technology. It is used widely for Personal Area Networking (PAN) with devices. It is particularly suitable for the transmission of small quantities of information between devices as it allows for different devices to talk to one another over a piconet or scatternet through easy interconnection of wireless devices such as mobile phones, computers, PDAs and a broad selection of other devices using a short-range wireless connection. The Bluetooth Special Interest Group (SIG) announced the formal adoption of the Bluetooth 4.0 Specification in July 2010. The first waves of smart-phones to include Bluetooth 4.0 technology as standard have started to come on stream, for example Apple's iPhone 4S & 5. Ultra-Low-Power (ULP) Bluetooth (Bluetooth 4.0) targets applications very similar to Radio Frequency Identification (RFID) and will be integrated in pretty much every mobile phone and consumer device in the next two years. It can be used for giving location-based information (Kalliola, 2011). This technology could be suitable for use in a navigation system at UCC. It offers many benefits over technologies such as RFID even though it is aimed at the same application-area.

Wireless Inertial Measurement Units (WIMU)

WIMUs typically provide data from accelerometer, gyroscope and magnetometer sensors. Pedestrian Dead Reckoning (PDR) alone is not sufficient for positioning, but it provides incremental, relative position information to complement any system that can provide absolute location estimates. WIMUs capture tilt, force and timings and represent an optical 3D motion capture system to provide a complete kinematic model of subjects (Hautefeuille et al., 2008). This technology can be combined with other technologies such as Bluetooth to provide an enhanced level of accuracy within a navigation system user through the gathering of inertial movement data (Walsh et al., 2011).

ASSESSING THE NEEDS OF CURRENT STUDENTS ON CAMPUS

Research was conducted to assess the needs and requirements of blind/vision-impaired students navigating campus at UCC. The study comprised a series of interviews and walkabouts with both students and the campus mobility trainer. The study provided a useful insight into the needs of students navigating the campus on a daily basis.

The participants represented both blind and vision-impaired students undertaking a number of undergraduate courses at UCC, from first year students (newly arrived) to final-year students. Therefore, those involved in the study had a varying level of familiarity with the campus. Both guide dog and white-cane users took part.

Participants were asked a series of questions regarding their own mobile device. These included questions regarding the brand of device and the normal interaction method associated with that device, for example, if the device had a touch-based interface. They were also asked if they had any experience with mobile-based navigation systems. In cases where students had used such navigation systems before, they were asked to comment on their experience.

During the interviews, the participants were asked to comment on the general issues that they faced while navigating campus. They were asked specifically to comment on

- General issues regarding navigation around campus.
- Issues associated with particular routes.
- Strengths and weaknesses of existing mobility training and support.

Students were also asked to describe both their thought processes during the walkabout and to indicate the cues and techniques they used to determine their position and orientation at each stage of the journey.

The results indicated that long-cane users and guide-dog users have significantly different needs. Long-cane users need to check and correct their position every few steps, whereas guide-dog users require less frequent position checks. However, guide-dog users must be able to identify route decision-points reliably and with great accuracy. This is particularly important where the dog is not familiar with the route being taken.

Both groups make use of environmental sounds as a primary means to determine position and as a confirmation of their location. Many of the participants indicated that they make use of manhole covers to mark their route. It is noted that in most cases participants had an aural ability to distinguish the degrees of water flow beneath the manhole to determine their location based on the level of silence, varying degrees of water flow, etc. The manholes are also used as place markers as they differ from solid ground. Participants also used other environmental cues. The flow of the campus river, wind direction, crowd noise and even the scent of coffee from the cafes, all represent environmental cues of importance to blind/vision-impaired users. Participants reported that open spaces, such as the Honan Plaza, cause them the most difficulty when navigating.

The review findings indicate that any navigation system developed for use by blind/vision-impaired students/staff and visitors on campus at UCC must meet the following requirements:

- Reliable operation both indoors and out.
- High levels of accuracy to within a few of meters at most, allowing effective use both indoors and also in large, open spaces such as the Honan Plaza.
- Operation on standard hardware (preferably user's existing smart-phone or similar device), with minimal network costs, battery-drain, etc.
- Simple but flexible user-interface offering choice of text, images, speech, non-speech sound and tactile feedback, selectable by user.
- Variable feedback to accommodate users from novice to expert output formats to suit both guide-dog and long-cane users.
- Optimized for use in conjunction with external sound and other environmental cues.
- Minimal mental/physical demand on the user.

Further details of the study are provided in Mehigan and Pitt (2012).

REVIEWING THE STATE-OF-THE-ART

Several solutions for the provision of an efficient and cost effective navigation system have been proposed in recent years. These include systems designed to increase the mobility of vision-impaired and blind pedestrians at exhibitions (Bellotti et al., 2006), museums (Santoro et al., 2007), and universities (Tee et al., 2007). There has been a particular focus in the literature on the provision of improved independence for vision-impaired pedestrians.

These systems have mainly been based on a combination of new technologies such as Near Frequency Communication (NFC) through the incorporation of Radio Frequency Identification Tags (RFID), Bluetooth, Dead-Reckoning systems using accelerometer devices and Global Positioning System (GPS). Other technologies have also been investigated for potential use by vision-impaired pedestrians: for example, Coughlan and Manduchi, (2007) have assessed the potential use of camera-phone-based navigation systems. All of these systems have been shown to offer benefits when compared with traditional approaches.

The aim of guide-based systems is to enable blind and low-vision students to walk around a university campus independently and safely. These systems use speech guidance to provide navigation information from the user's current location to a particular internal or external campus location.

Navigating at the Exhibition

The ELIOS Lab at the University of Genoa developed a multimedia guide for visually impaired people at an exhibition (Belotti et al., 2006). The system employs a location-aware tour-guide, based on RFID localization through handheld computers (Pocket PCs), dedicated but not exclusive to blind and vision-impaired users. The system uses 99 IP65 compliant self-powered RFID tags (water and dust resistant) installed at points of interest across an area of 30,000 square meters. UHF radio band (915 MHz to 868 MHz) was used for long range, high-speed communication for reliable data exchange. The Identec iCARD is incorporated to communicate with the RFID software module. The RFID software module an extension of MADE (Mobile Applications Development Environment) developed by the ELIOS lab to implement the localization algorithm, which is based on iterative scanning of tags. The iCARD is integrated into the handheld device to communicate with the RFID tags at distance up to 100 meters with an overall resolution of 5 meters.

The exhibition guide offers presentation based on two categories, general information regarding the exhibition, the guide itself, and services. This information is accessible to the user at any time and at any exhibition location. Descriptions of the selected interest points are provided. Every interest point is associated with an RFID tag, resulting in an event-driven system.

The User Interface provides a description of the points of interest, extended descriptions where required, and informs the user about, and guides them towards, exhibits. Each user is provided with information about service areas, for example restaurants, etc. and is supported in the development of a mental map of the surroundings through the inclusion of reference points within descriptions, for example, "you are near here." Identification of key crossroads (decision points) is indicated across the exhibition. Visual information is also provided for sighted users. Device control for the UI is conducted through hardware buttons as this solution is suitable for sighted and vision-impaired/blind users.

The guide was tested at "EuroFlora" 2006 with up to 100 blind users who used the system during the exhibition. Results indicated a high user acceptance rate. Issues initially reported by users included frustratingly long periods of silence between presentations where RFID tags had not been installed. The placing of tags in these locations overcame this. The ELIOS lab reported that positive comments were made regarding the system by the blind assistance experts whom they state, "highlighted the significant degree of independence the blind could reach through the guide" (Belotti et al., 2006, pp. 1-6). They also report on the opinion of one particular system user in their paper: "After always having been guided, for the first time I myself have been able to guide my wife and explain to her the exhibition!" (Belotti et al., 2006, pp. 1-6).

The Care-Giving System of Navigation

The School of Electrical and Electronics Engineering at the University of Nottingham, Malaysia developed the SmartGuide system (Tee et al., 2009), a care giving and monitoring system for the assistance of visually impaired students in a university environment. The system provides speech guidance based on current location, and navigation information on how to move toward a specific location.

The system consists of a wireless sensor network, speech synthesis chips and intelligent navigation software. The system uses RFID tags located at specific locations to create a 3D map for navigation purposes. It is based on the Drishti system (Helal et al., 2001), but the authors claim it offers advantages over Drishti. The system comprises four main layers, which are combined with a tag map; these are a speech module layer, a wireless sensor network (WSN) layer, a hardware layer and a user interface layer.

The speech module layer uses speech recognition on user input into a microphone, to provide speech based destination location information through speech synthesis, for navigation assistance. This information is then sent to the WSN layer. The WSN Layer is a centralized server that holds the software or user interface layer; it provides real time data transfer for the provision of navigation information to users. It handles user destination and current location data for path finding and monitoring. The hardware layer comprises two parts, the SmartGuide reader (an RFID reader), which is attached to the end of the user's long cane, and a SmartGuide tracker, which is attached to the user's belt or backpack. The two components are wirelessly connected via Bluetooth. The communication and navigation modules, part of the SmartGuide Tracker, use Bluetooth, RFID technology, GPS and Dead Reckoning to enhance the efficiency of the system. The user interface layer runs "Intelligent Navigation 2.0" software that works on any PC running the Windows platform. It is held centrally on a server.

The developers indicate that experiments were conducted using the system, but they do not specify the type of experiments conducted or the specifics of the associated results. They report that, "the experiments carried out indicate that the wireless sensor network has minimal delay in relaying even with walls and obstacles thus it is concluded the system would work reliably in indoor environments" (Helal et al., 2001, pp. 1-6). They state their intention to implement the system within all buildings of their university but it remains unclear whether this has been done.

Urban-Based Navigation

D'Atri et al., (2007) published a paper based on the RadioVirgilio/SesamoNet system. Currently in place in the city of Rome, the system uses mobile and wireless technology combined with RFID tags recovered from the beef trade to provide a cost effective mobility grid for the creation of improved mobility of vision-impaired civilians. The information is grid-based and accessible through the incorporation of an RFID reader into the vision-impaired user's long cane. The system relies on the use of passive RFID tags buried in the ground to 4cm. The tags originally served as identification tags for beef cattle and are recovered after the slaughter of trade animals. The re-use of these tags reduces the cost of tag implementation to the grid as they are already contained in plastic covers, making them waterproof. The tags are set to use the 134.2KHz channel to reduce environmental noise. The tags have a reading range of up to 15cm and require the RFID reader to be incorporated into the long cane. The reader then transmits data to a PDA or other mobile device via Bluetooth technology, now standard in most mobile phone devices. The reader is set to either operate in continuous read mode or on demand and is operated through an RFID controller button on the cane handle. An RFID antenna is incorporated into the tip of the cane. A Bluetooth terminal is also incorporated in the cane tip. The system was developed using a HP iPAQ hw6500 PDA with built in Bluetooth 1.2, 64MB memory and a Secure Digital (SD) slot. The system uses Text-to-Speech acoustic signals to orientate the user on a safe path.

The system's logical architecture comprises a Bluetooth Cane Connection Manager, a Navigation Data Interface and Navigation Logic. The software was initially developed for both the Microsoft .NET architecture and Java. The system's functionality facilitates the maintenance of a safe navigation path for the user and the provision of hazard information and directional turn information to the user. The user can access environmental information on demand through position tracking in tag grid. Online help and assistance (via GSM) is also provided.

To test the system, three cognitive walkthrough sessions were conducted on the initial development of the system. The purpose of the walkthroughs was to assess the system's functionality and discover where improvements could be made to the system. Walkthroughs 1 & 2 were conducted by the system developers and students; walkthrough 3 included the participation of a blind user. The blind student attempted to orientate his way along a path using the system to guide him. Some participants executed basic tasks within a relevant scenario. Other participants observed the session. Each participant was taught to recognize tones and the tag dispositions on the path. They were instructed on how to scan the path using the cane to access the tags. A path was constructed using grass-like carpets; two were positioned with RFID tags configured to a path with a starting point on the left, a straight section, and then a 900 turn. No tactile feedback was provided by the path. Currently no results are available from the testing process.

Systems in Place at Universities

Most universities do not have a specific navigation system in place other than the provision of PDF based map facilities. Some UK based universities, including the University of Southampton and Cambridge University, have employed the "DisabledGo" system ("DisabledGo", n.d.) as has Dublin City University (DCU) in Ireland.

"DisabledGo" at DCU

DCU uses the "DisabledGo" system to provide accessibility information for students and Web users. The system operates through the Disability Office Website. Three text-size variations are provided along with three screen-color contrast options. A "Listen to this Website" feature is also provided. These features are accessible via a menu located at the top of the screen.

When the Website is first accessed, a welcome screen offering a search facility is displayed to the user. The facility requires that a name or venue type is entered into a search box, and this can be further refined based on type classification, for example, libraries, teaching etc. Symbols for each classification are provided. Once the search is complete, the display changes to provide a disabled access guide for the student based on the building requested.

The disabled guide to the library represents just one of the searches that can be accessed by the student. The access guide initially offers the student the facility to filter the information based on their personal requirements, such as wheelchair user, mobility-impaired walker, adapted WC, Braille, Assistance Dog, etc. The user is then directed to the access guide, which provides textual information on opening times during semesters and off-season. The location of the building is also provided, as are the entrances. Lift locations, information desks, etc. are highlighted. The guide then progresses to describe certain aspects of the services and facilities within the building at an individual level. For example, information is provided regarding the reception desk: it is 12 m (13 yards) from the main entrance, there is level access to the reception from the entrance, and the reception desk is of medium height. Information is also given based on the environment, for example: lighting levels are medium, there are turnstiles, 56cm wide, and a gate, 104 cm wide, to the left of this reception giving access to the main library area. Additional information is also available, for example, that the staff has received disability training. Documents are available in Braille, large print, etc. An option to email the disability support office regarding the venue is also provided.

No directional information is provided between buildings on campus. A link to Google maps is provided, but the maps are generic Google city maps and are not specific to DCU. Therefore the maps provide insufficient direction to specific locations on campus. Information provided is building-specific and not related to campus layout. The implementation of this system at UCC would raise similar issues and would not be sufficient to meet the user requirements review findings.

"Here2There" at Temple University

Temple University in the USA, have employed the "Here2There" Media system since 2009. The system provides "To Campus, On-Campus and Inter-Campus way-finding" for everyone ("Temple", n.d.). The system offers an interactive Web-based interface to facilitate indoor and outdoor directions to and from any point of interest on campus. It indicates transport modes and provides route information on the location of accessible entrances. Users can instantly identify their current location through the graphical representation, which indicates "You Are Here". The system provides navigation information to facilitate travel between departments and buildings using directions, which are mapped with clear lines, and supporting text. Landmarks are included alongside images of final destinations. The system can be updated by the university as necessary. The system's main features include information on general inter-campus routes from department to department (both for the external campus and the interior of buildings), specific route information, for example, parking routes and emergency routes, and accessible routes for mobility and vision-impaired people.

This information is conveyed via textual directions from building to building and visual map-based directions with indication of here-to-there type directions.

To facilitate access by vision-impaired users, the Web site incorporates "BrowseAloud" ("BrowseAloud", n.d.) software tools. This is the only provision for vision-impaired students. The student is not provided with a facility to download the directions as an mp3 to use while they are traversing their chosen route.

Currently there is no support for mobile access to the existing system via mobile devices, such as phones, and consequently no support for access when a student or visitor is lost on campus. The current system requires user to pre-plan their route or gains access to the Internet while on campus. Temple University plans to incorporate a mobile system as part of the overall Here2There system to be available via personal information devices in the near future. The system is currently in development for use with iPhone and Android devices, and is awaiting certification by the NFB (National Federation of the Blind) and the AFB (American Foundation of the Blind). The main features will include general accessibility and specific accessibility routes, information and news. Specific features for vision-impaired users will include TTS directions, semi turn-by-turn directions, and directions to accessible entrances. The mobile system will be operated via GPS.

System issues are evident in relation to vision-impaired users. The system is Web-based and not currently supportive of mobile Web access via mobile/personal devices. Users have to pre-plan routes; this prevents users, particularly vision-impaired students, from having access to a system that allows them to plan en-route. The system does not enable vision-impaired students to re-gain their orientation should they become disoriented en-route. The system does not provide the facility to record directions for specific routes in audio format for re-use, for example, as an mp3 file. The system should provide sufficient access to visitors and non-disabled students on campus.

The campus at Temple University has a regular, grid-based layout. The incorporation of such a system at UCC would raise issues due to the non-linear layout and undulating topography of the campus. The ever-increasing spread of campus from Brookfield to WGB to Enterprise Centre would also raise issues. The planned mobile system at Temple University should provide a solution to some of these issues.

Commercial Systems

Commercial systems are also available for navigation purposes. For example, the HumanWare Trekker Breeze system ("Humanware", n.d.), is a lightweight system that is styled like a remote control. It's a handheld device that provides GPS based navigation to vision-impaired people. Users can control the device with one hand via buttons and can combine its use with their normal mobility aid, for example, their long cane. The system uses speech-based feedback to highlight a user's position based on street names, which are announced to the user as they pass intersections and specific landmarks. The main system features include large distinct buttons, quick volume adjustment, TTS feedback and maps available covering most western countries.

Outcome of Review

The review findings indicated that there is currently no off-the-shelf solution that meets the requirements of UCC. The systems that exist do not provide a sufficient level of accuracy to meet the needs of vision-impaired and blind system users. The reliability of the localization information provided to the user by existing systems is also an issue: UCC requires a system that provides a high level of accuracy and reliability. It also requires that any system installed for navigation purposes provides a seamless transmission between internal building spaces and the outdoor routes. A further drawback with most existing systems is that the feedback options are limited to TTS and visual content.

WAYFINDER: DESIGNING THE SYSTEM

In view of the review findings, it has been decided to develop a system that has two distinct layers - a navigation layer to obtain data on user position and orientation, and an interface layer that makes this information available to the user. Separation of the two layers allows development to proceed independently, and will also make it easier to upgrade or replace one layer at a later date without unnecessarily affecting the other. This will be of particular importance as new and more efficient technologies come on stream.

Designing a Campus-Appropriate Navigational Layer

The navigation layer will initially be based on ULP Bluetooth 4.0 combined with WIMUs. As noted before, Bluetooth 4.0 offers many benefits over technologies such as RFID even though it is aimed at the same application-area. It offers low energy demands with an ability to run for years on standard coin-cell batteries. It is a low cost solution that offers an enhanced range, particularly when combined with WIMUs. It provides multi-vendor interoperability. Thus ULP Bluetooth is an appropriate choice for the development of a low cost and highly accurate navigation system. It also offers the opportunity to develop a system that will be accessible via any smart-phone device.

WIMUs, capture tilt and force and facilitate the concurrent measurement of timing to represent an optical 3D capture of motion. They enhance a system to provide a complete kinematic model of a subject (Walsh et al., n.d.).

It became clear during the review that positioning technology is developing at a rapid pace, and that while ULP Bluetooth 4.0 currently appears the most appropriate choice, in the near future other technologies might overtake it. For example, GPS accuracy levels are increasing. This observation resulted in our decision to develop the navigation system in two distinct layers, separating the navigation layer from the user interface. The focus of this chapter is the development of the user interface layer.

Designing and Developing Wayfinder's User Interface

For the user interface, the aim of the design has been to address issues apparent in other systems while at the same time retaining beneficial elements evident on review of those systems. Therefore, while existing systems only provide TTS and visual feedback, we have extended the feedback facility to include optional audio and haptic components. Consideration has also been given to user control. In most existing systems, control has been facilitated through button control. Modern devices (for example the iPhone) use touch-screens and gesture-based system controls. Therefore we wish to ensure that this type of interaction is facilitated.

Attention has been paid to needs of users with differing levels of experience with technology. For example, some users will not have had experience with mobile devices and/or navigation based systems. This will be the case particularly for elderly visitors to campus. The system must also be designed to suit novice users, expert users, and casual/one-off users. Consideration must also be given to the differing requirements of guide-dog users and long-cane users.

As some vision-impaired users rely on echolocation techniques to gather information on their surroundings through environmental sounds, care must be taken to ensure that any audio interaction facility is optional and can be selected by the system user. Any audio facility must not totally rely on the use of headphones.

Based on the considerations outlined previously, the user interface incorporates the following key components:

- **A "You are Here" Facility:** To inform the system user of their exact location on campus at the time requested.
- **A "Route" Marking Facility:** To explain to the user the best route from their current location or a particular building to their required destination.
- **A "Near Me" Facility:** To provide information on services available on campus.
- **An "Orientation" Facility:** To aid users should they become disorientated while on campus; it provides feedback on the user's current location, the direction in which they are facing, and the nearest building in that direction and/or the nearest building/service in any direction requested.

The key components normally use speech feedback (Figure 2), but a ext option is available upon request. The font size and color of the text can be set by the user. Visual feedback, in the form of map-based information, is available for sighted users. Users can select any combination of these feedback mechanisms, and switch off any they find intrusive or frustrating.

The haptic feedback provides varying levels of pulsed vibration. This is used to indicate external key features, such as buildings. A device vibrates twice as the user approaches a building, for example. The haptic components are also used to convey a level of terrain information. For example, various levels of pulsed vibration are used to indicate stepped/ramped entrances (the device vibrates twice for steps, three times for ramps, etc.). This will be extended to include other potential obstacles for example road crossings and areas of high traffic.

The interface also offers three audio options: speech, simple audio and spatial audio. Simple audio uses a beeping sound with a repetition-rate proportional to landmark distance, while spatial sound uses directional cues that appear to come from the direction of the landmark. Orientation information, whereby the audio response varies as the device is turned relative to a landmark, is also available.

Other components make use of the compass within the user's mobile device to provide speech-based orientation information on request, for example, "You are Facing…". Decision Points provide information to the user to enable them to make informed route navigation decisions based, for example, on directional information, e.g. "Turn right to building 1/turn left to building 2".

The development of a beginner's level interface was an important objective of this project. This interface is aimed at users with little exposure to electronic systems, and thus it must offer a short learning curve. On completion of the user testing the system will be extended to support more advanced technology users.

Development of the prototype system was carried out using Eclipse. This is an open source tool for developing Java enterprise applications. It was installed with the Android SDK manager to allow Android development work to be carried out for mobile applications. The development was targeted at the Android 2.3 (Gingerbread) mobile operating system. This means smartphones running Android 2.3 or higher support the application. The User Interface (UI) is designed to be simple and intuitive in nature. All objects and elements on the UI are labeled and care has been taken to ensure that the interface is not cluttered with unnecessary elements (See Figures 1, 2, and 3).

TESTING WAYFINDER'S USER INTERFACE

User testing was iterative and continued throughout the development process. However, post-development testing comprised two "Wizard of Oz" studies, an initial pilot study followed by a larger study.

Initial Pilot Study

A small pilot study was conducted as the first stage in testing the WayFinder system. It involved two subjects who navigated a particular route across

Figure 1. The WayFinder prototype: a) WayFinder icon displayed on device widget screen, b) initial screen application display on selection of widget

Figure 2. The WayFinder system, level 1 data flow diagram

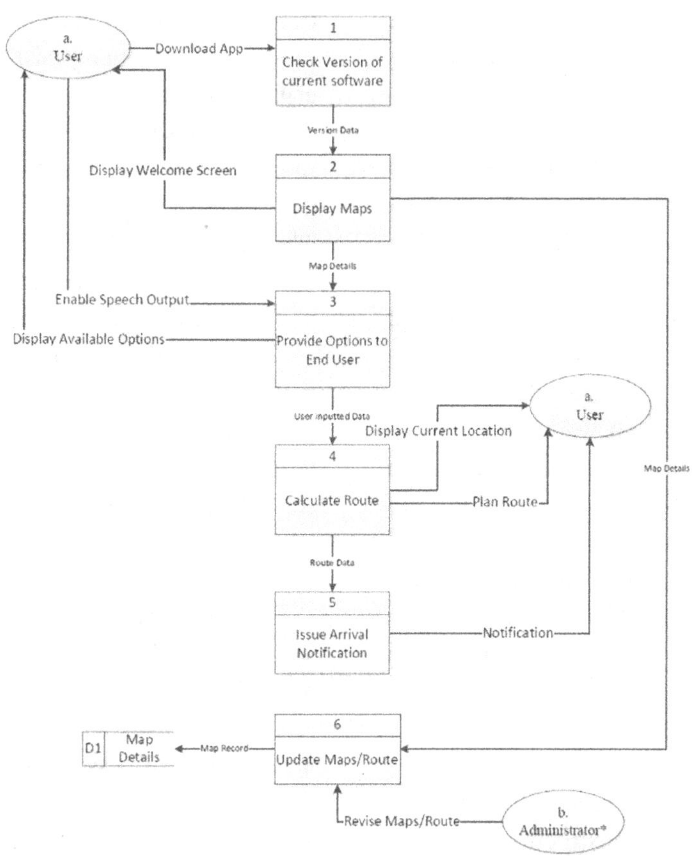

Figure 3. The WayFinder prototype, map-based information screen for sighted users

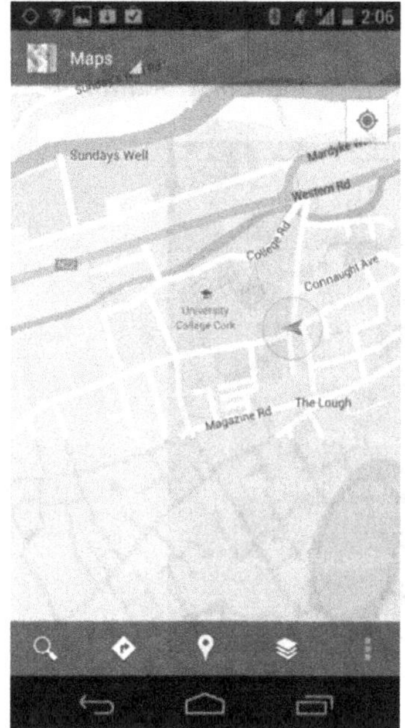

campus, and used a "Wizard-of-Oz" approach in which an experimenter remotely controlled the mobile devices used by the subjects. The testing involved the use of Android smartphones and Bluetooth connectivity to assess user actions in relation to information received via their mobile devices. Three Hypotheses were tested as part of the study:

1. Subjects will accurately react to instructions posed to them via the UI.
2. Subjects will complete the route independently based on information received from their interaction with the UI running on an Android device.
3. Based on their interaction with the system subjects (Both guide-dog and long cane users) will exhibit a high level of satisfaction with the system.

The two subjects were blind students enrolled on undergraduate programs at UCC. They were contacted through the Disability Support Services (DSS) Office at UCC. Both were experienced in using smartphones, in particular touch-screen devices. They varied in their level of familiarity with the UCC campus. One was a guide-dog user and the other a long-cane user (See Figure 4).

The subjects were asked to navigate a short route across campus, from the Clock Tower at UCC's North Wing to the entrance of the Boole Library, using WayFinder. Each subject was instructed on how the system works and how to interact with it. They were asked to use the system in combination with their existing mobility aid (guide dog or white cane), partly for safety purposes. On completion of the route participants were asked to answer a number of questions relating to the usability of the WayFinder user interface.

Pilot Study Analysis

Data was analyzed in order to determine how well users were able to interpret instructions received from the device (measured through observation by the test coordinator), their satisfaction with the application, and the usability of the system (measured via responses to the questionnaire). The questionnaire comprised two sets of questions, one aimed at gathering information on the individual participant, and the other aimed at gathering information about the system. For example, questions posed to the user concerned the type of mobile device they normally use and whether or not that device requires touch-screen interaction. The participants were also asked which mobility aid they use, and whether or not they had any prior experience with navigation-based systems.

The second set of questions was designed to gather information on the usability of the system and user satisfaction. This section required participants to give feedback on the following on a set of Likert scales:

- Their level of reliance on their usual mobility whilst navigating the route.
- The timeliness of the device feedback.
- Their ability to concentrate on environmental cues while using the system.
- The usefulness of the different feedback options (for example, the haptic components).
- The level of mental concentration/demand needed to use the system.
- Their extent to which they felt secure/insecure while using the system.

Figure 4. The WayFinder prototype, information screens for university building location

Participants were also encouraged to provide open comments regarding what they liked/disliked most about the system, what additional features might improve the system, etc.

The two participants in the pilot study had very different reactions to the system. One participant, a fourth year student and guide-dog user, reported that he had no difficulty in using the system. He expressed an interest in using the system in the future as a navigation aid on campus. The other participant, a first year student and long-cane user, indicated that her previous knowledge of the route made it difficult for her to use the system successfully. She found that she was anticipating the feedback from the system and as a result the feedback was redundant and distracting.

The problems reported by the second participant may result in part from the short time available to the subject in which to become familiar with the system. These problems might not have arisen had she had more time to experiment with the various forms of feedback available and to learn how to select appropriate feedback options to suit her needs. However, the results also highlight the differing needs of those who are new to a particular route and those who are more familiar with it. Those who are familiar with a particular route require less feedback than those who are not familiar with it. These issues were taken into account in the design of the next phase of testing.

The Main Study

A "Wizard of Oz" study was conducted to further test the viability of the pilot user interface for use by vision-impaired users when navigating campus.

Six blind/vision-impaired UCC students took part in the study. They were contacted through the DSS office at UCC and included three guide-dog users (one new) and three long-cane users. The group included one first-year undergraduate, four other students (both undergraduate and postgraduate) and one former student. Thus they had varying levels of familiarity with the campus. All participants were experienced in using smartphones.

The study was designed to test three hypotheses as follows:

1. Subjects will accurately react to instructions presented to them via the UI.
2. Subjects will complete the route independently based on information received from their interaction with the UI running on an Android device.
3. Based on their interaction with the system subjects (both guide-dog and long-cane users will exhibit a high level of satisfaction with the system.

The subjects were asked to use the Wayfinder system, in conjunction with their normal mobility aid, to navigate between the Boole Library to the Clock Tower. Subjects were given instruction on the planned route, on how the system works and on how to select appropriate system feedback options. After instruction, no further help was given to the subjects by the test coordinator. Any further instruction the participants received was route specific and provided via the system. For safety reasons, participants were accompanied while navigating the route. The test coordinator walked behind each participant to observe his/her reactions to the device's instructions during the test, but did not communicate with the participant until the test route was complete.. On completion of the route, subjects were asked a number of questions in order to assess usability and user satisfaction with the system.

Gathering Questionnaire-Based Data

In order to assess the usability of the WayFinder system and user satisfaction with it, subjects were asked to complete a questionnaire. It comprised two sections, one to be completed before attempting the route and the other upon completion of the route.

Before attempting the route, participants were asked five questions concerning their level of familiarity with smartphones, navigation-based systems and touch-screen interaction. Participants were also asked to indicate their normal mobility aid.

The post-navigation section of the questionnaire contained fifteen questions. Six of these were open questions, while the remaining nine required participants to rate their level of agreement with a statement on a scale from 3 (strongly agree) to -3 (strongly disagree). They included questions on:

- Participants' level of reliance on their regular mobility aid during navigation of the route.
- Timeliness of the feedback provided by the system.
- Level of distraction from normal environmental cues.
- Usefulness of the feedback tools.
- Level of mental demand placed on the subject in completing tasks.
- Level of security felt by the participant during the completion of tasks.

The open questions provided participants with an opportunity to give their own opinion of the system. For example, subjects were asked what they liked best and least about the system, and what features/feedback they felt could be added to improve the system. Participants were also asked if they would use the system again should the opportunity arise.

Evaluating the Results of the Main Study

Data gathered during the testing process was of two types: observations concerning users' reactions to information received from the smartphone device, and data concerning user satisfaction with the system obtained from the questionnaires.

All participants were observed as they navigated the route, the purpose being to assess users' reactions to the feedback provided by the system. If the system asked the user to turn left, for example, the user was observed to determine if their reaction was timely and appropriate. It was observed that all participants responded both appropriately and in a timely fashion to the instructions provided by the system.

The results from the questionnaire were also generally positive. The mean scores recorded for the main questions are listed in Table 2.

The results suggest that users relied mainly on their normal mobility aid rather than the WayFinder whilst navigating the route. While disappointing, this was perhaps to be expected as subjects were interacting with an unfamiliar system for the first time. Further consideration must be given to the level of reliance on mobility aids to ensure that students using the system have a higher level of confidence when using the system in the future.

The most popular features were the haptic feedback, the compass feature, and the "near-me" feature. Participants disliked the lack of control available over the audio feedback. This included the volume level, with some participants indicat-

Table 2. Usability study questionnaire: mean scores

Question	Mean Score
During the route navigation, I relied mostly on my normal mobility aid	2.66
The feedback provided by the system was timely (it gave appropriate instruction at each point of the route)	2
The system feedback made it difficult to concentrate on the sounds and other environmental cues that I normally rely on for the route	-1.66
I found the vibration-based feedback useful	2.33
I found the audio-based feedback useful	1
I found the task mentally demanding	-1.66
The pace of the task was rushed	-2.166
I felt insecure during the task	-1
I found the compass feature useful	2.4
The Near feature was useful I would use it again to regain my orientation	2.2

ing that the level was generally too low. Most participants indicated that a facility to switch on or off audio and/or vibration features would be beneficial.

The inclusion of optional features such as speech-based and/or spatial audio were also seen by participants as potentially beneficial inclusions in future versions of the system. Participants also suggested that the ability to identify buildings by letter as well as by name would be useful; this would make the system easier to use in conjunction with student timetables, on which buildings are indicated by letter only. This would enable more accurate identification of building entrances.

All participants said that they would use the system again should the opportunity arise.

FUTURE DIRECTION: EXTENDING THE STUDY

Work is currently underway to address the issues identified through the evaluation outlined previously. The user interface design is being expanded to include features such as spatial sound feedback and other features identified as potentially beneficial by the study participants.

The navigation layer is being developed using Bluetooth 4.0 and WIMU technologies to gather specific user location data based on inertial movement. It is planned that a pilot system will be in place on a specific route by mid-2013. This will include the temporary installation of Bluetooth beacons on route.

Further testing will be conducted with the aim of having the facility in place for use by incoming students in autumn 2013, at which stage further testing will be conducted. The study participants will include both new students (to assess the system with those using it on unfamiliar routes) and existing students (familiar with the route). If successful, further routes will be introduced.

CONCLUSION

Navigating a university campus can be difficult for visitors and incoming students/staff, particularly those who are blind or vision-impaired. A campus navigation system that makes use of mobile and wireless technologies could provide a more accessible and flexible navigation system than those currently available, enabling blind and vision-impaired people to navigate the space safely and independently.

This chapter has examined the difficulties faced by those navigating university campuses with a particular focus on blind and vision-impaired users. A review of relevant technologies and the state-of-the-art has been outlined. A study conducted to assess the needs of vision-impaired and blind users on campus at UCC has been described and the results discussed. This study comprised interviews and campus walkabouts with blind and vision-impaired students, and identified specific difficulties faced by such students.

A review of relevant technologies has been presented. The purpose of the review was to identify technologies capable of providing highly accurate positional information both indoors and out, an essential requirement of a navigation system for the vision-impaired. The review findings suggest that Bluetooth 4.0 is an appropriate technology. It offers high levels of accuracy without the problems associated with many other technologies, such as signal loss and high power demand. Since it is widely supported on smartphones, Bluetooth 4.0 allows the development of a system that runs on users' own mobile devices. This opens make it possible to develop a system that can be used by all, both sighted and vision-impaired, reducing the risk of isolation often faced by those with vision-impairments.

The design of a navigation system using mobile and wireless technologies has been described, along with the development of a prototype user interface based on data gathered in the study. The design includes a number of features necessary for safe, independent navigation around a university campus by vision-impaired students and others. Studies conducted to assess the usability of the prototype have been described and the results presented. This work is ongoing, and it is hoped that it will lead to the development of a campus navigation system that fully meets the needs of all users including those with vision-impairments.

REFERENCES

Ariadne GPS: Mobility and Map Exploration for All. (n.d.), Retrieved from http://www.ariadnegps.eu

Bellotti, F., Berta, R., DeGloria, A., & Margarone, M. (2006). Guiding visually impaired people in the exhibition. In *Proceedings of the Virtuality, Mobile Guide Workshop*. Retrieved February 26, 2011, from http://hcilab.uniud.it/sigchi/doc/Virtuality06/Bellotti&al.pdf

Bluetooth SIG. (n.d.), Retrieved from http://www.bluetooth.com/Pages/Low-Energy.aspx

BrowseAloud Software. (n.d.), Retrieved from http://www.browsealoud.com

Coughlan, J., & Manduchi, R. (2007). *Functional assessment of a camera phone-based wayfinding system operated by blind and visually-impaired users*. Paper presented at the IEEE-BAIS Symposium on Research on Assistive Technology. Dayton, OH.

D'Atri, E., Medaglia, C. M., Serbanati, A., & Ceipidor, U. (2007). *A system to aid blind people in mobility: A usability test and results*. Paper presented at the Second International Conference on Systems (ICONS). Sainte-Luce, Martinique.

Dixon, C. S., & Morrisson, R. (2008). A pseudolite-based maritime navigation system: Concept through to demonstration. *Journal of Global Positioning Systems*, *7*(1), 9–17. doi:10.5081/jgps.7.1.9.

Floerkemeier, C., & Lampe, M. (2004). Issues with RFID usage in ubiquitous computing applications. *Lecture Notes in Computer Science*, *3001*, 188–193. doi:10.1007/978-3-540-24646-6_13.

Gerst, M., Bunduchi, R., & Graham, I. (2005). Current Issues with RFID standardization. *Interop*. Retrieved January 2012, from http://www.york.ac.uk/res/e-society/projects/24/interop2005.pdf

Hautefeuille, M., O'Mahony, C., O'Flynn, B., Khalil, K., & Peters, F. (2008). A MEMS-based wireless multisensor module for environmental monitoring. *Microelectronics and Reliability*, *48*(1), 906–910. doi:10.1016/j.microrel.2008.03.007.

Helal, A., Moore, S., & Ramachandran, B. (2001). *Drishti: An integrated navigation system for visually impaired and disabled*. Paper presented at the Fifth International Symposium on Wearable Computers. Zurich, Switzerland.

Humanware. (n.d.). Retrieved from http://www.humanware.com/en-usa/products/blindness/talking_gps/trekker/_details/id_88/trekker_talking_gps.html

Kalliola, K. (2011). *High accuracy indoor positioning system on BLE*. Retrieved October 14th, 2011, from http://hermia-fi-bin.directo.fi/@Bin/b85924038a216224b987d81d837f0959/1327586874/application/pdf/865170/HighAccuracyIndoorPositioningBasedOnBLE_Kalliola_270411.pdf

Mau, S., Melchior, N. A., Makatchev, M. M., & Steinfield, A. (n.d.). *BlindAid: An electronic travel aid for the blind* (Technical Report Ref CMU-RI-TR-07-39). Pittsburgh, PA: The Robotics Institute, Carnegie Mellon University.

Mehigan, T., & Pitt, I. (2012). *Harnessing wireless technologies for campus navigation by blind students and visitors.* Paper presented at the 13th International Conference on Computers Helping People with Special Needs (ICCHP). Linz, Austria.

Mobile Pocket Speak. (n.d.). Retrieved from http://www.codefactory.es/en/products.asp?id=336

RFID Issues. (n.d.). Retrieved from http://www.slais.ubc.ca/courses/libr500/04-05-wt2/www/T_Gnissios/problems.htm

RFID Journal. (n.d.). Retrieved from http://www.rfidjournal.com/article/view/7224

Santoro, C., Paterno, F., Ricci, G., & Leporini, B. (2007). *A multimodal mobile museum guide for all.* Paper presented at Mobile Interaction with the Real World (MIRW). Singapore.

Tee, Z., Ang, L. M., Seng, K. P., Kong, J. H., Lo, R., & Khor, M. Y. (2009). *SmartGuide system to assist visually impaired people in a university environment.* Paper presented at the Third International Convention on Rehabilitation Engineering & Assistive Technology (ICREAT). Singapore.

Temple Go Mobile. (n.d.). Retrieved from http://www.here2there.com/pdf/H2T%20University%20PDF.pdf

UCC President's Report 2010. (n.d.). Retrieved from http://www.ucc.ie/en/news/newsarchieve/2010PressRelease/fullstory-91963-en.html

Walsh, M., Gaffney, M., Barton, J., O'Flynn, B., O' Mathuna, C., Hickey, A., & Kellett, J. A. (2011). *Medical study on wireless inertial measurement technology as a tool for identifying patients at risk of death or imminent clinical deterioration.* Paper presented at Pervasive Health. Dublin, Ireland.

ADDITIONAL READING

Al-Ammer, M., Al-Khalifa, M., & Abdulmalic, S. (2011). *A proposed indoor navigation system for blind individuals.* Paper presented at the Thirteenth International Conference on Information and Web-Based Applications and Services. Ho-Chi-Ming-City, Vietnam.

Ando, B., Graziani, S., Orazio Lombardo, C., & Pitrone, N. (2004). Development of a smart clear path indicator. In *Proceedings of the Instrumentation and Measurement Technology Conference,* (vol. 1, pp. 492-497). IEEE. Retrieved from http://0-ieeexplore.ieee.org.library.ucc.ie/stamp/stamp.jsp?tp=&arnumber=1351096&isnumber=29626

Belloni, F., Ranki, V., Kainulainen, A., & Richter, A. (2009). Angle-based indoor positioning system for open indoor environments. In *Proceedings of the IEEE 6th Workshop on Positioning, Navigation and Communication,* (vol. 1, pp. 261-265). IEEE. Retrieved from http://ieeexplore.ieee.org/stamp/stamp.jsp?tp=&arnumber=4907836&isnumber=4907789

El-Koka, A., Hwang, G.-H., & Kang, D. K. (2012). Advanced electronics based smart mobility aid for the visually impaired society. [ICACT]. *Proceedings of Advanced Communication Technology, 1,* 257–261.

Emerson, R., Naghshineh, K., Hapeman, J., & Wiener, W. (2011). A pilot study of pedestrians with visual impairments detecting traffic gaps and surges containing hybrid vehicles. *Transportation Research Part F: Traffic Psychology and Behaviour, 14*(2), 117–127. doi:10.1016/j.trf.2010.11.007 PMID:21379367.

Huang, H., & Gartner, G. (2009). A survey of mobile indoor navigation systems. *Cartography in Central and Eastern Europe, 1*, 305–319. doi:10.1007/978-3-642-03294-3_20.

Ifukube, T., Sasaki, T., & Peng, C. (1991). A blind mobility aid modeled after echolocation of bats. *IEEE Transactions on Bio-Medical Engineering, 38*(5), 461–465. Retrieved from http://0-ieeexplore.ieee.org.library.ucc.ie/stamp/stamp.jsp?tp=&arnumber=81565&isnumber=2674 doi:10.1109/10.81565 PMID:1874528.

Ivanov, R. (2010). *Indoor navigation system for the visually impaired.* Paper presented at the Eleventh International Conference on Computer Systems and Technology. Sofia, Bulgaria.

Ivanov, R. (2012). *RSNAVI: An RFID-based context aware indoor navigation system for the blind.* Paper presented at the Thirteenth International Conference on Computer Systems and Technology. Ruse, Bulgaria.

Jacob, R., Zheng, J., Ciepłuch, B., Mooney, P., & Winstanley, A. C. (2009). Campus guidance system for international conferences based on OpenStreetMap. *Lecture Notes in Computer Science, 5886*, 187–198. doi:10.1007/978-3-642-10601-9_13.

Jin, J., Lu, X., & Park, M. (2006). An indoor localization mechanism using active RFID tag. In *Proceedings of the IEEE Conference on Sensor Networks, Ubiquitous, and Trustworthy Computing,* (vol. 1, pp. 1-4). IEEE. Retrieved from http://ieeexplore.ieee.org/stamp/stamp.jsp?tp=&arnumber=1636157&isnumber=34301

Mehta, P., Kant, P., Shah, P., & Roy, A. (2011). *VI-navi: A novel indoor navigation system for visually-impaired people.* Paper presented at the Twelfth International Conference on Computer Systems and Technology. Vienna, Austria.

Mobile RFID. (n.d.). Retrieved from http://www.visionid.ie/t-low-frequency-mobile-rfid-readers.aspx

Pissaloux, E. E. (2002). A vision system design for blinds mobility assistance. *Engineering in Medicine and Biology, 3*, 2349-2350. Retrieved from http://0-ieeexplore.ieee.org.library.ucc.ie/stamp/stamp.jsp?tp=&arnumber=1053316&isnumber=22474

Ram, S., & Sharf, J. (1998). The people sensor: A mobility aid for the visually impaired. In *Proceedings of Wearable Computers,* (vol. 1, pp. 166-167). IEEE. Retrieved from http://0-ieeexplore.ieee.org.library.ucc.ie/stamp/stamp.jsp?tp=&arnumber=729548&isnumber=15725

Walsh, M. (2001). *A benchmark comparison between reconfigurable, intelligent and autonomous wireless inertial measurement and photonic technologies in rehabilitation.* Paper presented at the 3rd European Conference on Technically Assisted Rehabilitation (TAR). Berlin, Germany.

Willis, S., & Helal, S. (2005). *RFID information grid for blind navigation and wayfinding.* Paper presented at the 9th IEEE International Symposium on Wearable Computers. Osaka, Japan.

Yang, Y., Xu, J., Zheng, J., & Lin, S. (2009). Design and implementation of campus spatial information service based on Google maps. In *Proceedings of the IEEE International Conference on Management and Service Science,* (vol. 1, pp. 1-4). Retrieved from http://ieeexplore.ieee.org/stamp/stamp.jsp?tp=&arnumber=5301393&isnumber=5300803

Zhang, J., Dog, S., & Nee, A. (2008). *Navigation systems for individuals with visual impairments: A survey.* Paper presented at 2nd International Conference on Rehabilitation, Engineering and Assistive Technology. Bangkok, Thailand.

KEY TERMS AND DEFINITIONS

Bluetooth 4.0: Can be used to provide location-based information and offers a low energy and low cost wireless solution with an enhanced range over earlier Bluetooth specifications.

Mobility Aid: An aid used by a disabled person to facilitate mobility and independent navigation, for example, a guide dog used by a blind pedestrian.

Navigation: Involves movement from one location to another.

Navigation System: Involves the facilitation of movement (electronically or otherwise) from one location to another through localization of the subject's current position and direction in relation to their destination.

Prototype: A prototype is generally designed and developed to facilitate testing of a new product to allow it to be enhanced/improved.

User Interface: A user interface is a layer that facilitates the interaction between humans and machines.

Wireless Inertial Measurement Unit (WIMU): Typically provide data from accelerometer, gyroscope and magnetometer sensors, they capture tilt, force and timings to represent an optical 3D motion capture for the provision of a complete kinematic model of their subject.

Section 4
Health and Environment Monitoring

This section provides some emerging mobile-multimedia applications for health and environment monitoring.

Chapter 11
Ubiquitous Multimedia Data Access in Electronic Health Care Systems

Muhammad H. Aboelfotoh
Queen's University of Kingston, Canada

Patrick Martin
Queen's University of Kingston, Canada

Hossam Hassanein
Queen's University of Kingston, Canada

ABSTRACT

Advances in Information and Communication Technology (ICT) have enabled the provisioning of more cost-efficient means of delivering healthcare services through electronic healthcare systems (e-health). However, these solutions have constrained the mobility of medical professionals as well as patients. Mobile devices have been sought as a potential solution to free medical professionals and patients from mobility constraints. This chapter discusses the literature proposed in multimedia data transfer and retrieval, utilizing mobile devices and a multitude of wireless access technologies. A background section presents the different software technologies utilized by the proposed work, as well as a literature review. Following that, the authors compare these proposed systems and discuss issues and controversies found in these proposed systems, as well as propose means to address some of these issues. They conclude with an overall conclusion and outline future directions in this field.

INTRODUCTION

Telemedicine, literally meaning distant medicine, refers to the provisioning of medical services to distant locations via Information and Communication Technologies (ICT) (Moore, 1999). It includes sharing medical information, including multimedia data such as image and video, for consultation and diagnosis. Throughout the past two to three decades, countries have been exploiting advances in ICT in order to deploy more complex and resource-demanding telemedicine-based

DOI: 10.4018/978-1-4666-4054-2.ch011

services, such as those requiring high quality multimedia data, to deliver health care services to remote and underserved areas. This has enabled governments to deliver health care services to a wider audience in a more cost-effective manner.

ISSUES AND CHALLENGES

The adoption of electronic health care systems (e-health) has given the rise to the use of telemedicine Intensive Care Units (ICU), where intensive care patients are remotely managed from a central monitoring station that houses ICU medical specialists (Berenson, Grossman, & November, 2009). Functions of the central monitoring station include clinical multimedia data viewing, such as viewing diagnostic images and lab results (Cummings, Krsek, Vermoch, Matuszewski, & University HealthSystem Consortium ICU Telemedicine Task Force, 2007). Studies, such as Berenson *et al.* (Berenson et al., 2009), show that central monitoring stations constrain the mobility of the medical specialists, whom tend to perform more than one function and at times have to directly take care of patients.

There are currently mobile applications that have been developed to allow remote access of medical images, such as MRI and CT scans of patients (DigiSoft, n.d.; RemotEye, n.d.). Such applications may help enable medical professionals to remotely diagnose patients, as well as allow for better collaboration between peers, providing the ability to perform image manipulation operations such as cropping and annotating. However, these applications are typically agnostic of the dynamics of the underlying system and network. Wireless networks are in general unreliable with respect to their wired counterparts, and mobile access terminals tend to suffer from intermittent connectivity. In addition, image manipulation operations are known to be among the most resource-intensive tasks. System resources, such as processing (CPU, memory) and communicative power on mobile devices are limited. The speed at which medical images and other data is processed, sent and received is mainly at the mercy of the system load on the mobile device, as well as the network conditions such as load and variations in network bandwidth.

In some domains, constraining the mobility of the patients themselves can be an issue. With the rise in chronic illnesses worldwide, such as heart disease, the financial burden on long-term care facilities have reached a point to which they have either become unsustainable, or in the near future will become unsustainable (Social Security Advisory Board, 2009). To reduce this burden, some patients may live outside the long-term care facility, and some form of remote monitoring system would enable a medical facility to monitor their health conditions. Such a system would need to allow the patient to be mobile, and facilitate communication and feedback from the patient using multimedia such as sensor data, and perhaps audio and video streams. Most widely-adopted telemedicine solutions tend to be composed of stationary devices and only allow for in-door monitoring of the patient, constraining the patient's freedom of movement.

The cases previously presented mainly talk about issues in two categories of systems, namely *information retrieval systems* and *remote patient health monitoring systems*. It has become apparent that the lack of mobility and the inability to communicate multimedia data effectively has been major contributing factors to the issues and inefficiencies described before. Mobile devices are an increasingly attractive platform that can be used to overcome the mobility constraints described earlier. Coupled with the advancements made in wireless sensors, they have become the basis for a significant portion of the prototype systems and frameworks in the literature that target these issues.

However, just as the adoption of electronic health care systems introduced new issues, the use of wireless sensors, mobile devices and the wireless network environments that they operate in do not come without their issues and challenges. As a result, there has been a significant amount of research and development in enabling widespread utilization of mobile devices in the health care sector. In order to determine the extent to which research is closing the gap between research prototypes and widespread adoption of proposed systems by the health care sector, it is essential that one would derive a set of requirements based on the issues currently being faced, the issues that the adoption of mobile devices would introduce, and the goals that have to be achieved by these systems.

This chapter introduces a set of requirements that the proposed systems should satisfy. One of these requirements is efficiency and robustness, which includes the ability to transfer multimedia data such as medical images and sensor data with minimal resource consumption, in the face of stressful operating conditions. Other requirements include the ability to monitor vital signs and interoperability. It will also provide an overview of the different software technologies that these systems employ. In addition, this chapter provides a review of medical systems such as those used for multimedia data retrieval on mobile devices, as well as mobile patient health monitoring systems. Note that the term "mobile device" here in this chapter refers to any device that is generic in terms of functionality, portable, designed to be operational when mobile and has communication and processing capabilities. This does not include wireless sensors but includes devices that range from wireless handheld devices (e.g. feature phones, PDAs, smart phones) to laptops. Following that, issues and shortcomings in the reviewed systems are discussed. A set of criteria, based on the requirements, is used to provide a comparison between these systems. In that regard, we point out how different features in the software technologies used by these systems can affect the goal of satisfying one or more requirements. This discussion is followed by recommendations to how these issues and shortcomings could be addressed. Future trends are then highlighted and conclusions are presented.

REQUIREMENTS

The state of currently deployed systems and the issues and challenges faced, coupled with issues and challenges that would be introduced through the use of mobile devices and wireless networks, lead to the set of generic and function-specific requirements described next.

Efficient, Robust, and Reliable

Medical professionals need timely and accurate data for successful treatment of patients. The acquisition and the subsequent analysis of vital signs should occur in real-time so that any anomalies that occur must be immediately automatically reported to medical care. For example, Goni et al. (2009) state that after consulting with cardiologists, the maximum latency that a remote patient health monitoring system should be allowed for notifying the medical center that a patient has experienced a high-risk cardiac abnormality (symptom of a heart attack) is 13 seconds.

Systems with different functions may have different requirements for the duration for which they should be available. A mobile device used to forward vital signs to a medical center in a remote patient health monitoring system should be available for most of the day, where it can be assumed that at the end of the day the patient would be at home and can recharge the mobile device battery for use the next day. This can be mainly impacted by a combination of the efficiency and robustness mechanisms that a system uses. In addition, the reliability of the system software components also plays a role, but is beyond the scope of this discussion. For example, Goni et al. (2009) set the minimum required duration for the autonomy of the mobile device to 5 hours (duration without being connected to a charger), stating that it would be reasonable to assume that a patient utilizing a remote patient health monitoring system would be able to recharge the battery by then. This, obviously, may be subjective and may have to be determined by the health care provider, possibly based on the patient's schedule.

Battery life for mobile wireless sensors and mobile devices in general is crucial to the availability (reliability) of the system running on the device. Sensor duty cycles and communication are factors that affect power consumption on sensors. In general, there are many factors affecting power consumption on a mobile device, such as wireless communication, CPU cycles, and memory access. The magnitudes of these factors depend on the software running on the mobile device, and so it is imperative that the use of efficient algorithms in software can result in substantial savings in energy. In general, systems should strive to minimize power consumption by optimizing factors such as wireless communication and processing.

Systems operating on mobile platforms and in wireless environments face several stressful conditions, such as constrained resources on the mobile device, and network unreliability. Systems must be able to perform their required functions in such environments. In addition to enabling the retrieval of medical images in a timely manner, mobile medical image access systems should retain sufficient quality of these images to facilitate accurate diagnosis by the medical professional. For example, Park and Nam (Park, & Nam, 2009) report that images in the JPEG and JPEG2000 formats that have been compressed by up to 50% of their original size were acceptable for neurologists to read.

Unlike Internet access using Wi-Fi which is usually free of charge, the cost of using cellular network services varies from one provider and one region, to another. Charges that arise from mobile medical data access through cellular networks mainly depend on the communication protocols used, as well as the amount of data accessed. Data compression techniques may aid in reducing the size of data transmitted, hence reduce the monetary cost of communication, but would increase latency due to compression. Systems should be able to capture monetary cost requirements and allow for adjustment of parameters such as the use of compression, in order to satisfy all other requirements.

Maintainable

Since remote patient health monitoring systems deploy medical equipment which is outside the physical administrative boundary of medical centers, or "off campus", there needs to be means to remotely maintain the hardware and software at the patient's end. Maintenance tasks may include upgrading the software on the mobile device that is used to process sensor readings. It may also include updating an inference engine or a database of inference rules used in analyzing sensor data, which may require frequent updates as doctors refine these rules. New rules may need to be remotely added by a physician "on the fly" as the state of the medical condition changes. In addition, as mentioned before, some system parameters may be subjective as in patient health monitoring systems.

Interoperable

It is not uncommon for medical professionals to perform work at more than one hospital. For example, for a doctor to carry more than one wireless handheld device (or even a pager) in order to be able to communicate with different hospitals, because they deploy different interfaces that are incompatible with one another, is inconvenient and inefficient. Providing one mobile device that would allow the medical professional to communicate with different hospitals is more convenient and allows for more efficient sharing of information.

In addition, since one of the main goals of the health care industry is to reduce the cost of delivering health care services, it is imperative that the systems used to achieve this goal would be built from components that would be integrated and maintained at a low cost. This requires vendors of system components, such as sensors, mobile devices and medical software, to develop solutions that would allow *integration as a commodity*. This can be achieved with the help of employing standardized interfaces for these components.

Multiple Vital Signs

Remote patient health monitoring systems should support monitoring multiple vital signs simultaneously. Caring for patients with chronic illnesses usually involves monitoring and diagnosing multiple vital signs.

Secure

Information retrieval systems access sensitive personal information that may include the patient's medical history, which may contain a history of illness, medications, and allergies. It is imperative that this information must be communicated in a secure manner. In addition, remote patient health monitoring systems can be used to remotely alert a medical care center of an emergency condition. Tampering with the system might affect the ability of the system to send such alerts, possibly putting the life of the patient at risk. In general, the required security services for systems which involve the handling of medical data are authentication, privacy (e.g. using encryption), and data integrity. Important information that pertains to medical professionals, such as system access credentials, must also be protected.

Scalable

Systems need to sustain the same expected performance as the number of mobile devices and sensors in the network increase. The scalability of information retrieval systems and remote patient health monitoring systems mainly depends on the underlying network capacity, the scalability of the servers providing the services, and any mechanisms that application server software employ to aid in supporting a greater number of simultaneous users, and so providing a qualitative system analysis for this requirement is beyond the scope of this chapter.

COMPARISON CRITERIA

Based on the requirements mentioned earlier and the purpose of this chapter, a set of criteria is derived to allow comparison of different systems, and is presented in Table 1. The requirements efficient, robust, and reliable apply to both system categories, namely *information retrieval systems* and *remote patient health monitoring*, and so *efficiency, robustness, and reliability* can be used as a comparison criterion. Systems in both categories are required to be secure, therefore *security and privacy* is used as a comparison criterion. Definitions for efficiency, robustness, reliability, interoperability, and maintainability used in the description column in Table 1 and Table 2 were

Table 1. Generic comparison criteria derived from requirements

Criterion	Description
Efficiency, robustness, and reliability	Efficiency: the ability to perform the required functions with minimal cost, where cost refers to time (latency), monetary cost, and energy cost.
	Robustness: the ability to perform the required functions under stressful environmental conditions and invalid input.
	Reliability: the ability of a system or component to perform its required functions under stated conditions for a specified period of time. Synonymous with availability.
Interoperability	The ability of two or more systems or components to exchange information and to use the information that has been exchanged.
Security and privacy	Providing confidentiality and integrity to data being accessed.

Table 2. Additional comparison criteria specific to RPHMS

Criterion	Description
Vital signs monitored	The more vital signs that the system is able to monitor, the better. Not only that, but the type of vital sign monitored is also of importance; a system that monitors cardiac function to check for abnormal heart rhythms would prove valuable as it would assist in "offloading" patients with heart disease from chronic care facilities.
Maintainability	The ease to which a component can be modified to correct faults and improve performance.

derived from the IEEE Standards Glossary of Software Engineering (The Institute of Electrical and Electronics Engineers, 1990).

In addition, there are additional comparison criteria for remote patient health monitoring systems (RPHMS), which are outlined in Table 2.

BACKGROUND

This section first introduces some of the underlying software technologies used by the medical systems discussed in this chapter. These technologies are mainly deployed as a communication interface between the mobile device and a medical center. Then these technologies are compared in terms of efficiency, robustness, and reliability. In addition, these software technologies are discussed in terms of how they would impact the interoperability of software systems. Following that, the section briefly discusses the scalability, security, and privacy that can be provided by these technologies.

Software Technologies

The first two software technologies described in this subsection employ a Service-Oriented Architecture (SOA, defined shortly). From Sillitti et al. (2002), and the description of SOA found in McGovern et al. (2003), one can loosely define SOA as the following: constructing software using reusable components with a common, standard interface that facilitates interoperability within the system of a business and between systems of different business entities. An application built on a SOA is typically represented as a graph that represents services composed together, where more complex services are composed of simpler services.

JINI (Java Intelligent Network Infrastructure) (The Apache Software Foundation, 2007) is a Java-based implementation of SOA, where clients (service requesters) can use published services distributed over the network. Available services from service providers are published through *lookup* services. A service provider first searches for a lookup service, typically by sending out multicast messages. If a lookup service identifies itself, the service provider sends its *service object* to that lookup service. A service object is the instance of the Java class that implements the service code. A client can then *lookup* any required services from the lookup service. Once located, the client receives the service object, and loads that object, in order to invoke the service. Communication between the client and the service is typically achieved using the Remote Method Invocation (RMI) protocol (Oracle, n.d.). RMI allows for communication between two hosts running Java Virtual Machines, by facilitating invocation of methods of a Java object that resides on a remote host, by sending the method call and parameters, which are typically instances of Java classes, over the wire to the remote host.

In settings where it is not possible for a device to be part of a JINI network for reasons such as resource-constraints, or operating in a non-Java environment, the *JINI surrogate architecture* provides the means to connect to a JINI network, through a proxy host, which is capable of connecting to a JINI network. This proxy host, known as a *surrogate host*, executes code that interacts with the JINI network on behalf of the device, known as the *surrogate*. Any communication protocol between the device and the surrogate can be implemented.

In the discovery phase, the host sends out a multicast announcement to indicate its presence. A device, which needs a surrogate, has to listen for announcements, and upon receiving an announcement, sends a registration request to the surrogate host. Alternatively, the device would send out a multicast request for a host, and the host would reply back. Once the host accepts the request it retrieves the surrogate for execution, either from the device or from a URL specified by the device. The use of multicasting is usually not possible over the Internet as, generally speaking, the Internet is a non-multicast network; one would either require tunneling, or the use of a specific network which supports multicasting.

Web Services is a W3C standard that defines a language, the Web Service Description Language (WSDL), for defining and publishing service interfaces. Web Service technologies have defined one or more specifications that can optionally be deployed to enhance service delivery, such as Web Services Reliability and Web Services Security (discussed later) (W3C Web Services Architecture, n.d.).

SOAP (Simple Object Access Protocol) is a standardized simple XML-based protocol for remote method invocation. A SOAP message consists of a root XML element, known as a *SOAP envelope*, which signifies that the message is a SOAP message. A SOAP message can have zero or more optional *header* elements, and at least one body element containing the message data. In general, the lifetime of a SOAP message is composed of the following stages: serialization of data into a SOAP-XML message, transmission of the message over the network and de-serialization of the SOAP-XML message by the receiver (W3C SOAP Version 1.2 Part 1, n.d.).

SOAP supports two models, known as *bindings*, which describe how SOAP messages should be structured: Remote Procedure Call (RPC) style and document style. A method parameter XML element in a SOAP message specifies the parameter name and value (specified in a WSDL file by the attribute *use="literal"*), or can, in addition, specify the parameter type (specified in a WSDL file by the attribute *use="encoded"*). With RPC-style messages, the method name and its parameters are included in the message. With document-style messages, only the parameters are included. Matching the parameters to the corresponding method at the service provider's end is achieved by the use of a schema (part of the WSDL file), which defines aliases for parameters of each method in the WSDL file. It is these aliases that are used to specify method parameters in a document-style SOAP message. Using a document-literal style can result in a larger initial set-up time for Web Service invocation, but would result in greater messaging efficiency, as the size of a single SOAP message will be reduced.

Java Messaging Service (JMS) is an asynchronous messaging system API for applications implemented in the Java programming language. JMS does not define any particular network protocol to use; the implementation of the JMS is left to the JMS provider. The Java 2 Enterprise Edition (J2EE) platform includes an implementation for a JMS provider (Mahmoud, 2004).

COMPARISON OF DIFFERENT SOFTWARE TECHNOLOGIES

Systems that utilize widespread software technologies and standards have the potential to be adopted by health care institutions. However, it is necessary to examine these technologies and standards in order to determine their impact on a system in terms of issues such as performance. In this subsection, different software technologies and standards used by some of the surveyed systems are compared in terms of the generic criteria described previously.

Efficiency, Robustness, and Reliability

JINI

Wilkinson et al. (2001) note that JINI provides no mechanism to deal with unreliability in the underlying network. JINI allows a mobile service consumer to request a service by querying a lookup service. The service provider may either upload the entire service to the lookup service, or if operating conditions are prohibiting (e.g. network latency), upload a proxy to that service. In addition, relying on a single lookup service introduces a single point of failure. Increasing the number of lookup services can enhance reliability.

Web Service Technologies

In terms of the efficiency of using SOAP, one can clearly see that using an RPC-style binding instead of a document-style binding would yield a smaller WSDL file (no schema required), but can overall result in larger amounts of XML data being transmitted as a result of including the parameter types in a message. For XML messaging over slow networks, a document-style binding may be more beneficial, as, despite downloading the WSDL file would take more time, messaging throughput would be increased.

In terms of robustness, the use of document-style allows for the use of a streaming parser, which minimizes memory requirements for XML de-serialization (McCarthy, 2002). The schema can be used to validate the SOAP message i.e. check that the XML message is well-formed and conforms to the rules defined in the schema.

One way of providing standards-based reliability for Web Services is by adopting the Web Services Reliable messaging (WS-Reliability) standard. WS-Reliability was published by the Organization for the Advancement of Structured Information Standards (OASIS) in order to provide a "way to guarantee message delivery to applications or Web Services". As defined by the specification (Iwasa, 2004): "Web Services Reliability (WS-Reliability) is a SOAP-based protocol for exchanging SOAP messages with guaranteed delivery, no duplicates, and guaranteed message ordering. WS-Reliability is defined as SOAP header extensions and is independent of the underlying protocol".

Experiments, Analyses, and Comparisons from the Literature

Weibel (2002) conducted a set of experiments in order to identify strengths and weaknesses of SOAP Web Services and JINI in a wired environment. The main points discussed were initial set up time for a service, the actual service time, and the scalability of the service (discussed later). In order to model a highly distributed environment, a service that implements the exponentiation function, was created. In addition, a multiplication service and an addition service were also implemented. The exponentiation service used instances of the multiplication service, and instances of the multiplication services used instances of the addition services. The authors report that the service setup time for SOAP Web Services was almost 4 times larger as for JINI. The author states that JINI's exchange of whole Java objects, results in high network traffic, which is not at all suitable for slower networks such as mobile networks. In addition, the author pointed out that SOAP implementations spend a lot of time encoding/decoding XML messages, making SOAP slower than JINI. One must note that in this work only the RPC-style binding for SOAP was touched upon.

Eggen and Sunku (2003) conducted a study on the efficiency of SOAP-based communication, versus the use of JMS. In an experiment with the same scenario for each of SOAP and JMS, a client distributes an equal amount of data to worker servers, and then each server performs a sort operation on the data, and returns the results back to the client. Overall, JMS was found to be slightly more efficient.

Interoperability

This subsection explores how the utilization of Web Services or JINI can impact the interoperability of a system with other systems. In addition, the effect on how a system component can be integrated with components from other systems is discussed.

Web Services

The use of Web Services standards (WSDL, SOAP, WS-Addressing) would allow for the implementation of services in any programming language, providing different system providers with the freedom to use a programming language and environment which would best suite the hardware deployed in the system, or for other reasons such as exploiting existing developer skills in one particular technology or another. Having said that, the use of Web Services does not protect against changes in a Web Service interface (WSDL file) published by the system. With good software engineering practices such as the proper implementation of versioning (Lublinsky, n.d.; Brown, & Ellis, 2004; Endrei, Gaon, Graham, Hogg, & Mulholland, 2006) this should not be a major issue.

JINI

JINI enables communication between the service consumer and the service provider by means of a service proxy. This proxy is supplied by the service provider, which the service consumer uses to communicate with the service provider. This allows for "plug-and-play" protocols, and so allows for the service provider to adapt to different constraints (e.g. characteristics of underlying network) and use different protocol implementations accordingly. Having said that, using JINI as the underlying infrastructure would constrain the choice of the underlying platform, limiting the implementation to Java-based solutions (an interface layer would have to be added to facilitate interaction with non-Java code). It may not be desirable to enforce such a limitation on part of the system which interacts with mobile clients and clients outside the administrative border of the medical information system, for the reasons stated earlier regarding Web services. In addition, use of a non-standardized binary format to represent data would take us a step back from what technologies such as Web Services have achieved in interoperability.

Scalability

Scalability can be defined as the ability to sustain service requirements as workload increases. A remote patient health monitoring system should be able to sustain a growing number of remotely monitored patients, as well as the possible tremendous increase in location requests possibly emitted from mobile devices attached to these patients. The inability of a health care service provider to scale can result in increased delays when requesting patient status, and receiving alarm notifications of patients in dangerous conditions, which can have an adverse effect on the performance of the system as a whole. As part of the set of experiments described earlier, Weibel (2002) assessed the scalability of SOAP and JINI-based services. Weibel concludes that JINI was found to be more scalable than SOAP. This somewhat supports the idea, also hinted by Eggen and Sunku (2003), that SOAP achieves flexibility at the expense of scalability and performance. With regards to JMS, Eggen and Sunku concluded that both SOAP and JMS scale in a similar manner.

Security and Privacy

Systems used in health care such as remote patient monitoring systems, transmit vital signs data to a care centre to monitor a patient's health. This data is also used to determine whether the patient requires immediate attention. It is therefore imperative to secure the system at the patient, the care centre, as well as the communication between them. In addition, the access to the data being transmitted should be controlled, in order to prevent unwanted access of a patient's data.

JINI

JINI implements restrictions on access to a particular service by means of an access control list that is associated with the object that implements the service ("JINI Architecture Specification".2007). JINI services or service proxies may want to communicate with one another using non-standard protocols, or using non-standard ports, or ports which may not be allowed by firewall policies for some networks. The system should be designed in a way in which JINI services can communicate with each other over these networks, without compromising or reducing the security of any network. Poorly designed JINI services with buggy code may consume excess resources locally, or may incur a significant communication overhead when communicating with each other over the network. This unwanted behavior may have an adverse effect on the reliability of the system as a whole.

Web Services

One way of providing standard-based security is by using Web Services Security (WS-Security). As described by the specification, WS-Security "describes enhancements to SOAP messaging to provide message integrity and confidentiality". The specification "can be used to accommodate a wide variety of security models and encryption technologies" (OASIS Open, 2006). The SOAP XML structure itself is preserved but the data inside the SOAP message is encrypted. WS-Security does not define any security protocols, but provides tools and a framework for doing so. This allows for a flexible set of security protocols to be deployed. These protocols can then be activated in a "plug-and-play" manner. This is analogous to the flexibility that JINI provides in enabling a JINI service to dynamically specify the communication protocol via the use of service proxies. Despite the need for interface compatibility when using either technology (WS or JINI), the use of a service proxy by a JINI service consumer, may result in system policies being bypassed by the service proxy, and possibly exposing the system to undesired behavior due to buggy code in the service proxy. The use of Web Services, however, would protect the system from such risks.

Medical Systems

In the literature, different systems have been proposed to address the needs of one or more sub-domains in health care, such as cardiology or radiology, and have been created to serve different purposes such as information retrieval, or remote patient health monitoring. At the lower level, different parts of the system may utilize one or more communication protocols, as well as one or more wireless access technologies, such as Bluetooth (Bluetooth SIG, Inc, n.d.), ZigBee (ZigBee Alliance, n.d), Wi-Fi (WiFi Alliance, n.d), 3G (ITU, n.d), and CDMA (Telecommunications Industry Association, n.d). At the higher level, these systems may use different software architectural patterns, such as client-server or peer-to-peer. This section then presents a review of selected papers from the literature, and touches on a few commercial products. The systems examined fall into two categories depending on their functionality; namely *information retrieval* and *remote patient health monitoring*.

In the literature, there are several surveys that discuss different aspects of these approaches, and classify them in different ways. In addition to remote patient health monitoring systems, Orwat et al. (2008) also discuss systems geared towards the general welfare of patients (e.g. daily exercise routines), and covers systems in regular operation, but provide no technical in-depth analysis and comparison of these systems. This is partly due to the lack of in-depth technical information provided by the publications in which these systems are presented. Surveys, such as Rashvand (2008) discuss hardware technologies. Varshney (2009) discusses requirements in general, and networking issues in more depth.

Information Retrieval

These systems facilitate the retrieval of a patient's medical information, such as a patient's Electronic Health Record (EHR), or medical images for remote diagnosis. For instance, in an emergency situation, a radiologist may be contacted to help in diagnosing a stroke patient, in order to carry out the right treatment in time. A radiologist at that time may only have access to the health care provider through a handheld device. A typical diagnosis for a single stroke patient may involve examining tens of medical images for a single patient, and may go over a hundred images (Internet Stroke Center, n.d.). Radiologists examine several patients, and so the number of images required to be examined by the radiologist can easily be in the hundreds (Park, & Nam, 2009). These images have to be of sufficient quality, in order to enable the radiologist to make accurate and timely decisions.

Medical images in a traditional in-hospital setting tend to be of high quality and of a large size. Displaying such images on a wireless handheld for a radiologist to view can be challenging. This is due to the varying network bandwidth, the relatively smaller handheld display, and the limited resources on the handheld. It is imperative that a system can enable a radiologist or a medical professional in general, to perform their required tasks, in the face of these stressful technical operational conditions. In addition, the medical professional might want to perform image manipulation tasks like annotations, cropping, etc., in order to mark and share findings with other medical professionals. These image manipulation tasks are resource-intensive and may be a burden on a resource-constrained device such as a wireless handheld device. In order to relieve some of the burden on these resource-constrained devices, as well as provide a smooth user experience, some systems employ capability augmentation. The term "capability augmentation" refers to the system opportunistically offloading some of the workload from the resource-constrained device to a more powerful device. This is also commonly known in the literature as *cyber foraging* (Satyanarayanan, 2001). The Locusts cyber foraging framework (Kristensen, & Bouvin, 2008) is an example of a framework that supports performing resource-intensive image manipulation tasks by augmenting the capabilities of the mobile device.

Aspects Considered for an Information Retrieval System

Different systems are compared using the criteria defined in the Introduction section, in terms of the following aspects:

Protocols and Data Formats

The setting assumed here is that each medical professional's mobile device would be connected to the medical center where a particular operation is being performed on a patient. Since the main application here is the retrieval of medical images, the communication protocols used to retrieve those images can have a major impact on the overall performance. The data formats and the compression mechanisms they employ can provide robustness in environments where network bandwidth is constrained. Protocols and data formats should be able to cope with the different variations of bandwidth and reliability found in different wireless access technologies, and so it is essential that proposed systems would support and have its performance assessed against different wireless access technologies.

Resource Augmentation

Being aware of the amount of resources available on a mobile device and the amount of resources required by applications running on a mobile device can help in achieving better user experience by augmenting the resources of the mobile device with that of more powerful devices, such as clusters of servers. This is useful for resource-intensive applications such as those that involve image manipulation.

Comparing different systems in terms of the aspects *protocols and data formats* and *resource augmentation* gives a glimpse on how these systems deal with the two major challenges with the use of mobile devices, namely a device's resource constraints and the constraints imposed by the operating wireless environment.

Systems

A low-cost solution for accessing a patient's medical record can be implemented through the use of the ubiquitous WAP technology, as presented by Hung and Zhang (2003). The prototype system supports the following functionality: (1) ECG browsing and heart-rate estimation, (2) Blood pressure browsing, (3) Patient record browsing, (4) Clinic and hospital information inquiry (requested by patient), (5) Doctor's appointment browsing (requested by patient). A simple information retrieval experiment was carried out. This experiment involved no medical personnel, and so no information was given on whether or not the images of ECG graphs and information retrieved using WAP was of sufficient quality for medical diagnosis tasks. Such a system would be beneficial in a developing country, where only feature phones can be afforded, and where the communication infrastructure is not advanced. Such a system is monetary cost-efficient as data communications through cellular networks are relatively more expensive. The system is not reliable as there is no means of guaranteeing a certain level of quality of the information retrieved. The utilization of WAP technology allows use of the system across a diverse range of mobile devices. The scalability of the system depends on the scalability of the underlying cellular network. No security or privacy mechanisms were discussed.

Ivetic and Dragan (2009) present a prototype system for viewing medical images on a PDA. In this system, a mobile device retrieves medical images from a hospital information system server where medical images are stored in the standardized and widely adopted DICOM medical image format (NEMA, n.d.) (DICOM server). Mobile devices connect to an image server, which acts as an intermediary between the mobile device and the DICOM server. The image server is responsible for employing "scalable" image transmission, where image quality is compromised for the sake of performance; stationary devices receive high resolution, high quality images whereas handheld devices receive low resolution, low quality images. This can help make the system more robust against variations in resource availability, such as available bandwidth. The communication protocol used for image request and retrieval from the image server is JPIP (JPEG2000 Interactive Protocol), where JPEG2000 is the image format used by the protocol. JPEG2000 offers both lossy and lossless compression (JPEG, n.d.). The resolution, quality and region-of-interest can be defined in a DICOM request. Effects of different levels of compression on network transmission time and processing requirements were presented but means as to how to adjust the compression level to adapt to variations in network and mobile device resources were not discussed.

Park and Nam (2009) propose a framework for medical data access. A proof-of-concept system was developed to demonstrate the feasibility of the approach. The system allows a medical professional to use an application on his/her mobile device, to connect to a medical information system using a WiMAX wireless connection, in order to access medical images of patients (e.g. CT scans). The framework employs a service-oriented architecture, where an application is composed of one or more reusable software components, or *tasks*, and is typically represented as a graph. When an application residing on a mobile device invokes a particular service, a *context and service profile* is sent to a server acting as the *planning agent*, which is responsible for delegating tasks to distributed server farms, in order to allocate resources for the service. The profile contains application-specific parameters such as the time to execute the application and the duration for which the application should run. The profile also contains parameters that describe attributes of the mobile device such as the set of allowable display sizes and system resources available, as well as attributes that describe the network conditions such as the disconnection rate of the wireless network and the available network bandwidth. In addition, the profile contains a set of Quality-of-Service (QoS) levels, with each level corresponding to a particular average consumption rate of system resources.

The planning agent tries to assign tasks to servers in order to satisfy the QoS requirements. If the resources on the servers are insufficient, then the agent invokes a QoS control algorithm. The QoS control algorithm works as follows: from the set of applications running on the mobile device, select the application that yields the maximum resource savings if its QoS requirements are degraded, and degrade its QoS requirements. This process is repeated until the QoS requirements for each application are satisfied. One would like to note that there is an implicit assumption here, which is that all the applications running on the mobile device are of equal priority. The other issue is that there is no mention of how the attributes and QoS parameters are set or determined. With that said, the features described prior make the system robust against variations in the availability of resources on the mobile device (e.g. CPU, memory), and variations in network bandwidth. In addition, the system enables the mobile device to augment its computational resources with resources from clusters of servers. This may be beneficial if the application user, in this case a medical professional, performs a resource-intensive task on a medical image, such as cropping, annotating, or marking.

Sample applications were developed to demonstrate the effect of QoS adaptation on the image size and transmission time. Images stored using the medical image format DICOM can be converted to JPEG and JPEG2000 image formats before being sent to the mobile device. As part of the mechanism used to adapt to the mobile device and wireless environment conditions, the image quality (through different levels of compression) and resolution can be reduced, which helps making the system robust against severe resource constraints on the mobile device and in the network. The system uses the JPEG2000 Interactive Protocol (JPIP) over HTTP for image transmission to the mobile device.

Maglogiannis et al. (2009) propose the use of a wavelet-based compression algorithm for medical images accessed using mobile devices. Wavelet-based compression is different from the compression used in JPEG (JPEG, n.d.). JPEG2000 also employs wavelet-based compression. However, the authors state that a major drawback of the JPEG2000 standard is that it does not support lossy-to-lossless compression for a particular Region-of-Interest (ROI) in the image, a feature proposed by the authors. A mobile device user, such as a medical professional, remotely accesses a database of medical images via a Web-based system, using a Java-based client application that runs within a Java-enabled Web browser. The client application uses HTTP to access the

Web server and RMI for data exchange. The user is authenticated using a username and password pair, and data encryption is provided using a symmetric key of 128 bits length. The authors present a comparison of their experimental results for the response time for retrieving three types of medical images over GPRS and WLAN, when using JPEG and wavelet-based compression. Their results show that wavelet-based compression outperforms JPEG for all three types of medical images for both GPRS and WLAN.

Ninos et al. (2010) discuss the feasibility of using a PDA to connect to a hospital information system, to retrieve medical images in the DICOM format. The authors argue that research previous to their work, does not take into consideration the effects of reducing the quality of medical images on diagnosis results, and that no one considered transferring the medical images as they are (in DICOM format). One would strongly disagree, as previous work like Park and Nam (2009) have conducted experiments involving both transferring the image in its original DICOM format to the mobile device, and transferring a quality-degraded version of the image. The other issue is that the experiment presented took place in an "in-hospital" setting (Euromedica Medical Centre), where bandwidth resources may not be as scarce, whereas the framework presented by Park and Nam (2009) was not limited to an "in-hospital" setting and therefore would have to deal with variations in bandwidth resources, such as the difference between operating inside and outside the medical centre. Obviously such a system would not be robust in such varying operating conditions. Ninos et al. (2010) claim to have implemented CPU-efficient and time-efficient algorithms to enhance image quality, with no further details or references to these algorithms. No security issues were discussed.

Comparison

In Table 3 and Table 4, we define how we assess each of the proposed systems in Table 5. The properties in the table are based on the requirements defined in the Introduction section. The ranks defined in Table 4 are used to rank the different systems presented in the comparison Table 5.

The various criteria used to compare the systems presented in Table 4 show that each system combines one or more approaches to address one or more challenges defined by the criteria, but are either limited, or do not address all of the criteria mentioned before. Hung and Zhang (2003) propose a WAP-based solution for browsing ECG blood pressure information for patients. The WAP technology exists on almost all, if not all cellular phones, and supported by most, if not all cellular network operators, and so can handle the heterogeneity of cellular phones and networks. All the systems presented prior except for Hung and Zhang and Ninos et al. (2010) exploit the idea of sacrificing image quality for the sake of efficiency, robustness and reliability in one way or another. Ninos et al. limited medical information retrieval to an "in-hospital" setting, where DICOM images are retrieved onto the PDA with no compression whatsoever. Such approach is inefficient since

Table 3. Information retrieval system ideal properties

Criterion	System properties in an ideal system/ setting
Efficiency, robustness, and reliability	(a) Study effect of different compression levels (b) Observe network conditions (c) Observe mobile device resources (CPU, memory) (d) Protocol which dynamically adapts compression level based on observed conditions
Interoperability	(a) Standard-based interfaces (b) Standard-based data formats
Security and privacy	(a) Access control (b) Data encryption

Table 4. Basis for assessment based on opinion

Assessment	Efficiency, robustness, and reliability	Interoperability	Security and privacy
Very good	All	All	All
Good	(a), (b) and (c)	All	All
Fair	(a)	(a) or (b)	(a) or (b)
Poor	None	None	None

Table 5. Information retrieval systems comparison

Reference	Setting	Efficiency, robustness, and reliability	Interoperability	Security and privacy
Hung and Zhang (2003)	Outdoor	**Poor.** No mechanisms proposed	**Fair.** Uses WAP for information retrieval	**Poor.** N/A
Ivetic and Dragan (2009)	Outdoor	**Fair.** Propose using JPIP to provide different image quality levels; stationary devices receive high quality images, wireless handheld devices receive low quality images	**Fair.** Uses DICOM/JPEG as the data format	**Poor.** N/A
Park and Nam (2009)	Outdoor	**Good.** QoS control; mobile device resources augmented with grid resources. Proposes use of JPIP for adjusting compression level	**Fair.** Uses DICOM/JPEG as the data format	**Poor.** N/A
Maglogiannis et al. (2009)	Outdoor	**Good.** Proposes the use of a wavelet-based compression that outperforms the compression used in JPEG	**Fair.** Uses RMI for data exchange	**Good.** User authentication using username/password. Data encryption using symmetric key with 128bits length.
Ninos et al. (2010)	Indoor	**Poor.** No mechanisms proposed	**Fair.** Uses DICOM as the data format	**Poor.** N/A

the system would face difficulties in operating reliably outside hospital settings, where network bandwidth would vary and at many instances may be scarce.

Ivetic and Dragan (2009), Park and Nam (2009) and Maglogiannis et al. (2009) all present more practical means of dealing with a wireless environment. Ivetic and Dragan present the effect of using different compression levels on the image size, the network transmission time, as well as processing requirements. Park and Nam propose a SOA-based framework for medical data access, with QoS guarantees making the system more reliable. The framework uses a pool of servers to provide additional resources to the service being accessed from the mobile device, in order to alleviate the load on the mobile device. The effect of using different compression levels on the image size and the network transmission time, but not processing requirements, were presented. Maglogiannis et al. proposed the use of an image compression algorithm that has shown to be more efficient than JPEG, in reducing the size of medical images. However, no mechanisms to help alleviate the load on the resources of the mobile device were proposed.

None of the work discussed previously, however, described what protocols or interfaces the mobile device uses to access the information system of a medical center, in order to allow a medical professional to remotely access medical data. This would have provided a hint as to how different medical professionals with mobile devices would collaborate. This would have also hinted at the degree of interoperability that the system would allow between different health care institutions. In addition, all of the systems, which proposed the use of adjustable compression, did not demonstrate how and when compression levels would be adjusted, but merely demonstrated the effect of the compression level used. A simple example of dynamically adjusting the compression level to deal with variations in bandwidth and mobile device resources is to monitor the mobile device battery lifetime as well as the available network bandwidth, measure these parameters at regular intervals, compare the values against a set of rules with thresholds, which determine whether or not the compression level should be changed. Finally, despite the fact that some of the work discussed has studied mobile access using different wireless access technologies, none of the work proposed has mentioned or cited work that addresses means to utilize these different wireless access technologies together in a seamless manner.

Remote Patient Health Monitoring

Remote patient health monitoring systems are used to remotely monitor a patient's health status, by attaching or implanting biological sensors to the patient, and forwarding the sensor readings to a health care provider. Since these sensors typically have a short transmission range, a more powerful gateway is used to forward the readings to the provider. A system may continuously transmit sensor readings to a medical care center, or transmit sensor readings in specific situations (context-based). The gateway may in some systems also perform analysis of sensor data, in order to detect anomalies and determine if the patient requires immediate attention.

Aspects Considered for a Remote Patient Health Monitoring System

Different systems are compared using the criteria defined in the Introduction section, in terms of the following aspects:

Observed Operating Parameters

In order for a remote patient health monitoring system to be efficient, robust, and reliable, it has to be aware of the operating environment and conditions. The system should observe parameters including but not limited to: available network bandwidth, mobile device battery lifetime, as well as parameters pertaining to the patient's context.

Controlled Parameters

A system should be able to make the most out of configuration parameters in order to optimize its performance. Such controlled parameters include but are not limited to: transmission rate, compression rate, and location of sensor data analysis (mobile device vs. medical center).

Protocols and Data Formats

The nature of the protocols and data formats used to forward the vital signs from the mobile device to the medical center can provide insight into the communication cost associated with transmitting these vital signs.

Adaptation Techniques

A Remote Patient Health Monitoring System (RPHMS) should be able to cope with variations in capabilities of different wireless access technologies. Therefore, efficient algorithms and techniques should be developed to make the best use of observed and controlled parameters, as well as make efficient use of the wireless link.

As previously illustrated, a Remote Patient Health Monitoring System (RPHMS) mainly consists of the following components: sensors, a mobile device, and associated software at the medical center. The benefits provided by using standard-based protocols and interfaces for components of a RPHMS are: easier integration with components from other vendors (better maintainability), and the freedom of health care institutions to choose components from different vendors to build a RPHMS, thereby reducing monetary costs. There are mainly two aspects behind which the choice of using standards versus in-house or proprietary protocols and interfaces is made, a business aspect, and a technical aspect. The business aspect is related to the business strategy and business model adopted by the vendor and is beyond the scope of this chapter. The technical aspect is related to the performance impact of using standard-based protocols, interfaces and data formats in a RPHMS.

Systems

Doukas et al. (2007) present a patient health-monitoring framework where vital signs are sent from sensors to a PDA, which sends patient status and control data to a hospital information system server using SOAP-XML. The system also supports real-time video transmission from the patient's PDA. Messages are transformed into the widely adopted and standardized Health Level 7 (HL7) message format (Health Level Seven International, 2012) before they are sent to a Hospital Information System (HIS), which facilitates interoperability between other systems, as HL7 is a standard that is widely adopted in health care. The system can adapt the use of encryption (e.g. encryption versus no encryption) and quality of data sent (no compression, low compression, high compression implies lowest quality), depending on the network transmission rate, and whether the patient's status is normal or urgent. The measurement data levels, which are used to indicate an urgent state, are fixed value ranges. The same goes for the transmission rate thresholds, which, along with the patient's status, determine the form in which data is sent.

Goni et al. (2009) present a cost model to help increase efficiency for patient health monitoring systems. The architecture of the monitoring system assumed here is one where sensors attached to a patient send biological readings to a mobile device. The mobile device, depending on the configuration, forwards those readings to a medical care center. No data transmission protocols are discussed. In one type of configuration, data analysis is performed on a mobile device. When an anomaly (medical condition requiring attention) is detected, an alarm message is sent to the care center, containing the sensor data that caused the alarm. In another type of configuration, the mobile device continuously forwards the sensor data to the care center (without performing analysis). The cost model relates three configuration variables, namely granularity, compression, and analysis. The granularity parameter refers to the amount of time spent collecting sensor readings before performing an analysis on those readings. Compression refers to whether or not to compress data before transmission. Analysis refers to the choice of either analyzing sensor data on the mobile device or at the medical center. The objectives are to minimize the rate of data transmitted (to reduce monetary cost), minimize the latency in receiving an alarm, and minimizing the rate of battery consumption. The formulas used to describe these objectives capture different parameters, such as the time it takes to compress sensor data. For each of these objectives, a threshold is applied. These objectives and thresholds are formulated into a multi-objective optimization problem which can be solved to obtain the optimal set of values for granularity, compression, and analysis. A prototype system was developed and experiments were conducted. Two medical conditions were considered. These medical conditions are arrhythmia (irregular heart rhythm) and apnea (sleep disorder). For each medical condition, the optimal set of configuration values was obtained.

Both optimal configurations suggested analysis on the mobile device. However, the optimal configuration for the arrhythmia monitor suggested the use of compression, whereas the optimal configuration for the apnea monitor suggested not using compression.

Tahat (2009) presents a prototype patient health monitoring system, which utilizes the ubiquitous cellular network technologies SMS and MMS. An ECG sensor is connected to a microcontroller, which is connected to a Bluetooth transceiver, which wirelessly connects to a mobile phone. The mobile phone is responsible for collecting the ECG data. The mobile phone can graph ECG data and send the graph image as an MMS message, or, alternatively, send an SMS message that contains ECG samples. The data the mobile phone receives from the sensors is written continuously in blocks of 16kB to the phone's SD memory. A one-hour long single-channel ECG will occupy 900 kB of memory space. An ECG sample with 8-bit resolution allows loading of 140 samples (bytes) in an SMS message. At a sampling frequency of 125Hz (8 milliseconds), a single SMS message can contain 1.12 seconds of ECG data.

In order to facilitate the adoption of standards, Lee and Gatton (2010) propose a prototype patient health monitoring system that is based on standards. Sensors that read biological signals wirelessly connect to a mobile device using Bluetooth, through an interface that is based on the IEEE1451 standard (IEEE, 2007). The IEEE1451 standard defines an interface for data exchange with sensors. The mobile device sends the sensor data to a server at the care centre, and the server stores the data in an HL7-compliant format. The protocol used to communicate between the mobile device and the server is "in-house", and does not conform to any standard. The mobile device, upon initialization, searches for sensors, and verifies any detected sensors with an identification and password. The system supports commands that can be executed remotely on the mobile device, from the medical center. These commands allow resetting the mobile device state, as well as halting data transmission from the mobile device to the medical center. However, no robustness mechanisms were employed to deal with varying operating conditions. In their prototype implementation, sensors were simulated on a PDA running Windows mobile operating system.

The INHOME project presented by Vergados (2010) comprises a prototype system to provide *ambient assisted living* to patients in need of care. Even though it discusses other aspects of assisted living such as entertainment, only the patient health monitoring system is discussed here. In this proposed system, sensors are connected to an "INHOME terminal," which is a mobile device that connects using Wi-Fi to a residential gateway, such as a wireless-enabled desktop computer. Some sensors can also connect directly to the residential gateway using Bluetooth. Measurements of vital signs made by the sensors are forwarded to the residential gateway, where the data is analyzed to check for abnormalities; different measurements are checked against a pre-determined set of rules, and if rule conditions are satisfied, then a corresponding action is triggered. The conditions use pre-defined fixed value ranges. This residential gateway is connected to the home phone, and can initiate a phone call to alert a care centre, should an emergency be detected. An alert can also be sent via the HTTP protocol. The system employs a role-based access control scheme to control access to patient data, service management, and defining the rules used for analysis of vital signs.

Another proposed prototype health monitoring system (Neubert et al., 2010) utilizes a proprietary module, which comes attached to a chest belt, and has integrated sensors. The module collects different physiological measurements and transmits them to a mobile device via Bluetooth. The analysis of these measurements is performed on the server. The mobile device is continuously connected to the server using a TCP-based connection, and synchronizes and transmits the collected data to a server on the World Wide Web. The mobile device should be able to communicate with the server through Wireless LAN (Wi-Fi), or using cellular radio standards. If disruptions in the communication between the mobile device and the server occur, the mobile device would store the data in First-In-First-Out (FIFO) storage. Then the mobile device periodically tries to reconnect to the server to transfer the stored data. For every data set sent, the mobile device expects a confirmation acknowledgement (ACK) message from the server, to acknowledge correct data transmission (via a hash value), and to acknowledge a successful data update operation to the database. If the server sends back an error message (NAK), or no ACK message has been received after a certain time period, the communication between the mobile device and the server is restarted.

Chowdhury et al. (2010) present a prototype patient health monitoring system, where, as opposed to continuous streaming of biological sensor data, the data is only sent under specific situations. The system utilizes an Internet-connected mobile phone to transmit the biological sensor data to a medical care centre server. Different contextual states suggest different conditions such as the patient being disturbed at night, or may be depressed. These states are checked against a set of rules which are set by a medical professional. One of the main contributions in this work is that existing rules can be composed to form new rules, and new sensor data streams can be added and incorporated into these rules in order to monitor new medical conditions. The contextual states trigger actions such as the collection and transmission of one or more vital signs. Context and event processing is done by the client software, which resides on the mobile device.

Better system efficiency and reliability can be achieved by giving higher priority to the more important segments of a sensor data stream. This is demonstrated by Wang et al. (2010), where a framework for a patient health monitoring system is proposed. In this system, groups of wearable sensors are wired to *health nodes*. A health node processes the sensor data and wirelessly sends it to a mobile device. The mobile device then sends this data to hospitals, via a Wi-Fi or GSM connection to the Internet. The main focus of this work is implementing a health node that would maximize the quality of forwarded sensor data, with minimum energy consumption, in order to maintain a reliable data stream (i.e. increase availability as much as possible). In a way similar to the optimization problem described by Goni et al. (2009), the average total energy consumption is expressed as a function of the following parameters:

- Desirable bit error rate requirement (similar to thresholds used by Goni et al.)
- Frame length
- Control packet transmission rate
- Data packet transmission rate
- Transmission power
- Retransmission limit (in case of a transmission error)

The health node adjusts the sensor data quality by dividing the sensor data into blocks of more important and less important parts. For example, for an ECG sensor which measures heart rate, the signal can be divided in two "blocks" of data, where the more important block would be the QRS complex (the spike associated with a heart beat that you see on a heart beat monitor screen). Being able to adjust the number of more important blocks sent, versus the number of less important block sent, allows the health node to adjust the quality of the data being sent. Encryption is only applied to the more important blocks of a data stream. In their previous work that they have referred to in the paper, the Advanced Encryption Standard (AES) was selected as an example, but in theory any encryption algorithm can be used.

Another system that employs the use of cellular network technology for monitoring the biological signals from patients is proposed by Nikolidakis et al. (2010). In this prototype system, sensors that measure a patient's biological signal connect to a mobile device using Bluetooth. The mobile device connects to a hospital, through the cellular network, using the Session Initiation Protocol (SIP) (Rosenberg et al., 2002), and sends medical data collected by the sensors using the HL7 format. The sensors are first placed on the patient when the patient is first admitted to the hospital. The authors report using AES for encrypting the data sent from the sensors to the hospital. However, it is not clear if the encryption is applied from the sensors to the mobile device, or applied from the mobile device to the hospital, or both.

The MobiHealth project (Konstantas et al., 2004) is a patient health monitoring system that supports continuous real-time transmission of a patient's biological signals using cellular data networks. The sensors forward biological signal readings to the care centre through a mobile device. The sensors connect to the mobile device using Bluetooth. The mobile device connects to the server using the BANip protocol (Dokovsky et al., 2004), which is derived from the JINI IP Interconnect Protocol, which uses the HTTP protocol as the underlying transport protocol. The mobile device acts as an interface to the sensors. JINI supports the dynamic discovery of services. Therefore, it was decided that the system would be based on JINI, in order to solve the problem of dynamically discovering sensors in a Body Area Network (BAN). JINI requires the use of the Remote Method Invocation (RMI) protocol during the discovery phase. Rather than using RMI directly on the mobile device, the JINI surrogate architecture was used in order to allow for less resource consumption on the mobile device. Even though some preliminary tests of the system were conducted, there was no mention of testing the dynamic discovery of sensors. One would suspect that the dynamic discovery of sensors has failed, as JINI uses multicasting in order to search the network for servers which can act as surrogates. The problem is that multicasting is traditionally used on Local Area Networks (LANs), and so it cannot be assumed that multicasting would work over the Internet without some form of tunneling. This problem would not be encountered, if the surrogate is on the same LAN as the BAN. One other benefit from using JINI is that JINI scales well, relative to other Web service technologies. As one can infer, scalability is of paramount importance to patient health monitoring systems, as monitoring hundreds of patients simultaneously, with each patient wearing several sensors, can be a burden on Information and Communication Technology (ICT) infrastructures. No security mechanisms have been proposed to protect the BAN, or the system as a whole, from attacks.

Instead of solely relying on best effort methods in delivering remote patient health monitoring services, quality of service agreements can be made between the health care institution and the cellular network provider. For this to be achieved, one has to able to understand the relationship and tradeoffs between different service parameters. To this goal, Puming (2006) has developed a prototype system and proposed the use of a model, which "enables the mapping of the vital sign quality parameters to network service capacity needs that are able to transfer the vital signs." Such a model allows a stakeholder such as a medical care centre to set the desired vital sign to monitor (e.g. blood pressure, oxygen saturation), as well as the sample rate (how many vital sign readings are taken, per second), compression level, and resolution, which is the number of bits used to represent a sample reading. Increasing the resolution increases the quality, but is more demanding in terms of transmission bandwidth.

In the prototype system presented by Dağtas et al. (2008), body sensors are connected wirelessly to a mobile device, using ZigBee wireless technology. The mobile device sends raw ECG data to a health care centre through the cellular network. The health care centre can invoke three commands on the mobile device: *set* the sampling ECG frequency; *start* reporting ECG data to the server; *stop* reporting ECG data to the server. At the network level, the system invokes an ECG data call on the cellular network in order to establish a connection with the care centre and send the patient's ECG measurement data. When this call is made, the cellular network allocates bandwidth for this call (just like any regular cellular phone call). However, in the advent of an emergency, if bandwidth is not available to satisfy an incoming ECG data call, the system invokes a proposed "dynamic channel allocation algorithm", which is based on the idea of cellular base stations "borrowing" channels from one another. As for security, symmetric keys are used to protect communication from sensors to the mobile device, and between sensors. Data integrity is achieved using keyed-hash messages, somewhat aiding in making the system more robust, as it somewhat protects against invalid input.

A & D Medical (n.d.) has released a suite of products, known as Wellness Connected, for an in-home patient health monitoring system. A software application, the Wellness Connected Software, is installed on a patient's home computer, and can be used to track and display health-related information to the patient, such as the patient's blood pressure and weight. In addition, there is also an online service, ActiHealth (n.d.), which can automatically upload and record the data collected by the Wellness Connected Software installed on the patient's home computer. This enables peers such as medical professionals and relatives, to remotely track the patient's health. This system does not provide automated monitoring of other common medical conditions which are of major concern, such as cardiovascular diseases. In addition, the scope of monitoring is confined to the patient's home.

Biotronik Home Monitoring (n.d.) is a patient health monitoring system that uses the cellular network to connect to a care centre via the Internet. Implanted sensors monitor cardiac function and forward ECG measurements to a custom-made mobile device. The mobile device transmits the data to the Biotronik Home Monitoring Service Centre for analysis. In the event that an anomaly is detected, a notification is sent to physicians. An SMS or fax can also be sent as an additional notification by the service centre. Despite Biotronik reporting that "more than 150,000 BIOTRONIK Home Monitoring systems have been implanted, in more than 50 countries worldwide," clinical trials are still being conducted in order to evaluate the cost-effectiveness of the system (U.S National Library of Medicine, 2011). The sensor communicates with the mobile device using the standardized ULP-AMI (ultra low-power active medical implants) protocol (ETSI, 2011). One disadvantage of this home monitoring system is that external sensors such as blood pressure monitors cannot be integrated into the system (Müller et al., 2011). Biotronik employs an XML-based data format specification, the Biotronik IEEE 11073-10103 standard, for the exchange of patient information with other institutions.

MedApps Mobile Health Monitoring (MedApps, 2011) is, in its current state, a means to forward a patient's vital signs, currently blood sugar levels for diabetic patients, to a care centre's server, via a custom-made mobile device. In their disclaimer (can be found at the end of the demo at reference [MedApps, 2011]) they state that their system and their system-related products are not for emergency or time-critical patient monitoring. Blood sugar levels are measured from a sample of the patient's blood. Once the measurement is taken, the device is connected to the mobile device in order to transmit the measurements. The mobile device receives the data, and then attempts to acquire a cellular network signal. Once the signal is acquired, the data is sent to the server at the care centre. The mobile device currently supports Bluetooth, GSM, CDMA, and GPS. It also features a programmable reminder to assist in patient compliance with the health care program. The mobile device only attempts to acquire a wireless signal when it needs to send data to the care centre.

Comparison

The systems previously presented can be classified in terms of where the vital signs analysis is performed; be it on the mobile device, at a nearby stationary gateway, at the medical center, or a combination of these locations. These systems are compared in terms of how they achieve efficiency, robustness, and reliability, interoperability with different communication technologies, maintainability, and security. Monitoring a chronic health condition typically involves monitoring multiple vital signs, and so it is important for these systems to be able to support such requirement.

The bare minimum function of the mobile device is to forward the vital signs being read from the sensors. In addition to monitoring a patient's vital signs, additional context obtained from monitoring network conditions, the mobile device, and the patient's activities may help in enhancing the performance of the system. These contexts are found along with the mechanisms described under the *efficiency, robustness, reliability* column. The *interoperability* column denotes the protocols, data formats and software interfaces used between the mobile device and the medical center.

Table 6 presents the ideal cases for which a proposed system would satisfy all of the criteria. The properties in the table are based on the requirements defined in the Introduction section. The ranks defined in Table 7 are used to rank the different systems presented in the comparison Table 8.

Table 6. Remote patient health monitoring system ideal properties

Criterion	System properties in an ideal system/ setting
Efficiency, robustness, reliability	Observes network conditions Observes mobile device conditions Adapts use of encryption and data quality SLA with cellular network provider
Vital signs monitored	Cardiac function Oxygen saturation Other vital signs
Interoperability	Standard-based interface Standard-based data formats
Maintainability	Set sampling frequency Set rules used in sensor data analysis Use other contextual information such as patient's schedule Reuse existing analysis rules Set limits for rate of volume of sensor data transmitted to medical center Set limit for maximum allowable latency for receiving an alarm Set limit for minimum autonomy
Security and privacy	Role-based access control for system users (physicians, family members, etc.) Non-role-based access control for system users Access control for sensors Data encryption

Most of the remote patient health monitoring systems presented prior continuously forward vital signs data from sensors to a medical center, where analysis of that data is performed. Continuous wireless transmission can be taxing on the battery lifetime of the mobile device forwarding the vital signs, and can result in high monetary costs. Only Doukas et al. (2007) and Goni et al. (2009) propose to employ some form of automated analysis of vital signs on the mobile device. Experiments conducted by Goni et al. have shown that performing analysis on the mobile device rather than continuously transmitting vital signs to a medical center via Wi-Fi, is optimal for the battery lifetime of the mobile device, for the heart rhythm monitor and sleep disorder monitor applications that were tested. All systems from the previous literature employ best effort approaches in order to maintain availability except for the system proposed by Puming (2006), which employs resource reservation via a Service-Level Agreement (SLA) between the medical center and the cellular network operator. The system proposed by Dağtas et al. (2008) is the only other system where the underlying cellular network would recognize a request to establish a remote ECG monitoring session. However, unlike the system proposed by Puming, the cellular network

Table 7. Basis for assessment based on opinion

Opinion	Efficiency, robustness, reliability	Vital signs monitored	Interoperability	Maintainability	Security and privacy
Very good	All	All	All	All	All
Good	(a), (b), and (c)	(a) and optionally (c)	(a)	((b), (c), (d)) or ((e), (f), (g))	((a) or (b)) and (d)
Fair	(a), and (b)	(b)	(b)	One of (b), (c), (d) or (e), (f), (g)	(a) or (b) or (c) or (d)
Poor	(a) or (b)	(c)	-	(a)	-
Very poor	None	None	None	None	None

Table 8. Remote patient health monitoring systems comparison

Reference	Location of vital signs analysis	Efficiency, robustness, reliability	Vital signs monitored	Interoperability	Maintainability	Security and privacy
Doukas et al. (2007)	Mobile device	**Good.** Contexts observed: network conditions. Adapts use of encryption and data quality depending on network load and patient status	**Very good.** Cardiac function, blood pressure, pulse rate, heart rate, oxygen saturation	**Good.** HL7 over SOAP-XML	**Very poor.** N/A	**Fair.** SSL encryption
Dağtas et al. (2008)	Medical center	**Good.** Contexts observed: network conditions. Network resource allocation where resources are "borrowed" if not sufficient resources are available	**Good.** Cardiac function	**Very poor.** N/A	**Poor.** Set sampling ECG frequency	**Good.** Symmetric keys to protect communication from sensor to mobile device, integrity using keyed—hash messages
Goni et al. (2009)	Mobile device/ Medical center	**Good.** Contexts observed: N/A Cost model which relates data rate and energy consumption on the mobile device, deriving an optimal set of configuration parameters prior to starting the monitoring session	**Very good.** Cardiac function, oxygen saturation	**Very poor.** N/A	**Good.** For different symptoms, allows adjustment of limits for: rate of volume of information transmitted, maximum allowable latency for receiving an alarm, minimum autonomy (battery life time)	**Very poor.** N/A
Tahat (2009)	Medical center	**Very poor.** N/A	**Good.** Cardiac function	**Very poor.** N/A	**Very poor.** N/A	**Very poor.** N/A

continued on following page

Table 8. Continued

Reference	Location of vital signs analysis	Efficiency, robustness, reliability	Vital signs monitored	Interoperability	Maintainability	Security and privacy
Lee and Gatton (2010)	Medical center	**Very poor.** N/A	**Very poor.** N/A	**Very poor.** "in-house" protocol	**Very poor.** N/A	**Fair.** Verification of sensors connecting to mobile device using id and password
Vergados (2010)	Static residential gateway	**Very poor.** N/A	**Very poor.** N/A	**Very poor.** N/A	**Good.** Allows for setting rules for patients, used for sensor data analysis	**Fair.** Role-based access control scheme
Neubert et al. (2010)	Medical center	**Fair.** Contexts observed: N/A Disruptions in communication handled using FIFO storage	**Good.** Cardiac function, pulmonary function, skin temperature	**Very poor.** N/A	**Very poor.** N/A	**Very poor.** N/A
Chowdhury et al. (2010)	Medical center	**Good.** Contexts observed: mobile device, patient location, patient activity. Context-dependent sensor data collection and transmission	**Very poor.** N/A	**Very poor.** N/A	**Very good.** Rule-based data analysis engine can be update with new rules, existing rules can be reused to detect new symptoms	**Very poor.** N/A
Wang et al. (2010)	Medical center	**Good.** Contexts observed: mobile device. Cost model that relates parameters such as transmission rate and power	**Good.** Cardiac function	**Very poor.** N/A	**Very poor.** N/A	**Fair.** Parts of data are encrypted using AES
Nokilidakis et al. (2010)	Medical center	**Fair.** Contexts observed: N/A Data is cached for later transmission when unable to establish a network connection	**Very poor.** N/A	**Fair.** HL7	**Very poor.** N/A	**Good.** AES. Authentication, integrity and non repudiation achieved with X.509
Puming (2006)	Medical center	**Fair.** Contexts observed: not applicable (based on resource reservation). Cost model SLA-based	**Very good.** Cardiac function, oxygen saturation, blood pressure	**Very poor.** N/A	**Very poor.** N/A	**Very poor.** N/A

continued on following page

Table 8. Continued

Reference	Location of vital signs analysis	Efficiency, robustness, reliability	Vital signs monitored	Interoperability	Maintainability	Security and privacy
Konstantas et al. (2004)	Medical center	**Fair.** Contexts observed: N/A Multiple "sensor data" messages aggregated in a single HTTP request; HTTP chunking; "deflate" compression (Dokovsky et al., 2004)	**Good.** Breathing frequency, oxygen saturation, pulse rate, blood pressure, pupil size, reactions and amount of fluid infused	**Fair.** protocol derived from JINI IP Interconnect protocol	**Very poor.** N/A	**Very poor.** N/A
Biotronik (n.d.)	Medical center	No information available	**Good.** Cardiac function	**Very poor.** Proprietary	No information available	**Very poor.** N/A
MedApps (n.d.)	Medical center	**Fair.** Contexts observed: N/A Wireless signal only acquired upon sending data	**Poor.** Blood sugar levels	**Very poor.** Proprietary	No information available	**Very poor.** N/A

would strive to allocate sufficient network resources to satisfy the request.

All of the systems proposed target cellular networks except for the system proposed by Vergados (2010). Goni et al. state that their proposed cost model approach should also work in cellular network environments, despite only being tested in a Wi-Fi setting. Vergados (2010) targets an "in-home" setting, where Wi-Fi would be available to allow a mobile device to forward vital signs to a residential gateway which can be connected to the Internet using a high-speed connection, and so does not have to deal with the more stressful network conditions encountered in cellular networks. The system can alert the care center either through the Internet or by auto-dialing the care center.

The systems that use cellular networks to forward vital signs utilize a cellular data connection. However, Tahat (2009) instead utilizes the ubiquitous SMS and MMS technologies to forward ECG readings. In general, lightweight use of SMS and MMS is cost-efficient, especially in the developing world, where the communication infrastructure is not advanced. However, the cost of continuously sending an SMS message every few seconds may be of concern. The technologies SMS and MMS exist on almost all, if not all cellular phones, and supported by most, if not all cellular network operators, and so can handle the heterogeneity of cellular phones and networks.

There are two aspects to the maintainability criterion: maintaining the rules for sensor data analysis, and providing configuration parameters or "control knobs" for manually fine-tuning the system's performance. Dağtas et al. (2008) allow for setting the ECG sampling frequency for sensor data acquisition. Vergados (2010) allows for defining rules used for sensor data analysis. In addition to being able to add new rules or update existing rules, the approach proposed by Chowdhury et al. (2010) allows for more flexibility by proposing a set of operations that allow the dynamic creation of new rules from existing rules, via the use of a proposed set of operations that allow for the composition of these rules in one way or another.

Both Doukas et al. (2007) and Wang et al. (2010) propose sacrificing the use of encryption for the sake of performance, during harsh network conditions. Wang et al. employ a more fine-grained approach, which helps achieve better sensor data transmission reliability, rather than choosing between encryption versus no encryption as in Doukas et al.

GENERIC ISSUES

This section discusses major observations of the different proposed systems surveyed in this chapter, in order to shed some light on some of the deficiencies found in the literature.

Lack of Utilization of SOAP Features

As mentioned in the Background section, there are two ways of structuring a SOAP message, namely RPC-style and Document-style. In general, document-style is more efficient in terms of messaging, and would help in reducing the message size, as only unique aliases to parameters of method names are used in the message, as opposed to using both the method name as well as the names of its parameters in the message as in the RPC-style. Surprisingly, the systems reviewed did not report the use of such optimizations, which are not new (Butek, 2005).

XML Messaging for Resource-Constrained Devices and Networks

As noted by Kangasharju (2008): "XML as a highly redundant text-based format is not obviously suitable for mobile devices that need to avoid extraneous processing and communication". In order to tackle this issue, and provide a data exchange format that can be used with resource-constrained devices and networks, such as mobile devices and networks, the W3C Efficient XML Interchange (EXI) Working Group (WG) was formed. The EXI WG issued a call for contributions, and received several submissions, one of which was Efficient XML from AgileDelta (2012). After conducting a series of measurements on all submissions, Efficient XML, was deemed to be the basis for EXI. W3C EXI is based on applying several optimizations to XML structures, and uses a binary format. Kangasharju (2008) has conducted a study on the implementation and performance of XML-based formats for mobile devices, which included EXI. In comparing different existing XML-based data exchange formats, the work identifies four areas of improvement in XML messaging systems, which developers can use as guidelines when implementing XML-based messaging systems. Surprisingly, during the course of this survey, one has not come across a single proposed medical system that exploits the optimizations documented for EXI, or even experiment with the open source Xebu format (Kangasharju, 2008) that employs similar optimizations.

Software Portability

One way of addressing the issue of heterogeneity of software platforms (further discussed in the next section) on mobile devices, is by using the ubiquitous Java Connected Limited Device Configuration (CLDC) profile for resource-constrained devices, such as mobile devices. Using the CLDC profile helps in achieving a set of common (thus minimal) functions, or a "lowest common denominator", between different platforms, and thus provides a greater chance that an application that uses this profile would run on a diverse range of mobile devices without compatibility issues. This approach, however, comes at a cost. As reported by Konstantas et al. (2004), the use of the

CLDC profile for a Java-based implementation on a mobile device can be problematic, when using HTTP connections. This is because the CLDC profile does not allow persistent HTTP connections. This can disrupt the intended functionality of a system; the mobile device would have to establish a new TCP/IP connection whenever it wants to send data, making the communication process highly inefficient. This suggests that other approaches for addressing the heterogeneity issue may be required, such as the use of standard protocols and model-based application development. In model-based application development, a high-level modeling language may be used to describe the application logic, and software tools would be used to generate an application that can run natively on the mobile device. This would allow for portability, albeit at the model-level, without sacrificing functionality of a more capable device. The system utilizing the mobile device should then be flexible enough to adapt to the utilization of different capabilities of different devices.

Non-Standard Security Mechanisms

Previous research, such as that of Nabi et al. (2010), supports the idea that available security solutions for transmission of medical records mostly tend to be hybrid approaches (SOAP used with non-standard security mechanisms) using security mechanisms such as cryptography, digital signatures, and tokens, etc. The adoption of standards needs to be pushed for if we are to see widespread deployment of these proposed systems.

SOLUTIONS AND RECOMMENDATIONS

When developing a system for information retrieval, researchers should utilize the studies conducted on the effect of changing compression levels of medical images on latency, and deploy algorithms that achieve better efficiency in accessing those images. Furthermore, if dynamically reducing image quality is used as a tool to achieve better efficiency in downloading these images, then standards such as those from the American College of Radiology (2002) and the Canadian Association of Radiologists (2008) should be taken into consideration and adhered to. For example, the CAR Standards for Irreversible Compression in Digital Imaging within Radiology specify the maximum acceptable compression ratios for JPEG and JPEG2000. In addition, incorporating knowledge of imaging-based reference frameworks (Muir et al., 2006) that are applied by radiologists in diagnosis of conditions such as acute stroke would help the system achieve better efficiency.

Systems deployed for remote patient health monitoring should be aware of variations of network conditions throughout a monitoring session, and should be able to adapt accordingly. As discussed before, considering analyzing vital signs on the mobile device can yield resource savings for the mobile device, thereby attaining greater availability. However, these systems also need to adapt to variations in network bandwidth, which is commonly encountered when operating in wireless and mobile environments.

In addition, they should also be aware of the variations in resources on the mobile device that it is running on. This would help the system in determining better means to dynamically reconfigure the system accordingly, thereby increasing the efficiency, robustness, and reliability of the system. The system should also be able to utilize the multitude of wireless access technologies that exist nowadays to determine efficient means to forward sensor data as well as provide multimedia communication services with health care personnel.

Most interoperability solutions discussed from the literature were poor. Efforts and developments made by the Integrated Healthcare Enterprise (IHE) (IHE International, 2011) and IEEE standards should be utilized in order to facilitate ease of integration with existing electronic health care systems. On that note, security and privacy solutions should also integrate easily with existing systems.

As a potential solution to the interoperability problem, Elgazzar et al. (2012) propose a remote patient health monitoring system where vital signs data from sensors are exposed as Web Services on the mobile device. This system enables health care practitioners to access, with permission from the patient, the sensor data provided by the mobile device. This system provides better interoperability and ease of integration with existing systems deployed at medical facilities. The Web Services are deployed using the RESTful architectural style (Fielding, 2000). The system supports both the option to continuously forward vital signs to a medical facility, or employ event-based forwarding. Events may include exceeding certain defined thresholds.

Despite the existence of proprietary solutions and the lack of existence of standardized solutions, the efforts put in working standards by collaboration of health care professionals and industry is paving the way for widespread availability of standardized solutions that integrate easily with existing systems. Eventually, one would expect solutions developed from a combination of Commercial-Off-The-Shelf (COTS) components, where the software would be available for easy deployment on COTS devices.

FUTURE RESEARCH DIRECTIONS

From our discussion, we learned that research on medical data access in a wireless and mobile environment, coupled with research studies on the software technologies that may be used to develop these systems can be of detrimental value. To illustrate this let us discuss some of the open problems in these proposed systems.

Information Retrieval Systems

All of the information retrieval systems reviewed only present the effects of different levels of compression on image size, transmission time and processing requirements. No proposed system actually demonstrates how different compression levels would be dynamically adjusted. There needs to be more work into developing frameworks which actually utilize the proposed compression mechanisms and demonstrate how the system would adapt to variations in resources using these mechanisms. In addition, there needs to be more studies into how mobile devices would interface with medical information systems, and how collaboration would take place between different peers. Medical information systems must also be able to interoperate, and a study on the use of standards to achieve this interoperability can be useful. In addition, more work needs to be done on ensuring the security and privacy of medical information as well as the credentials of a medical professional.

Remote Patient Health Monitoring Systems

Most systems do not employ some form of analysis of vital signs on the mobile device, despite work in the literature which shows benefits of doing so. The systems which do employ analysis on the mobile device lack in one or more aspects such as robustness against variations in network bandwidth, dynamic readjustment of control parameters such as the rate of forwarding vital signs to the medical center, remote maintenance of the software running on the mobile device, or the lack of support for extensible means to add new contextual information such as a patient's schedule or location. These proposed systems do not consider quantifying the reliability of sensor data analysis on the mobile device. Quantifying the reliability of sensor data analysis on the mobile device versus the reliability of sensor data analysis at the medical service center, can help the monitoring system make better decisions as to whether or not it is more beneficial to analyze sensor data on the mobile device when providing a remote monitoring service. In addition, more work needs to be done with regards to security and privacy, and the plausibility of proposals to sacrifice encryption during harsh network conditions needs to be assessed. A few systems adopted standards for the protocols, data formats and software interfaces used between the mobile device and the medical center. For these systems to become more attractive to health care institutions, and provide the ability to build a system from components that may not necessarily be from one provider but are interoperable, these systems must be built from components with standardized interfaces.

Currently most solutions tend to be proprietary, where the vendor would provide an entire integrated system for the consumer. This puts the service consumer at the mercy of the vendor. We are presently seeing a greater push for standards-compliant systems and systems, which can interoperate. Efforts such as those from the IHE (IHE International, 2011), Bluetooth Continua Alliance (Continua Health Alliance, 2010), and the IEEE 11073 standards (IEEE-SA, 2012) pave the way to realizing such dream.

CONCLUSION

This chapter first provides an introduction to the issues and challenges in remote access to medical data in two different settings, and then provides a background on the different software technologies used in the medical systems that have been proposed for use in these settings. It presents a comprehensive overview and qualitative analysis of the functional aspects of these different medical systems. Mobile devices are increasingly gaining widespread adoption by medical professionals in the health care domain. They provide medical professionals with the convenient access they need to medical information systems, without constraining their freedom of movement, enabling them to conduct their duties in a more efficient manner. The adoption of proposed medical systems in the literature depends on addressing issues and challenges faced in the operating environments that mobile devices are deployed in, and solutions which provide flexible systems which adapt and evolve as technology advances. Developing system components which employ standardized interfaces would provide health care institutions with integration as a commodity, giving them the ability to build customized systems at a reduced cost, and providing them with control over their systems, allowing them to spend more on providing better medical services.

REFERENCES

Ableiter, D. (2008). *Smart caching for efficient information sharing in distributed information systems*. (M.S. thesis). Naval Postgraduate School, Monterey, CA. Retrieved from http://www.dtic.mil

Abuan, L. A. (2009). *Information sharing for medical triage tasking during mass casualty/ humanitarian operations*. (M.S. thesis). Naval Postgraduate School, Monterey, CA. Retrieved from http://www.dtic.mil

ActiHealth. (n.d.). Retrieved May 1, 2011, from http://www.actihealth.com

A&D Medical. (n.d.). *A&D medical*. Retrieved March 5, 2011, from http://www.andonline.com

AgileDelta. (2012). *AgileDelta technologies: Software and services for XML, web services and the mobile internet*. Retrieved June 19, 2012, from http://www.agiledelta.com

American College of Radiology. (2002). *ACR standard for teleradiology*. Retrieved from http://imaging.stryker.com/images/ACR_Standards-Teleradiology.pdf

Berenson, R. A., Grossman, J. M., & November, E. A. (2009). Does telemonitoring of patients--the eICU--improve intensive care? *Health Affairs*, *28*(5), 937–947. doi:10.1377/hlthaff.28.5.w937.

Biotronik. (n.d.). *Biotronik home monitoring*. Retrieved March 7, 2011, from http://www.biotronik.com/biohm/home

Bluetooth, S. I. G. (n.d.). *The official bluetooth® technology web site*. Retrieved April 18, 2011, from http://www.bluetooth.com

Breslow, M. J. (2005). *Building a better delivery system: The eICU solution: A technology-enabled care paradigm for ICU performance*. Washington, DC: National Academies Press.

Brown, K., & Ellis, M. (2004, January 30). *Best practices for web services versioning*. Retrieved August 6, 2011, from http://www.ibm.com/developerworks/Webservices/library/ws-version/

Butek, R. (2005). *Which style of WSDL should I use?* Retrieved May 1, 2011, from https://www.ibm.com/developerworks/Webservices/library/ws-whichwsdl/

Canadian Association of Radiologists. (2008). *CAR standards for irreversible compression in digital diagnostic imaging within radiology*. Retrieved from http://www.car.ca/uploads/standards%20guidelines/Standard_Lossy_Compression_EN.pdf

CDMA Development Group. (n.d.). *CDG*. Retrieved June 19, 2012, from http://www.cdg.org

Chowdhury, A. R., Falchuk, B., & Misra, A. (2010). MediAlly: A provenance-aware remote health monitoring middleware. In *Proceedings of the IEEE International Conference on Pervasive Computing and Communications* (pp. 125-134). IEEE.

Clotfelter, C. T., & Towl, J. E. (2007). *Twiddlenet: Metadata tagging and data dissemination in mobile device networks*. (M.S. thesis). Naval Postgraduate School, Monterey, CA. Retrieved from http://www.dtic.mil

Continua Health Alliance. (2010). *Homepage*. Retrieved June 27, 2012, from http://www.continuaalliance.org

Cummings, J., Krsek, C., Vermoch, K., & Matuszewski, K., & University HealthSystem Consortium ICU Telemedicine Task Force. (2007). Intensive care unit telemedicine: Review and consensus recommendations. *American Journal of Medical Quality*, *22*(4), 239–250. doi:10.1177/1062860607302777 PMID:17656728.

Dağtas, S., Pekhteryev, G., Sahinoğlu, Z., Çam, H., & Challa, N. (2008). Real-time and secure wireless health monitoring. *International Journal of Telemedicine and Applications*, (1): 1–10. doi:10.1155/2008/135808 PMID:18497866.

Department of Health and Human Services. (2002). *Standards for privacy of individually identifiable health information, final rule*. Retrieved from http://www.hhs.gov/ocr/privacy/hipaa/administrative/privacyrule/privrulepd.pdf

Department of Health and Human Services. (2003). *Health insurance reform: Security standards, final rule*. Retrieved from http://www.hhs.gov/ocr/privacy/hipaa/administrative/securityrule/securityrulepdf.pdf

DigiSoft. (n.d.). *DigiSoft: Innovative software solutions*. Retrieved August 8, 2011, from http://www.digisoftdirect.com

Dokovsky, N., van Halteren, A., & Widya, I. (2004). *BANip: Enabling remote healthcare monitoring with body area networks. Scientific Engineering of Distributed Java Applications*. Berlin, Germany: Springer.

Doukas, C., Maglogiannis, I., Anagnostopoulos, I., & Peraki, K. (2007). A context-aware telemedicine platform for monitoring patients in remote areas. *The Journal on Information Technology in Healthcare*, *5*(4), 255–262.

Eggen, R., & Sunku, S. (2003). Efficiency of SOAP versus JMS. In *Proceedings of the International Conference on Internet Computing* (pp. 99-105). IEEE.

Elgazzar, K., Aboelfotoh, M., Martin, P., & Hassanein, H. S. (2012). Ubiquitous health monitoring using mobile web services. In *Proceedings of The 3rd International Conference on Ambient Systems, Networks and Technologies (ANT)*. Niagara Falls, Canada: ANT.

EN 13606 Association. (n.d.). *The CEN/ISO 13606 association site*. Retrieved April 21, 2011, from http://www.en13606.org

Endrei, M., Gaon, M., Graham, J., Hogg, K., & Mulholland, N. (2006, May 1). *Moving forward with web services backward compatibility*. Retrieved August 6, 2011, from http://www.ibm.com/developerworks/java/library/ws-soa-backcomp/

ETSI. (2011). *ETSI medical*. Retrieved May 1, 2011, from http://www.etsi.org/WebSite/technologies/Medical.aspx

Fielding, R. T. (2000). *Architectural styles and the design of network-based software architectures*. (Doctoral Dissertation). Retrieved from http://www.ics.uci.edu/~fielding/pubs/dissertation/top.htm

Goni, A., Burgos, A., Dranca, L., Rodriguez, J., Illarramendi, A., & Bermudez, J. (2009). Architecture, cost-model and customization of real-time monitoring systems based on mobile biological sensor data-streams. *Computer Methods and Programs in Biomedicine*, *96*(2), 141–157. doi:10.1016/j.cmpb.2009.04.010 PMID:19481289.

Health Level Seven International. (2012). *Homepage*. Retrieved June 27, 2012, from http://www.hl7.org/

Hung, K., & Zhang, Y. (2003). Implementation of a WAP-based telemedicine system for patient monitoring. *Transactions on Information Technology in Biomedicine*, *7*(2), 101–107. doi:10.1109/TITB.2003.811870 PMID:12834165.

IEEE. (2007). *1451.0-2007 - IEEE standard for a smart transducer interface for sensors and actuators - Common functions, communication protocols, and transducer electronic data sheet (TEDS) formats*. Retrieved May 31, 2011, from http://standards.ieee.org/findstds/standard/1451.0-2007.html

IEEE-SA. (2011). *The IEEE standards association*. Retrieved June 27, 2012, from http://standards.ieee.org/

Indrajit, I., & Verma, B. (2007). DICOM, HL7 and IHE: A basic primer on healthcare standards for radiologists. *Indian Journal of Radiology and Imaging*, *17*(2), 66–68. doi:10.4103/0971-3026.33610.

International, I. H. E. (2011). *IHE.net home*. Retrieved June 27, 2012, from http://www.ihe.net/

Internet Stroke Center. (2011). *Radiology image library | internet stroke center*. Retrieved April 17, 2011, from http://www.strokecenter.org/radiology/

ITU. (n.d.). *What really is a third generation (3G) mobile technology*. Retrieved April 28, 2011, from http://www.itu.int/ITU-D/tech/FORMER_PAGE_IMT2000/DocumentsIMT2000/What_really_3G.pdf

Ivetic, D., & Dragan, D. (2009). Medical image on the go! *Journal of Medical Systems*, *35*(4), 499–516. doi:10.1007/s10916-009-9386-2 PMID:20703540.

Iwasa, K. (2004). *Web services reliable messaging (WS-reliability 1.1)*. Retrieved from http://docs.oasis-open.org/wsrm/ws-reliability/v1.1/wsrm-ws_reliability-1.1-spec-os.pdf

Jassal, S., Brissenden, J., Raisbeck, A., & Roscoe, J. (1998). Comparative cost-analysis of two different chronic care facilities for end-stage renal disease patients. *Geriatric Nephrology and Urology*, *8*(2), 69–76. doi:10.1023/A:1008378422292 PMID:9893214.

JPEG. (n.d.). *JPEG 2000*. Retrieved April 18, 2011, from http://www.jpeg.org/jpeg2000

Kangasharju, J. (2008). *XML messaging for mobile devices*. (Doctoral dissertation). Department of Computer Science, University of Helsinki, Helsinki, Finland. Retrieved from http://www.doria.fi/

Konstantas, D., Halteren, A. V., Bults, R., Wac, K., Jones, V., Widya, I., & Herzog, R. (2004). MobiHealth: Ambulant patient monitoring over next generation public wireless networks. In Demiris, G. (Ed.), *E-Health: Current Status and Future Trends* (pp. 107–122). Boca Raton, FL: IOS.

Kristensen, M. D., & Bouvin, N. O. (2008). Developing cyber foraging applications for portable devices. In *Proceedings of the IEEE International Conference on Portable Information Devices*. Garmisch-Partenkirchen, Germany: IEEE.

Kyriacou, E. C., Pattichis, C. S., & Pattichis, M. S. (2009). An overview of recent health care support systems for eEmergency and mHealth applications. In *Proceedings of the Annual International Conference of the IEEE Engineering in Medicine and Biology Society* (pp. 1246-1249). IEEE.

Lee, M., & Gatton, T. M. (2010). Wireless health data exchange for home healthcare monitoring systems. *Sensors (Basel, Switzerland)*, *10*(4), 3243–3260. doi:10.3390/s100403243 PMID:22319296.

Lublinsky, B. (n.d.). *Versioning in SOA*. Retrieved August 6, 2011, from http://msdn.microsoft.com/en-us/library/bb491124.aspx

Maglogiannis, I., Doukas, C., Kormentzas, G., & Pliakas, T. (2009). Wavelet-based compression with ROI coding support for mobile access to DICOM images over heterogeneous radio networks. *IEEE Transactions on Information Technology in Biomedicine*, *13*(4), 458–466. doi:10.1109/TITB.2008.903527 PMID:19586812.

Mahmoud, Q. (2004, November). *Java messaging service*. Retrieved March 5, 2011, from http://java.sun.com/developer/technicalArticles/Ecommerce/jms/index.html

McCarthy, J. (2002, June 1). *Reap the benefits of document style web services*. Retrieved March 5, 2011, from http://www.ibm.com/developerworks/Webservices/library/ws-docstyle.html

McGovern, J., Tyagi, S., Stevens, M., & Mathew, S. (2003). Service-oriented architecture. In *Java Web Services Architecture* (pp. 35–63). New York: Elsevier Science. doi:10.1016/B978-155860900-6/50005-1.

MedApps. (2011). *MedApps mobile health monitoring*. Retrieved March 7, 2011, from http://www.medapps.net/HealthPAL.html

Moore, M. (1999). The evolution of telemedicine. *Future Generation Computer Systems*, *15*(2), 245–254. doi:10.1016/S0167-739X(98)00067-3.

Muir, K., Buchan, A., von Kummer, R., Rother, J., & Baron, J. (2006). Imaging of acute stroke. *The Lancet Neurology*, *5*(9), 755–768. doi:10.1016/S1474-4422(06)70545-2 PMID:16914404.

Müller, A., Helms, T. M., Wildau, H., Schwab, J. O., & Zugck, C. (2011). Remote monitoring in patients with pacemakers and implantable cardioverter-defibrillators: New perspectives for complex therapeutic management. In M. Kumar Das (Ed.), *Modern Pacemakers - Present and Future*. Retrieved from http://www.intechopen.com/articles / show / title / remote - monitoring - in - patients - with - pacemakers - and - implantable -cardioverter-defibrillators-new-perspe

Nabi, M., Kiah, M., Zaidan, B., Zaidan, A., & Alam, G. (2010). Suitability of using SOAP protocol to secure electronic medical record databases transmission. *International Journal of Pharmacology*, *6*, 959–964. doi:10.3923/ijp.2010.959.964.

Nakata, K., Maeda, K., Umedu, T., Hiromori, A., Yamaguchi, H., & Higashino, T. (2009). Modeling and evaluation of rescue operations using mobile communication devices. In *Proceedings of the ACM/IEEE/SCS 23rd Workshop on Principles of Advanced and Distributed Simulation* (pp. 64-71). ACM/IEEE/SCS.

NEMA. (n.d.). *Digital imaging and communications in medicine*. Retrieved June 27, 2012, from http://medical.nema.org/

Neubert, S., Arndt, D., Thurow, K., & Stoll, R. (2010). Mobile real-time data acquisition system for application in preventive medicine. *Telemedicine and e-Health Journal, 16*(4), 504-509.

Niagara Health System. (n.d.). *Niagara health system – Services – Complex care (chronic care)*. Retrieved July 6, 2011, from http://www.niagarahealth.on.ca/services/chronic_care.html

Nikolidakis, S. A., Georgakakis, E., Giotsas, V., Vergados, D. D., & Douligeris, C. (2010). A secure ubiquitous healthcare system based on IMS and the HL7 standards. In *Proceedings of the 3rd International Conference on Pervasive Technologies Related to Assistive Environments*. Samos, Greece: IEEE.

Ninos, K., Spiros, K., Glotsos, D., Georgiadis, P., Sidiropoulos, K., & Dimitropoulos, N. et al. (2010). Development and evaluation of a PDA-based teleradiology terminal in thyroid nodule diagnosis. *Journal of Telemedicine and Telecare*, *16*(5), 232–236. doi:10.1258/jtt.2010.090512 PMID:20423934.

Open, O. A. S. I. S. (2006). *Web services security: SOAP message security 1.1 (WS-security 2004)*. Retrieved from http://www.oasis-open.org/committees/download.php/16790/wss-v1.1-spec-os-SOAPMessageSecurity.pdf

Oracle. (n.d.). *Remote method invocation home*. Retrieved May 31, 2011, from http://www.oracle.com/technetwork/java/javase/tech/index-jsp-136424.html

Orwat, C., Graefe, A., & Faulwasser, T. (2008). Towards pervasive computing in health care – A literature review. *BMC Medical Informatics and Decision Making*, *8*(26). PMID:18565221.

Park, E., & Nam, H. S. (2009). A service-oriented medical framework for fast and adaptive information delivery in mobile environment. *IEEE Transactions on Information Technology in Biomedicine*, *13*(6), 1049–1056. doi:10.1109/TITB.2009.2031495 PMID:19775976.

Puming, L. (2006). *Service agreements and facilities for m-health vital sign monitoring*. (M.S. thesis). Department of Computer Science, University of Twente, Enschede, Netherlands. Retrieved from http://essay.utwente.nl/

Rashvand, H. F., Salcedo, V. T., Sanchez, E. M., & Iliescu, D. (2008). Ubiquitous wireless telemedicine. *IET Communications*, *2*(2), 237–254. doi:10.1049/iet-com:20070361.

RemotEye. (n.d.). *DICOM viewer – RemotEye*. Retrieved August 8, 2011, from http://www.neologica.it/eng/RemotEye.php

Rosenberg, J., Schulzrinne, H., Camarillo, G., et al. (2002). *SIP: Session initiation protocol*. Retrieved November 18, 2011, from http://www.ietf.org/rfc/rfc3261.txt

Satyanarayanan, M. (2001). Pervasive computing: Vision and challenges. *IEEE Personal Communications*, *8*(4), 10–17. doi:10.1109/98.943998.

Schloeffel, P., Beale, T., Hayworth, G., Heard, S., & Lesli, H. (n.d.). *The relationship between CEN 13606, HL7, and openEHR*. Retrieved April 6, 2011, from http://www.oceaninformatics.com/Media/docs/Relationship-between-CEN-13606-HL7-CDA--openEHR-2ba3675f-2136-4069-ac5c-152139c70bd0.pdf

Shnayder, V., Chen, B., Lorincz, K., Jones, T. R. F. F., & Welsh, M. (2005). Sensor networks for medical care. In *Proceedings of the 3rd International Conference on Embedded Networked Sensor Systems*. San Diego, CA: IEEE.

Sillitti, A., Vernazza, T., & Succi, G. (2002). Service oriented programming: A new paradigm of software reuse. *Lecture Notes in Computer Science*, *2319*, 268–280. doi:10.1007/3-540-46020-9_19.

Social Security Advisory Board. (2009). *The unsustainable cost of health care*. Retrieved from http://www.ssab.gov/documents/TheUnsustainableCostofHealthCare_graphics.pdf

Subramanya, A. (2001). Image compression technique. *IEEE Potentials*, *20*(1), 19–23. doi:10.1109/45.913206.

Tahat, A. A. (2009). Mobile messaging services-based personal electrocardiogram monitoring system. *International Journal of Telemedicine and Applications*, 85929. PMID:19707531.

The Apache Software Foundation. (2009). *Jini_Architecture_Specification – River wiki*. Retrieved June 19, 2012, from http://wiki.apache.org/river/Jini_Architecture_Specification

The Institute of Electrical and Electronics Engineers. (1990). *IEEE standard glossary of software engineering terminology*. Retrieved from http://ieeexplore.ieee.org/stampPDF/getPDF.jsp?tp=&arnumber=159342

U.S National Library of Medicine. (2011). *Home-monitoring in implantable cardioverter defibrillator (ICD) patients (monitor-ICD)*. Retrieved April 28, 2011, from http://clinicaltrials.gov/ct2/show/NCT00787683

Varshney, U. (2009). Wireless health monitoring: State of the art. In *Pervasive Healthcare Computing* (pp. 119–146). New York: Springer US. doi:10.1007/978-1-4419-0215-3_6.

Vergados, D. D. (2010). Service personalization for assistive living in a mobile ambient healthcare-networked environment. *Personal and Ubiquitous Computing Journal*, *14*(6), 575–590. doi:10.1007/s00779-009-0278-8.

W3C. (n.d.a). *SOAP version 1.2 part 1: Messaging framework (2nd Ed.)*. Retrieved May 31, 2011, from http://www.w3.org/TR/soap12-part1/

W3C. (n.d.b). *Web services architecture*. Retrieved May 31, 2011, from http://www.w3.org/TR/ws-arch/

Wafa, T. (2010). How the lack of prescriptive technical granularity in HIPAA has compromised patient privacy. *North Illinois University Law Review*, *30*(3).

Wang, H., Peng, D., Wang, W., Sharif, H., Chen, H., & Khoynezhad, A. (2010). Resource-aware secure ECG healthcare monitoring through body sensor networks. *IEEE Wireless Communications*, *17*(1), 12–19. doi:10.1109/MWC.2010.5416345.

Weibel, N. (2002). *Web services technologies SOAP vs. JINI*. Retrieved from http://www.rudibelotti.com/doc/projects/Webservices/Webservices.pdf

Weiser, M. (1991). The computer for the 21st century. *Scientific American*, *265*(3), 66–75. doi:10.1038/scientificamerican0991-94 PMID:1754874.

Wi-Fi. (n.d.). *Wi-Fi alliance*. Retrieved April 18, 2011, from http://www.wi-fi.org

Wilkinson, T., Haines, S., & Williams, C. (2001). Jini in military system applications. In *Proceedings of the RTO Information Systems Technology Panel (IST) Symposium on Information Management Challenges in Achieving Coalition Interoperability*, (pp. 154-172). Quebec, Canada: IST.

WiMAX. (n.d.). *What is WiMAX? | general | WiMAX FAQ*. Retrieved April 18, 2011, from http://www.wimax.com/general/what-is-wimax

ZigBee. (n.d.). *ZigBee alliance*. Retrieved April 18, 2011, from http://www.zigbee.org

ADDITIONAL READING

Angius, G., Pani, D., Raffo, L., Randaccio, P., & Seruis, S. (2008). A tele-home care system exploiting the DVB-T technology and MHP. *Methods of Information in Medicine*, *47*(3), 223–228. PMID:18473088.

Ash, J. S., Berg, M., & Coiera, E. (2004). Some unintended consequences of information technology in health care: The nature of patient care information system-related errors. *Journal of the American Medical Informatics Association*, *11*(2), 104–112. doi:10.1197/jamia.M1471 PMID:14633936.

Asplund, M., Nadjm-Tehrani, S., & Sigholm, J. (2009). Emerging information infrastructures: Cooperation in disasters. In *Critical Information Infrastructure Security*. Berlin, Germany: Springer. doi:10.1007/978-3-642-03552-4_23.

Baker, S. D., & Hoglund, D. H. (2008). Medical-grade, mission-critical wireless networks (designing an enterprise mobility solution in the healthcare environment). *IEEE Engineering in Medicine and Biology Magazine*, *27*(2), 86–95. doi:10.1109/EMB.2008.915498 PMID:18463024.

Benson, T. (2010). Why interoperability is hard. In Hannah, K. J., & Ball, M. J. (Eds.), *Principles of Health Interoperability HL7 and SNOMED* (pp. 25–34). London, UK: Springer-Verlag. doi:10.1007/978-1-84882-803-2_2.

Christensen, E., Curbera, F., Meredith, G., & Weerawarana, S. (n.d.). *Web service description language (WSDL)*. Retrieved May 1, 2011, from http://www.w3.org/TR/wsdl

Dang, S., Dimmick, S., & Kelkar, G. (2009). Evaluating the evidence base for the use of home telehealth remote monitoring in elderly with heart failure. *Telemedicine and e-Health, 15*(8), 783-796.

Ekeland, A. G., Bowes, A., & Flottorp, S. (2010). Effectiveness of telemedicine: A systematic review of reviews. *International Journal of Medical Informatics*, *79*(11), 736–771. doi:10.1016/j.ijmedinf.2010.08.006 PMID:20884286.

Fergus, P., Haggerty, J., Taylor, M., & Bracegirdle, L. (2011). Towards a whole body sensing platform for healthcare applications. In England, D. (Ed.), *Whole Body Interaction* (pp. 135–149). London, UK: Springer London. doi:10.1007/978-0-85729-433-3_11.

Gao, T., Massey, T., Selavo, L., Crawford, D., Chen, B., Lorincz, K., Welsh, M. (2007). The advanced health and disaster aid network: A light-weight wireless medical system for triage. *IEEE Transactions on Biomedical Circuits and Systems, 1*(3), 203-216.

Gao, T., Pesto, C., Selavo, L., Chen, Y., Ko, J. G., & Lim, J. H. Welsh, M. (2008). Wireless medical sensor networks in emergency response: Implementation and pilot results. In *Proceedings of the IEEE Conference on Technologies for Homeland Security* (pp. 187-192). IEEE.

Gonzalez-Valenzuela, S., Chen, M., & Leung, V. C. M. (2011). Mobility support for health monitoring at home wearable sensors. *IEEE Transactions on Information Technology in Biomedicine*, *15*(4), 539–549. doi:10.1109/TITB.2010.2104326 PMID:21216718.

Huang, E., & Liou, D. (2007). Performance analysis of a medical record exchanges model. *IEEE Transactions on Information Technology in Biomedicine*, *11*(2), 153–160. doi:10.1109/TITB.2006.875681 PMID:17390985.

Jung, D. K., Kim, G. N., Kim, G. R., Shim, D. H., Ham, K. Y., & Kim, M. H. et al. (2006). A biosignal monitoring system for mobile telemedicine. *The Journal on Information Technology in Healthcare*, *4*(3), 173–183.

Koch, S., & Hagglund, M. (2009). Health informatics and the delivery of care to older people. *Maturitas*, *63*(3), 195–199. doi:10.1016/j.maturitas.2009.03.023 PMID:19487092.

Koff, D., Bak, P., Brownrigg, P., Hosseinzadeh, D., Khademi, A., & Kiss, A. et al. (2009). Pan-Canadian evaluation of irreversible compression ratios (lossy compression) for development of national guidelines. *Journal of Digital Imaging*, *22*(6), 569–578. doi:10.1007/s10278-008-9139-7 PMID:18931879.

Kohler, B. U., Hennig, C., & Orglmeister, R. (2002). The principles of software QRS detection. *IEEE Engineering in Medicine and Biology Magazine*, *21*(1), 42–57. doi:10.1109/51.993193 PMID:11935987.

Krupinski, E. A., Williams, M. B., Andriole, K., Strauss, K. J., Applegate, K., & Wyatt, M. et al. (2007). Paper. *Journal of the American College of Radiology*, *4*(6), 389–400. doi:10.1016/j.jacr.2007.02.001 PMID:17544140.

Kukawka, B., & Wilk, S. (2012). Indexing and retrieval of medical resources for a telemedical platform. In Pietka, E., & Kawa, J. (Eds.), *Information Technologies in Biomedicine (LNCS)* (pp. 603–614). Berlin, Germany: Springer. doi:10.1007/978-3-642-31196-3_60.

Laner, M., Svoboda, P., Hasenleithner, E., & Rupp, M. (2011). Dissecting 3G uplink delay by measuring in an operational HSPA network. In *Proceedings of the 12th International Conference on Passive and Active Measurement* (pp. 52-61). Berlin: Springer-Verlag.

Mamaghanian, H., Khaled, N., Atienza, D., & Vandergheynst, P. (2011). Compressed sensing for real-time energy-efficient ECG compression on wireless body sensor nodes. *IEEE Transactions on Bio-Medical Engineering*, *58*(9), 2456–2466. doi:10.1109/TBME.2011.2156795 PMID:21606019.

Martínez-Espronceda, M., Martínez, I., Escayola, J., Serrano, L., Trigo, J., Led, S., & García, J. (2009). Standard-based homecare challenge. In Yogesan, K., Bos, L., Brett, P., & Gibbons, M. C. (Eds.), *Handbook of Digital Homecare, Series in Biomedical Engineering* (pp. 179–202). Berlin, Germany: Springer. doi:10.1007/978-3-642-01387-4_9.

Prokkola, J., Perala, P. H. J., Hanski, M., & Piri, E. (2009). *3G/HSPA performance in live networks from the end user perspective*. Paper presented at the IEEE International Conference on Communications. Dresden, Germany.

Schrenker, R., & Cooper, T. (2001). Building the foundation for medical device plug-and-play interoperability. *Medical Electronics Manufacturing,* 10-16.

Singh, G. (2008). Information sharing for emergency response. In *Proceedings of the IEEE Conference on Technologies for Homeland Security* (pp. 421-425). IEEE.

Srinivasan, A., Mayank, G., Al Azri, F., & Lum, C. (2006). State-of-the-art imaging of acute stroke. *Radiographics, 26*(1), 75–95. doi:10.1148/rg.26si065501 PMID:17050521.

Tan, W. L., Lam, F., & Lau, W. C. (2008). An empirical study on the capacity and performance of 3G networks. *IEEE Transactions on Mobile Computing, 7*(6), 737–750. doi:10.1109/TMC.2007.70788.

Toomey, R., Ryan, J., McEntee, M., Evanoff, M., Chakraborty, D., & McNulty, J. et al. (2010). Diagnostic efficacy of handheld devices for emergency radiologic consultation. *AJR. American Journal of Roentgenology, 194*(2), 469–474. doi:10.2214/AJR.09.3418 PMID:20093611.

Walter, M., Eilebrecht, B., Wartzek, T., & Leonhardt, S. (2011). The smart car seat: personalized monitoring of vital signs in automotive applications. *Personal and Ubiquitous Computing, 15*(7), 707–715. doi:10.1007/s00779-010-0350-4.

Wilson, L. S., Stevenson, D. R., & Cregan, P. (2010). Telehealth on advanced networks. *Telemedicine and e-Health, 16*(1), 69-79.

KEY TERMS AND DEFINITIONS

BAN: Body Area Network, which comprises two or more nodes, most of which are biological sensors for measuring vital signs, interconnected through wires or wirelessly.

DICOM: The Digital Imaging and Communications In Medicine standard. This standard was created to address the needs for a standardized format for storing and distributing medical image data. These images are created from different imaging equipment, known as *modalities*, such as ultrasound, CT, MRI. In addition, DICOM also supports operations for the querying and transfer of DICOM images.

Information Retrieval System: A system which facilitates transmission and retrieval of medical images onto a medical professional's mobile device to allow remote diagnosis of patients while mobile.

Medical System: A system that utilizes information and communication technology to help deliver one or more medical services to health care practitioners or patients.

Mobile Device: A portable device with communication and processing capabilities.

Remote Patient Health Monitoring System: A system which utilizes a mobile device to forward data collected from sensors attached to a patient's body, to a medical facility and other stakeholders.

Ubiquitous Service: A service that is virtually available in an "anywhere, anytime, anyhow" fashion, by utilizing a multitude of wireless access technologies.

WSDL: Web Service Description Language, which is used to describe the interface of a Web Service (the functions provided by a service, and the means to invoke them). This interface description is stored in what is known as a WSDL file, which needs to be retrieved by the service consumer in order to use a particular service.

Chapter 12
BioCondition Assessment Tool:
A Mobile Application to Aid Vegetation Assessment

Chin Loong Law
Queensland University of Technology, Australia

Paul Roe
Queensland University of Technology, Australia

Jinglan Zhang
Queensland University of Technology, Australia

ABSTRACT

Environmental degradation has become increasingly aggressive in recent years due to rapid urban development and other land use pressures. This chapter looks at BioCondition, a newly developed vegetation assessment framework by Queensland Department of Resource Management (DERM) and how mobile technology can assist beginners in conducting the survey. Even though BioCondition is designed to be simple, it is still fairly inaccessible to beginners due to its complex, time consuming, and repetitive nature. A Windows Phone mobile application, BioCondition Assessment Tool, was developed to provide on-site guidance to beginners and document the assessment process for future revision and comparison. The application was tested in an experiment at Samford Conservation Park with 12 students studying ecology in Queensland University of Technology.

DOI: 10.4018/978-1-4666-4054-2.ch012

INTRODUCTION

The world is undergoing drastic changes due to increased population and rapid urban development. As environmental health is degrading at an alarming rate, ecologists have tried to tackle the problem by monitoring the composition and condition of the environment. This chapter will present the use of mobile technology to assist novice ecologists to conduct vegetation assessment surveys and also to document the survey process to increase validity and accuracy of the assessment result.

BioCondition is a vegetation assessment framework developed by Queensland Department of Resource Management in Australia. It uses surrogates to represent the health or condition of the environment. The framework is designed to be simple and accurate therefore users without ecological background can easily learn the process. However, as tested in an experiment, getting started with BioCondition is still a complex process that requires lengthy study. Beginners also find that they frequently need to refer to the assessment manual for information and thus requiring much longer time to complete the survey. Furthermore, an interview with experts of BioCondition discovers that due to the relaxed nature of BioCondition, it is difficult to validate the assessment result without re-conducting the survey.

Mobile technology has advanced at a tremendous rate in the past few years. Many field work researchers have started evaluating the use of commodity mobile devices. This research aims to take advantage of mobile platforms in order to greatly benefit beginners in learning and data collection and also increase the accuracy of the assessment result.

BioCondition Assessment Tool (BAT), a Windows Phone application, is designed to guide users while undertaking the survey and providing mechanisms to document the process for further validation. The application streamlines the process by adding additional guidelines, examples and interactivity to BioCondition survey. Currently, BAT uses two simple mechanisms to achieve the above goals. Firstly, BAT leverages cameras on mobile devices to take visual evidence of flora. BioCondition involves identification of species. Since beginners may not know the exact species name, keeping a visual record and associating a description can be beneficial in the future for further research or validation by expert ecologists. Secondly, BAT provides context appropriate guidelines and examples to users. Instead of constantly referring the assessment manual, users can interact with graphical elements displayed on a mobile phone which provide description and examples of the actions required to complete the task.

The BioCondition Assessment Tool is evaluated in an experiment conducted with a group of students studying ecology in Queensland University of Technology. The aim of this experiment is to verify that beginners can greatly benefit from mobile technology in the field in guiding them in a BioCondition survey and documentations can increase the accuracy and validity of the assessment result. Experimental results show that the BioCondition Assessment Tool achieves the goals of guiding users, documenting and simplifying the BioCondition survey process.

ENVIRONMENTAL MONITORING

BioCondition

BioCondition is a rapid vegetation assessment framework designed to capture the biodiversity of a terrestrial ecosystem. It is a site-based, quantitative and therefore repeatable assessment procedure that can be used in any vegetation state (Eyre et al., 2011). BioCondition produces a numeric score (1 to 4) that can be used as a condition rating, or functional to dysfunctional condition for biodiversity. The result is achieved by comparing the vegetation condition attributes of the site with a reference (or benchmark) site of the same type that represents the Best-on-Offer condition.

The assessment method for BioCondition is intended for use by assessors who have some knowledge of Regional Ecosystem (RE) mapping and vegetation assessment at a site scale. Primary users of BioCondition include ecologists, resource managers and landholders. Traditional or full-scale assessment requires assessors to have extensive ecological knowledge primarily because such method involves full-scale species sampling and identification. These methods are costly, time-consuming, and complicated. In comparison, BioCondition only requires assessors to have knowledge about assessment site selection and species differentiation. The procedure of BioCondition is relatively simple. It involves tasks such as counting number of large trees, measuring canopy height and length of coarse woody debris. These tasks, though time consuming, can be performed by assessors with no prior knowledge or ecological background.

The procedure to conduct a survey with BioCondition is divided into three steps:

- **Preparation:** Preparation is the first crucial step of BioCondition. This step involves defining an objective, searching for mapping data in order to choose appropriate assessment site and collecting benchmark data for the selected regional ecosystem. It directly affects the accuracy and outcome of the assessment.
- **Field Assessment:** Field assessment is the most time consuming part of BioCondition. It involves setting up the site and collecting data of the environment. Specifically:
    - Trees
    - Plants
    - Coarse Woody Debris
    - Ground Cover
    - Canopy Cover
- **Score Calculation:** The result of BioCondition (the conditional score) is determined by adding values of the collected attributes and apply the formula specified in the assessment manual.

Prior to visiting the assessment site, it is important that assessors gather existing mapping data of the regional ecosystem. These data contributes to the decision making of assessment site selection. Although the actual suitability of the selected area for BioCondition can only be determined when assessors physically visit the site, regional ecosystem mapping data will inform assessors the general condition of the area, and therefore differentiating the area to various assessment units and sites.

According to BioCondition Assessment Manual (Eyre et al., 2011, pp. 9-10), the components needed to perform the assessment include:

- 100 m and 50 m transect tapes
- 1 × 1 m quadrat for measuring ground cover
- Global Positioning System (GPS) for navigation
- Compass for direction
- Diameter tape or small measuring tape for tree diameter measurement
- Clinometer, hypsometer or rulers for measuring tree height
- Digital or print film camera

Optionally, the following components are suggested to be included:

- Star pickets for site relocation
- Assessment manual and data sheet
- Internet access to obtain information
- Benchmark documents

- Clipboard, pencils and erasers
- Flagging tape
- Plant identification books

As can be seen from the above bullet points, many of the resources can be emulated on a modern mobile device or implemented as part of the software system. Although necessary, these resources may be inaccessible or too costly for beginners or novices therefore reducing public participation.

MOBILE COMPUTING

Mobile Devices

Mobile computing describes a form of computing that emphasizes portability (Lane et al., 2010). In the early days, mobile computing was not widely adopted primarily due to high cost and power consumption and low device capabilities. However in recent years, the increased interests of mobile devices in general public have become a prime driving force to reduce the cost while increasing device capabilities. This trend has led to the expansion of ubiquitous computing, which focuses on the research of human-computer interaction where computing devices are used in human's everyday activities.

Previously, portable computing devices were categorized in terms of their processing power and capabilities. For example, smartphone and Personal Digital Assistant (PDA) were in different categories because their functionalities and processing power were significantly different. This concept has been redefined due to growing similarities. Smartphones nowadays are equipped with sophisticated components, such as Global Positioning Unit (GPS), digital compass and networking chipsets, which can be utilized in not just personal entertainment but also scientific research (Lane et al., 2010).

The basic functionality of these devices is the ability to communicate via one or many network interfaces, for example 3G or Wi-Fi. Laptops are usually better suited for heavyweight tasks because of the powerful integrated general and graphic processors. Although laptops are portable, they are however mainly used in stationary situation because of weight and physical size of the device. On the other hand, smartphones are exceptionally versatile in any use-case scenario due to its small physical dimension. With the increased processing power, modern smartphones are not only used for personal communication. For instance the tracking components in smartphones can be used for collecting data of an environment. This allows a large scale, voluntary survey to be conducted: Participatory Sensing (Burke et al., 2006).

Mobile Hardware and Software

Modern mobile devices are equipped with advance hardware components, for example high quality still and video camera, accelerometer, global positioning system (GPS), graphic processing unit (GPU), digital compass and gyroscope. These components can be utilized to achieve augmented reality by providing geographical data and additional information it needs, for example snapshots of an environment with camera direction and position information can be reconstructed as 3D models and stored according to the timeline. A significant application of this is to monitor changes of environment.

Mobile Hardware Components

Compare to early days, modern mobile devices are equipped with high quality imaging sensor. For example, University of Cambridge and Oldenburg both conducted researches that use camera on mobile device as input to recognise natural features. The tests aimed to exercise several stripped-down versions of popular object recognition algorithms using camera on mobile phone. They both concluded that modern smartphones are equipped with central processing unit (CPU) powerful enough to perform sophisticated algorithms for object tracking and recognition (Pielot et al., 2009; Wagner, Reitmayr, Mulloni, Drummond & Schmalstieg, 2008). Both tests successfully detected natural features using modern smartphones and overlaid 2D virtual objects on display.

In addition to imaging sensor, many modern mobile devices are also equipped with tracking sensors such as accelerometer, digital compass, and gyroscope. These components are used to provide device orientation and acceleration changes, and compliment raw data reading from each other. Table 1 summarises each component:

Accelerometer measures the magnitude and direction of acceleration (e.g. left, right, up, down, forward and backward), digital compass measures direction where the device is pointing at (e.g. north or south), and gyroscope measures angular velocity of the device (e.g. pitch, roll and yaw rotations). The combined readings of all three components can be used to track local device movement accurately. For example, a robot with accelerometer is able to move forward and backward but it cannot rotate itself, however with the assistance of gyroscope it will be able to turn around and bend over. These two components give a total of six-degree of freedom (assuming each component measures three-axis). With the addition of digital compass, in the example the robot can now move towards a specific direction as instructed by the user. Even though the combination of digital compass and accelerometer is able to provide position and orientation information of a device in 3D space, however it does not offer the accuracy and fluency of a gyroscope. The distinctive difference is accelerometer and digital compass measure the rotation using Earth's magnetic fields. However, gyroscope measures the rotation relative to the device itself and therefore it is faster and better at picking up subtle motion. Furthermore, gyroscope does not take into account any external interference (Qiang et al., 2009). The distinctive enhancement and integration of sensors to environmental monitoring is discussed in mobile software systems section.

Table 1. Tracking sensors in modern mobile devices

Component	Description
Accelerometer	Accelerometer is a device that measure proper acceleration experienced relative to free-fall. It can be used to sense orientation, acceleration, vibration shock and falling (Keir, Hann, Chase & Chen, 2007).
Digital compass (magnetometer)	Magnetometer, or more commonly known as digital compass, is a scientific instrument used to measure strength and direction of magnetic field relative to the component.
Gyroscope	Gyroscope is a device for measuring or maintaining orientation, it does not measure the linear acceleration of the device (Qiang et al., 2009).

Global Positioning System (GPS) is an outer space-based global navigation satellite system that provides reliable location and time information. Modern mobile devices usually contain a GPS receiver to retrieve GPS satellite broadcast signals from outer space and calculate its three-dimensional location (longitude, latitude and altitude). Many augmented reality projects use the combined reading of GPS, accelerometer, digital compass and gyroscope to track location and movement of the user in real-world to provide location-sensitive digital content, for example augmented reality real-world browser such as Layar Reality Browser and Wikitude (Wither et al., 2009). This type of browsers is typically used on portable devices, it displays digital content relative to the real-world on a device's screen using camera view. In relation to vegetation assessment, it is possible to take advantage of such technology to aid users keeping track of tasks in the field, for example using augmented reality to keep track of identified trees in the assessment area.

Besides the addition of new components, the most significant change on mobile hardware is the processing capability. Over the past decade,

Central Processing Unit (CPU) and graphic processing unit (GPU) on a mobile device are also getting more powerful. Most noticeably, the graphic performance of modern mobile phones is significantly better than low-to-mid range netbooks (or sub-notebooks) and tablets. The performance difference is easily noticeable when playing a high definition video, performing computing intensive tasks (e.g. algorithms and video games), and browsing multiple Webpages. One of the reasons why mobile device, specifically smartphone, needs better processing capability is because of increased complexity and development of mobile operating systems, for example Apple iOS and Google Android.

Mobile Software Systems

General consumer's decision-making on purchasing a new mobile device has shifted from pure hardware capabilities to software and ecosystem provided by the vendor. A great example of such transition is the decreased awareness of Nokia and the rising popularity of Apple iPhone and Samsung Galaxy smartphones. Since the release of first Apple iPhone in June 2007, the strategy of mobile operating systems has drastically changed to focusing on the wide variety of Application Programmable Interfaces (APIs) and user friendliness of the Graphical User Interface (GUI). Aside from marketing strategy, these two features are the main reasons for attracting a broad range of programmers, companies and consumers to invest in the platform.

Advance mobile hardware and operating systems expose the possibility of using mobile device for scientific purpose, for examples the research of using mobile device to preserve historical objects and sites using 3D reconstruction method (Thormählen and Siedel, 2008; Ziegler et al., 2003), and sound pollution monitoring in urban environment (Kanjo et al., 2008). Portability of mobile devices makes it the prospective platform for fieldwork. Previously, field workers often carry a customised Personal Digital Assistant (PDA) to collect data on-site. However, the trend has changed towards using smartphones because of processing power and programmability.

In relation to BioCondition, modern mobile devices provide many advantages over traditional pen and paper approach. Combined with software capabilities, mobile device can improve many aspects of the assessment process. The list below summarises a few improvements:

- **Imaging Sensor:** Digital (or print film) camera is one of the required resources in BioCondition. It is used to keep a visual record of environment to monitor changes over time. Camera on a mobile device can be used instead of carrying a separate camera. Furthermore, it allows us to coordinate both software and hardware to achieve the goal of documenting the assessment process. BioCondition does not enforce the rule of keeping a visual record of all natural objects in the environment therefore it is not possible to justify the assessment outcome. A mobile application can then be used as a data collector to require photo evidence when inputting data, thus improving accuracy and allowing detail environmental monitoring over time.
- **GPS and Sensors:** Similar to camera, GPS is one of the required resources for acquiring location data of the assessment site. Modern mobile operating systems often include Geographical Information Solutions (GIS) or mapping system such as Google Maps and ArcGIS. We can take advantage of this to create an interactive map showing location of user, the assessment site and also overlaying critical information on the map to help user navigating the area and aid them in conducting the assessment, for example overlaying a push pin for each counted tree to eliminate the possibility of counting the same tree more than once.

Furthermore, these sensors also allow the emulation of a digital range finder on a smartphone. Using simple trigonometry and embedded sensors, it is possible to triangulate the distance from smartphone to an object, and height, area, and width of the object. There is no scientific research of such methodology as of the publication of the book chapter. The implementation of such measurement and emulation is discussed in BioCondition Assessment Tool (BAT) section.

- **Networking Capabilities:** Networking capabilities on a mobile device, for example 3G and Wi-Fi, can also be used to aid user in obtaining information. For example, instead of carrying the assessment manual and searching for related field guides before the assessment, the mobile application can retrieve such information over Internet.
- **Multimedia Capabilities and Interactivity:** Modern mobile devices are capable of processing multimedia content and engaging user due to high graphical processing power. It introduces several possibilities such as using video playback for training or demonstrating examples using video or images instead of text. This is particularly beneficial to beginners because the mobile application can be used as a digital teaching assistant to increase their proficiency over time at performing the tasks. Interactivity provided by mobile platform also makes the tasks more engaging to users, for example showing user's location and boundary of the assessment area on interactive map.

However, the improvements shown on the bullet points above may suffer from some constrains and limitations. A great example of such disadvantages is battery life. Since the mobile device will be using tracking sensors and computational power at all times, battery consumption may increase tremendously over time. Moreover, GPS, tracking sensors and networking capabilities may be obstructed depending on the environment setting. For example a very dense forest may drastically decrease or obstruct network coverage and GPS accuracy thus disabling some crucial features of the mobile application. Furthermore, display on mobile device may be too reflective under direct sun light therefore obstructing the viewing capability. However, these hardware specific issues are out of reach of our project's scope.

CITIZEN SCIENCE

With reduced cost and increased capabilities, most people in the world, even in the poorest countries, have mobile phones (International Telecommunication Union, 2011). Today's smartphones are powerful networked computing devices with an array of sensors. These capabilities make smartphones powerful data collection devices, which can be used for collecting environmental data. Their ubiquity enables large scale, voluntary environmental monitoring to be conducted by the public utilizing mobile phones as sensing device. This is an example of Participatory Sensing (Burke et al., 2006), which explores community-based monitoring. Two closely related concepts are citizen science and crowd sourcing.

Crowd Sourcing

Crowd sourcing takes a task traditionally performed by an employee or contractor and outsourcing it to an undefined, large group of people in the form of an open call (David, 2011). Since it broadcasts the problem solving and production tasks to the crowd, the tasks can often be completed more effectively at comparatively little cost. As many organisations and communities have gained the benefits from crowd sourcing, many researchers have applied it into the science area, leading to the notion of citizen science. Although citizen scientists have no adequate professional training and scientific expertise, they still can perform or manage some research related tasks, such as measurement and observation. Citizen science has the potential to broaden the research scope and strengthen the ability to gather scientific data. Galaxy Zoo is a classic example of this approach, with over 250,000 active users helping to manually classify galaxy types according to their shapes. Galaxy Zoo provides users with initial identification training and testing and then provides an interface for classifying galaxies, deferring the final complex analysis task to humans. Identification of the same galaxy by multiple users ensures consistency and accuracy. The astronomers and members of Galaxy Zoo can manually classify the morphology of one million galaxies in less than three weeks.

Participatory Sensing

Participatory sensing is a type of citizen science focusing on data collection. In participatory sensing, citizens voluntarily carry their mobile phones as sensing devices to gather information about important resources and share this information through Internet communication infrastructure. As mobile phone prices are reducing quickly, it can be used to inexpensively obtain measurement of the environment on a regular basis and at large scale.

Major Projects

In practise, many participatory sensing systems have been successfully deployed and utilised for community-based monitoring of important resources. Coral Watch, developed at the University of Queensland, Australia (Abdulmonem & Hunter, 2010), is used for coral reef monitoring. Other popular systems are Zooniverse, which hosts one of the largest and most popular citizen science projects that primarily focuses on space, weather, and humanities. The site is a portal to individual project sites that all retain a very similar layout and design features, which helps to reinforce to the user that they are within the wider Zooniverse scope of projects. The site is supported by the Citizen Science Alliance (Smith et al., 2011).

Another important site is Project Noah, which makes excellent use of Web 2.0 tools and social media to support and nurture the volunteer community. The presence of active users is very visible across the site, and they have adopted a model that not only allows users to contribute to other's submissions (e.g. by helping to name organisms photographed by contributors), but also to 'follow' other users, tag photos as favourites, and also rewards users for the increasing number and kind of contributions they make with a 'patch' program. The site focuses on engaging participation rather than quality data collection.

The Centre for Embedded Networked Sensing (CENS), established by University of California, Los Angeles (UCLA), has envisioned and researched on how communities can use every day mobile phones to gather data about things that are important to them. They have applied this concept to critical scientific and societal pursuits such as Participatory Urban Sensing in the area of Public Health and Urban Planning.

eBird is an online tool for global birders to collect critical bird data for science. It aims to maximise the utility and accessibility of the vast numbers of bird observations made each year by recreational and professional bird watchers. Launched in 2002 by the Cornell Lab of Ornithology and National Audubon Society, it is growing rapidly and has approached a new benchmark in eBird – 100 millionth bird observation by July 2012. There are eBird applications for smartphones.

As citizen science approach becomes more popular, online tools have been developed which support the simple creation of some citizen science Websites. FieldData provides primarily the functionality required to record, manage and review observational data about species. It can be used as a plug-in for providing data collection and management capabilities to an existing Website or used for creating a simple Website for collecting and managing data.

Most existing citizen science projects provide a means (e.g. online forms) to allow citizen scientists to contribute observations, photos and other files to a project and to view and edit their records. Many of them also provide some online training materials, often in the form of a dedicated Webpage containing related texts, videos, and links to other Websites. However, few assist in the reliable collection of data and combine the following key features:

- Dynamic data collection guidance.
- Collection and use of provenance data to aid data cleansing and interpretation.
- Context-sensitive training materials that can be used while sensing in the field and provide just-in-time learning.

BIOCONDITION ENVIRONMENTAL ASSESSMENT

Vegetation monitoring generally consists of sampling and identifying all species in order to determine the condition or health of the environment. However, a full scale survey usually takes a substantial amount of time to complete. As the world is facing critical environmental degradation, this technique is highly undesirable. Therefore, a simple yet accurate approach is preferable as it facilitates the uptake of resources managers (Eyre et al., 2011). Various states in Australia have developed condition assessment frameworks that utilise key attributes or surrogates of biodiversity values that can be rapidly measured in field (Eyre et al., 2011), for examples BioCondition in Queensland and Habitat Hectares in Victoria.

Since the introduction of BioCondition in 2006, Queensland Department of Resource Management (DERM) has been actively encouraging ecologists, resource managers, and landholders to use BioCondition to assess or monitor the vegetation state of their property. To an extent, Queensland universities and organisations also provide opportunities for learning and on-site training, for example Condamine Alliance and Burmett Mary Regional Group.

Even though BioCondition is designed to be simple, fast, and approachable, however it still poses a significant challenge to beginners. A great example of this is the reference manual. The only reliable source of information is the assessment manual published by Queensland government. There are not many examples, in written form or as interactive media, accessible to the general public. Therefore, it is difficult for people without ecological background to understand what the requirements are and how to effectively perform the associated tasks.

As previous section discussed, many of the required resources can be found on a mobile device or implemented as part of the software system. Mobile technology has advanced to the point where it is being actively used in scientific research, for example in public sector (Kuznetsov and Paulos, 2010) and environmental monitoring (Nurminen, Kruijff and Veas, 2011). However, there is no existing application or smart tool to assist and guide users in performing the tasks. This can be a significant roadblock to beginners due to the following reasons:

- Novices (e.g. students and landholders) are to gather the required equipment, such as transect tape, range finder and GPS, and learn its usage in addition to the procedure of BioCondition. For some of the users the assessment may be one-time only thus increasing the cost.
- Users need to fully understand the procedure of BioCondition and the only source of information is the assessment manual. There is no interactive guide to assist users in the field.

- Some associated tasks are repetitive and time consuming, which may be significantly easier to perform with technological assistance.
- There is no way to instantly analyse and compare data on-site.

Furthermore, there is no way to justify or revise the assessment result without revisiting the assessment site because BioCondition does not enforce the rule of documenting the environment. This may significantly affect accuracy of the result if the assessor is a beginner with no ecological background.

The portability of mobile device makes it the perfect candidate for field workers. Modern mobile devices are equipped with components and processing capability powerful enough to assist, guide and automate some tasks in BioCondition. Introducing an interactive mobile application to assist users on-site can mitigate issues above. Table 2 summarises the issues and solutions using a mobile device (detail discussions can be found in later section):

Table 2. Issues and possible solutions of BioCondition using mobile device

Issue	Solutions
No on-site interactive guidance	Context-sensitive mobile application that provides step-by-step guidance for users on-site
Costly compulsory resources to perform the assessment (e.g. GPS, compass and camera)	Modern smartphones are capable of emulating the tools as software features using Application Programmable Interfaces (APIs) available on the mobile chosen platform
Time consuming and repetitive tasks (e.g. layout assessment site and tree counting)	Automate the tasks using provided APIs on the platform
No evidence or artifacts of the assessment to justify or revise the assessment result	Use camera on the mobile phone to capture photo evidence of the process

BIOCONDITION ASSESSMENT TOOL (BAT)

BioCondition Assessment Tool (BAT) is designed to assist beginners in conducting BioCondition in the field. The initial version of the BioCondition Assessment Tool uses modern smartphone features including GPS, camera, sensors and interactive touch-sensitive screens to assist the assessment of certain on-site attributes (See Figure 1).

The BioCondition Assessment Tool (BAT) has been implemented on Windows Phone operating systems and deployed to HTC Titan. The application is developed in XAML, Silverlight and SQLite. Functional aspects of BAT include:

- Full data collection for BioCondition.
- Capability to take photo evidence for each assessment task.
- Context-sensitive interactive guidance.
- Mapping system for ease of on-site navigation.

BAT features full data collection for BioCondition. The assessment tasks are grouped as following:

- **Trees:** In 100 × 50 m area, find median height of ecologically dominant layer, count the number of large trees and differentiate different tree species.
- **Coarse Woody Debris:** In 50 × 20 m area, measure length of fallen logs and dead trees that are more than 0.5 m in length, 10 cm in diameter and 80% contact with ground.
- **Plants:** In 50 × 10 m area, count number of different plant species, including shrub, grass, forbs and other species, and non-native plants.
- **Ground Cover:** Along the 100 m transect line, estimate ground cover components for five 1 × 1 m quadrats. Specifically, the percentage of native perennial grass cover and organic litter.

Figure 1. Overview and dataflow

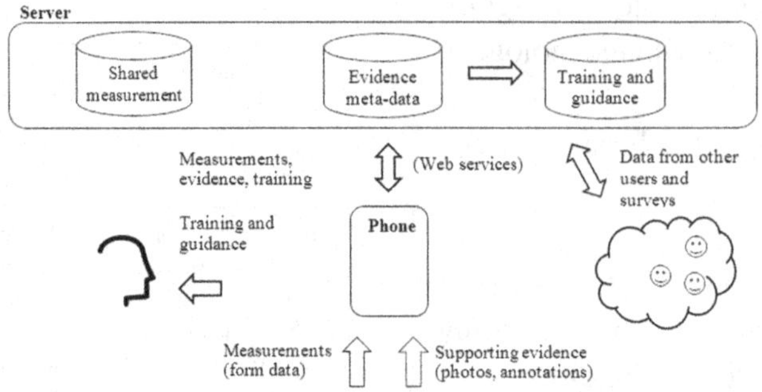

- **Canopy Cover:** Walk along the 100 m transect line and measure length of tree and shrub canopy cover using line-intercept method.

User Interface Design

The design of BioCondition Assessment Tool (BAT) focuses on engaging user with their tasks at hand while still providing relevant guidance without obstructing or interrupting user's workflow. BAT takes full advantage of Windows Phone Modern (or previously known as Metro) design language and adheres to its design principles and best practices. Panoramic experience is used in the main page and assessment user interface to provide a quick access to logically grouped tasks or information. For example, data on the assessment sheet is logically grouped and divided vertically into sections and horizontally into its related sub-sections on a panorama user interface (See Figure 2). We believe such intuitive user interface enables fast and fluid navigation to minimise the idle time when conducting the assessment. Most noticeably, it gives users more constructive information compare to the assessment sheet. (See Figure 3)

User interface elements on each section contain context-sensitive information required to perform the task, for example interacting with each section header (e.g. trees, debris and subplots) will provide user with overview and brief description of tasks related to the section, however interacting with sub-section header (e.g. define large threshold) will introduce user specific instruction of the task (See Figure 4). Furthermore, data input on BAT is not simply a text box, majority of input elements are custom built to include guidance and validations of input data.

Figure 2. Panoramic user experience for logically grouped assessment data

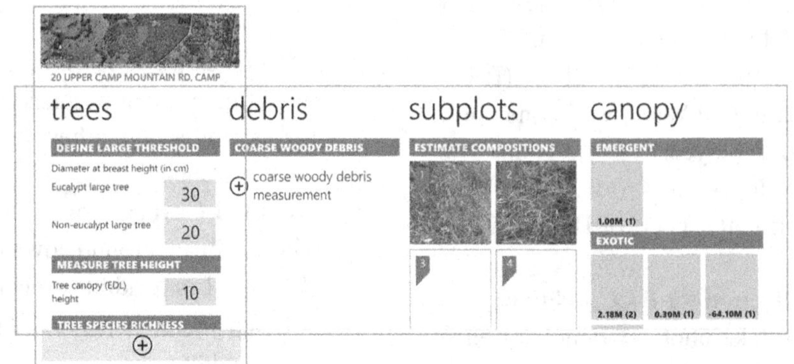

Figure 3. Screenshot of BAT showing the tree section for BioCondition and the same section on the assessment sheet

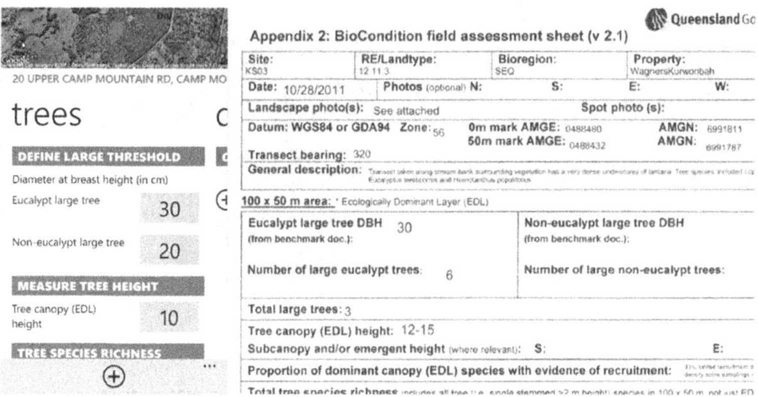

Figure 4. Screenshots of the context-sensitive digital guide on BAT

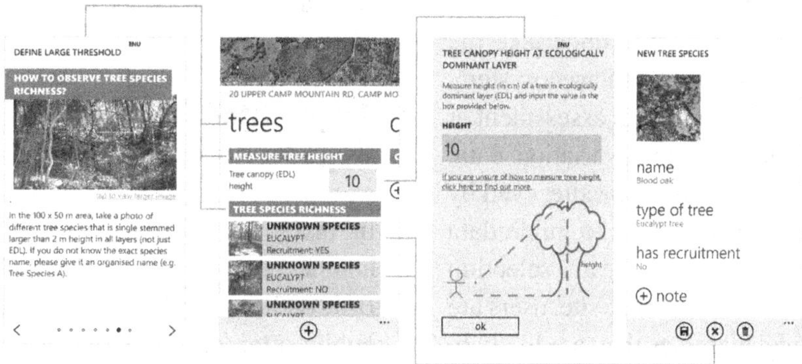

Training Materials, Guided Data Collection, and Data Validation

Novice users often need to refer to the assessment manual to confirm details concerning procedures. Therefore training content, especially the assessment manual, are stored in the application as indexed and searchable learning materials. We also attach to each data collection activity instructional videos and photos depicting how to undertake measurements. Examples from reference meta-data are also provided, showing how to undertake particular measurement and collection activities (See Figure 5).

Guided data collection simplifies the collection of data through BAT providing guidance on where, when and how to collect data and take measurements. Geo-tagging and location information ensure that measurements are taken in the right place. A simple form of Augmented Reality (AR) shows the location of the main measurement transects and the different sized areas for undertaking different tasks. All data and measurements are timed and geo-tagged. A list of survey tasks is shown to the user, the progress is measured automatically by BAT. This enables user to keep track, which tasks have been undertaken and which remain.

Environmental assessment is typically a relative measurement of environmental condition. BioCondition compares sites to a reference (or benchmark) site. Reference sites are chosen and surveyed by experts. Experts use the same tool to collect data, meta-data and training materials.

Figure 5. Training information, virtual transect, and task list

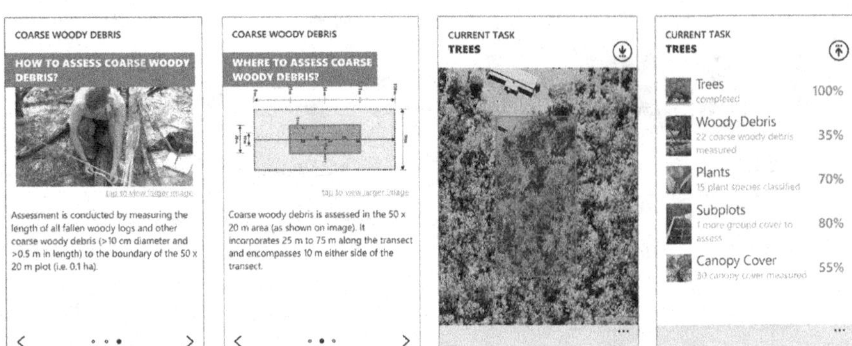

BioCondition entails the comparison of data. The meta-data collected by experts is used to aid novice users in their data collection. By looking at photos and videos of data collected at the reference site, users get a better idea of how to undertake the survey and what to measure in a similar ecosystem. Furthermore, since environmental assessment at its core concerns spatial-temporal variation, data, and meta-data collected at the same site, even by different people can also be used to guide data collection. For example, by looking at a selection of grasses collected at the reference site, users are given an idea of which grasses they are likely to find at the target site. Reference sites are by design similar (same regional ecosystem) to their comparative sites. This aids objective measurement across sites and between different users.

In addition to the time and geo-stamping of measurements, the survey tasks request user to document the collection of activities by taking photos of the measurement activities. Thus, the application collects provenance meta-data which support the simple quantitative measurements required by the paper based survey. This additional meta-data can be used by data consumers to help vet, clean, and interpret data.

Furthermore, photos may also be annotated. For example, some measurements require the length of canopy cover to be calculated; photo of a group of trees can be annotated in this case to differentiate between closely located trees, and also encourage further discussion of the result among peers. Similarly, a photo of ground cover can be annotated with identified perennial grass and organic litter, which must be measured. Such annotated photos can be used for verification and guidance.

Augmented Reality and Sensors Integration

In many occasions, the measurement of tree height and object length is required to complete BioCondition. Enthusiasts, beginners, and general public often do not have the required hardware tools to complete the survey. Thus resulting in additional cost to users and often preventing them from adopting new survey framework.

With the addition of sensors, GPS and increased processing power, modern smartphones are capable of emulating measurement instruments with little compromise. The recent version of BAT includes features such as height and length measurements that facilitate embedded sensors to simplify tree height measurement and coarse woody debris length measurement. It provides users with an augmented reality user interface that guides them to complete the task (See Figure 6).

However, BAT requires user to input an estimate phone height (distance from ground to the phone) in order to achieve fast and simple procedure but still emulate the actual measurement instruments. Once the phone height is provided, BAT will first require user to measure distance to

Figure 6. Augmented reality distance and height measurement

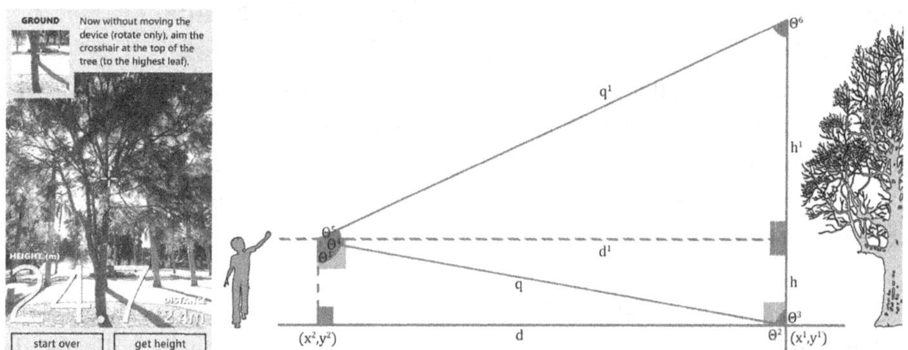

the tree by pointing the device at the bottom of the tree, and then pointing it at the top of tree to get an estimated tree height. The calculation is achieved by manipulating sensor readings and applying trigonometry as illustrated on Figure 6.

Benefits of the application of augmented reality and sensors integration are obvious in the case of BAT. With little compromise, users are now able to complete BioCondition survey with no additional hardware tools and prior learning. Furthermore, scientifically mobile sensing technology provides scientists and researchers with new approaches and instruments to conduct study. For example, marker-less augmented reality object tracking achieved significant advancement recently due to increased maturity of tracking sensors embedded in Microsoft Kinect (Oikonomidis et al., 2011).

The formal usability, accuracy, and effectiveness research studies of such features have not yet been conducted due to project scope.

FIELD EXPERIMENT

An experiment was conducted to test guidance, documentation, and usability aspects of BioCondition Assessment Tool (BAT). The experiment was conducted in an open forest in Samford Conservation Park with 12 students. All 12 students participated in the experiment did not have experience with BioCondition. They were students of Dr. Susan Fuller in Queensland University of Technology studying Bachelor of Ecology. In the experiment, two students were assigned as a group with one student performing the task using pen and paper and another student using the application. The assessment site was laid out by all students with supervision of Dr. Susan Fuller and her associate David Tucker. Each group was instructed to perform a different task, including tree, plant, coarse woody debris, ground cover and canopy cover.

The students were given a survey after the field trip asking the following groups of questions:

- Self-Assessment
    - As a beginner, how difficult is BioCondition?
    - Do you think BioCondition is time-consuming?
    - How often do you find yourself referring to the assessment manual?
    - How tech savvy are you?
- For Each Task
    - Was the task difficult to perform?
    - Was the task time consuming to perform?
    - Did you lose track when performing the task?

- ◦ Did the application help you in performing the task?
- Mobile Application
  - ◦ Will you prefer using the mobile application next time when you conduct a BioCondition assessment?

Result of Experiment

For self-assessment questions, all of the students thought BioCondition was easy to perform. None of them thought BioCondition was time-consuming. They did not find themselves refer to the assessment manual too much and majority of them are neutral to technology.

Interestingly, my expected outcome of this series of questions was tailored to BioCondition being difficult and time consuming because of repetitive tasks and constant needs to refer to assessment manual. The reason for such drastically different outcome is because the tasks were already broken down and instructions were clearly given to each group, therefore they completely understood what they needed to perform and thus there was no need to consult the assessment manual or the application. Secondly, each group only required to perform one task; all groups performed the assigned tasks simultaneously therefore they did not think BioCondition was time-consuming in general.

However, for questions designed specifically for each task, the result was quite different. In average, they thought the assigned task was repetitive and somewhat time consuming but easy to perform. Majority of the students agreed that the application was useful-to-very useful in assisting them. Compare to pen and paper approach, they also agreed that photo taking capability of the application was simple but very useful. Furthermore, all students that used the application also said that they will choose to use the application next time they need to conduct a BioCondition assessment.

Field Observation and Student Feedback

During the experiment, interviews were randomly conducted with the students about their thought of BioCondition in general and usefulness of the application. Feedback for the application was generally positive, specifically:

- Since each group performed different task, they had to transfer collected data to an assessment sheet. With the application there was no need to transfer data. Collected data can be transfer from device to device.
- Photo taking capability is implemented for each task. They thought it was very useful in the aspect that they could ask expert's advice in the future or have a peer discussion of the species that they could not identify on-site.
- Instead of sampling each plant, photo taking is easier for them to perform comparison.
- Pen and paper method uses tally approach to keep count of identified species. The students thought the application was more accurate due to photo evidence.

However, the students also pointed out a few negativities of the application. Majority of the negativities were due to unfamiliarity of Windows Phone operating system, some students were confused by the user interface at the beginning but they overcame the problem easily.

Observation in the field also yielded interesting results. One group of students were classifying tree species in the 100 × 50 m area, I followed them closely during the assessment and instead of only one student using the application, in the middle of performing the task both students had started to rely on the application because the student using pen and paper approach lost track of which tree species had been classified. Furthermore, navigation in the assessment area also proved to be an issue to beginner. Due to the reason that the assessment site was not set up properly, they also often lost track of the boundary or extent of the 100 × 50 m assessment area. One of the features of the application is mapping system that accurately shows the assessment area and also tracking position of the user. This feature was tested but failed due to the density of forest obstructed GPS signal. However, this feature is highly desirable and receives positive feedback from the participants because it saves time from having to constantly searching for a visual evidence of the boundary (See Figure 7).

FUTURE RESEARCH DIRECTIONS

Large-Scale User Study

The result of initial experiment is generally positive, however, our research can benefit from user study in a larger scale. In addition, formal study of the inclusion of height and length measurement features also needs to be conducted, particularly accuracy of the algorithm and sensor performance on multiple devices.

Due to time and project scope constraints, previous experiment lacks some considerations such as accuracy comparison with an expert and time spent compare to manual approach. We plan to include these measurements in the next experiment with larger user participation and different cohorts of users to understand issues for different type of users. A formal evaluation is currently being studied.

Furthermore, a big issue we have not addressed is how to keep users motivated. In order to achieve wider public adoptions, it would be interesting and beneficial to understand if the idea of 'gamification' could be used to keep users interested and motivated in the assessment.

Complete Survey Framework

BioCondition Assessment Tool (BAT) currently does not support advance BioCondition survey techniques, such as ecological layer differentiation (e.g. ecologically dominant layer and emergent) and post-assessment survey score calculation. A full transfer of the survey work will be included in the next release of BAT, including preparation such as site selection and an automated benchmark site comparison framework.

In addition, accurate species identification can greatly contribute to the study of species richness. BioCondition does not require user to identify species type, however the geo-tagged visual evidence can be used to conduct complex analysis, such as distribution network of species, and monitor

Figure 7. Participant taking photo of a tree and its leaf to determine its species family

changes of environment overtime. The survey framework will also include a complete server infrastructure to allow community participation, real-time result analysis, remote collaboration, and data sharing.

Crowd Sourcing and Data Analysis

Currently the research project is heading towards the direction of crowd sourcing. We plan to expand the opportunity to other mobile platforms, such as Apple iOS and Google Android, to increase the participation and awareness of BioCondition. However, crowd sourcing also attracts many challenges, such as maintaining data quality and application scalability. A formal study of adopting crowd sourcing, mitigations and approaches to deal with these challenges is needed. Furthermore, BAT encapsulates many additional environmental features using images. We can take advantage of the artefacts to increase and justify accuracy of the result. Furthermore, a suitable image processing framework can also be utilised to understand user's actions on-site and its relation to accuracy of the result.

Augmented Reality and 3D Visualisation

BAT can benefit from taking advantage of Augmented Reality to overlay environmental information on the screen. It has been proven that the immersive and interactivity of 3D visualisation approach is more beneficial in fieldwork because it constantly keeps users engaged at the tasks. As such, this new experience can substantially improve the learning ability of beginners. These will be explored in the future.

CONCLUSION

This chapter introduced BioCondition, a vegetation assessment framework developed by Queensland Department of Resource Management (DERM). Although the framework is designed to be simple and accurate, it has yet to be widely adopted by users. Some of the required resources can be found in modern mobile devices, such as GPS, compass and camera. Mobile technology has made significant advancement in the past decade; many researchers have started to exploit the use of mobile technology especially ecological science due to its capabilities and portability. We introduced a new mobile application—BioCondition Assessment Tool (BAT)—to aid beginners in conducting BioCondition survey in aspects such as providing context appropriate guidance and photo documenting the survey process. The experiment conducted at Samford Conservation Park with 12 students studying ecology in Queensland University of Technology yielded very positive result. Observations in the field and feedback from the students have proven that the application helped the students in providing guidance, simplifying the survey process, and improving accuracy by documenting the assessment.

ACKNOWLEDGMENT

We thank Queensland University of Technology, Microsoft QUT eResearch and Microsoft Research for their support and funding of this research project. We also thank Dr. Susan Fuller, Mr. David Tucker, and all participants in the experiment.

REFERENCES

Burke, J., Estrin, D., Hansen, M., Parker, A., Ramanathan, N., Reddy, S., et al. (2006). Participatory sensing. In *Proceedings of the Workshop on World-Sensor-Web: Mobile Device Centric Sensor Networks and Applications* (pp. 117-134). Boulder, CO: ACM.

Eyre, T. J., Kelly, A. L., Neldner, V. J., Wilson, B. A., Ferguson, D. J., Laidlaw, M. J., & Franks, A. J. (Eds.). (2011). *BioCondition: A condition assessment framework for terrestrial biodiversity in Queensland: Assessment manual*. Brisbane, Australia: Biodiveristy and Ecosystem Sciences.

Henze, N., Schinke, T., & Boll, S. (2009). What is that? Object recognition from natural features on a mobile phone. In *Proceedings of the Workshop on Mobile Interaction with the Real World.* Retrieved October 30, 2011 from http://mirw09.offis.de/paper/

International Telecommunication Union. (2011). *Key global telecom indicators for the world telecommunication service sector.* Retrieved July 6, 2012, from www.itu.int/ITU-D/ict/statistics/at_glance/KeyTelecom.html

Kanjo, M., Benford, S., Paxton, M., Camberlain, A., Fraser, D., & Woodgate, D. et al. (2003). MobGeoSen: Facilitating personal geosensor data collection and visualization using mobile phones. *Personal and Ubiquitous Computing*, *12*(8), 599–697. doi:10.1007/s00779-007-0180-1.

Keir, M., Hann, C., Chase, G., & Chen, X. (2007). A new approach to accelerometer-based head tracking for augmented reality & other applications. In M. Keir (Ed.), *IEEE International Conference on Automation Science and Engineering* (pp. 603-608). Scottsdale, AZ: IEEE Computer Press.

Kuznetsov, S., & Eric, P. (2010). *Participatory sensing in public spaces: Activating urban surfaces with sensor probes*. Paper presented at the 8th ACM Conference on Designing Interactive Systems. Aarhus, Denmark.

Lane, N. D., Miluzzo, E., Hong, L., Peebles, D., Choudhury, T., & Campbell, A. T. (2010). A survey of mobile phone sensing. *IEEE Communications Magazine*, *48*(9), 140–150. doi:10.1109/MCOM.2010.5560598.

Nurminen, A., Ernest, K., & Eduardo, V. (2011). *HYDROSIS: A mixed reality platform for on-site visualization of environmental data*. Paper presented at the 10th International Conference on Web and Wireless Geographical Information System. Kyoto, Japan.

Oikonomidis, I., Kyriazis, N., & Argyros, A. (2011). Efficient model-based 3D tracking of hand articulations using Kinect. In *Proceedings of the British Machine Vision Conference* (pp. 101.1-101.11). BMVA Press.

Pielot, M., Henze, N., Nickel, C., Menke, C., Samadi, S., & Boll, S. (2009). Evaluation of camera phone based interaction to access information related to posters. In *Mobile Interaction with the Real World* (pp. 61–72). Citeseer.

Qiang, L., Stankovic, J. A., Hanson, M. A., Barth, A. T., Lach, J., & Zhou, G. (2009). *Accurate, fast fall detection using gyroscopes and accelerometer-derived posture information*. Paper presented at the 6th International Workshop on Wearable and Implantable Body Sensor Networks. Berkeley, CA.

Thormählen, T., & Seidel, H. (2008). 3D-modeling by ortho-image generation from image sequences. *ACM Transactions on Graphics*, *27*(3), 1–5. doi:10.1145/1360612.1360685.

Wagner, D., Reitmayr, G., Mulloni, A., Drummond, T., & Schmalstieg, D. (2008). *Pose tracking from natural features on mobile phones*. Paper presented at the 7th IEEE/ACM International Symposium on Mixed and Augmented Reality. Cambridge, UK.

Wither, J., DiVerdi, S., & Höllerer, T. (2009). Annotation in outdoor augmented reality. *Computers & Graphics*, *33*(6), 679–689. doi:10.1016/j.cag.2009.06.001.

Ziegler, R., Wojciech, M., Hanspeter, P., & Leonard, M. (2003). *3D reconstruction using labeled image regions*. Paper presented at Eurographics Symposium on Geometry Processing. Aachen, Germany.

ADDITIONAL READING

Chambers, R. (2006). Participatory mapping and geographic information systems: Whose map? Who is empowered and who disempowered? Who gains and who loses? *The Electronic Journal of Information Systems in Developing Countries*, *25*(2), 1–11.

Foote, K. E., & Lynch, M. (2009). *Geographic information systems as an integrating technology: Context, concepts, and definitions*. Retrieved July 25, 2011 from http://www.colorado.edu/geography/gcraft/notes/intro/intro_f.html

Halpern, S. (2010). The iPad revolution. *New York Review of Books*, *57*(10), 1-3. Retrieved September 6, 2011, from http://www.nybooks.com/articles/archives/2010/jun/10/ipad-revolution/

Hu, S., Qiao, J., Zhang, A., & Huang, Q. (2008). *3D reconstruction from image sequence taken with a handheld camera*. Paper presented at the International Society for Photogrammetry and Remote Sensing. Beijing, China.

Keir, M., Hann, C., Chase, G., & Chen, X. (2007). A new approach to accelerometer-based head tracking for augmented reality & other applications. In M. Keir (Ed.), *Proceedings of the IEEE International Conference on Automation Science and Engineering* (pp. 603-608). Scottsdale, AZ: IEEE Computer Press.

Lane, N. D., Eisenman, S. B., Musolesi, M., Miluzzo, E., & Campbell, A. T. (2008). *Urban sensing systems: Opportunistic or participatory?* Paper presented at the 9th Workshop on Mobile Computing Systems and Applications. Napa Valley, CA.

Lenders, V., Koukoumidis, E., Zhang, P., & Martonosi, M. (2008). *Location-based trust for mobile user-generated content: Applications, challenges and implementations*. Paper presented at the 9th Workshop on Mobile Computing Systems and Applications. Napa Valley, CA.

Lutz, W., Sanderson, W., & Scherbov, S. (2008). The coming acceleration of global population ageing. *Nature*, *451*(7179), 716–719. doi:10.1038/nature06516 PMID:18204438.

Mohan, P., Padmanabhan, V. N., & Ramjee, R. (2008). *Nericell: Rich monitoring of road and traffic conditions using mobile smartphones*. Paper presented at the 6th ACM Conference on Embedded Network Sensor Systems. Raleigh, NC.

Mwanundu, S. (2010). *Participatory mapping and communication*. Retrieved September 8, 2011, from http://www.iapad.org/publications/ppgis/jon_corbett_ifad_pm-communication.pdf

Reddy, S., Samanta, V., Burke, J., Estrin, D., Hansen, M., & Srivastava, M. (2009). MobiSense—Mobile network services for coordinated participatory sensing. In *Proceedings of the International Symposium on Autonomous Decentralized Systems* (pp. 1-6). Athens, Greece: IEEE.

Reddy, S., Shilton, K., Burke, J., Estrin, D., Hansen, M., & Srivastava, M. B. (2008). Evaluating participation and performance in participatory sensing. In *Proceedings of the International Workshop on Urban, Community, and Social Applications of Networked Sensing Systems* (pp. 7-11). Raleigh, NC: IEEE.

St-Onge, B., Vega, C., Fournier, R. A., & Hu, Y. (2008). Mapping canopy height using a combination of digital stereo-photogrammetry and lidar. *International Journal of Remote Sensing*, *29*(11), 3343–3364. doi:10.1080/01431160701469040.

Thau, D., Morris, R. A., & White, S. (2009). Contemporary challenges in ambient data integration for biodiversity informatics. In *Proceedings of the International Workshops and Posters on the Move to Meaningful Internet Systems*. Vilamoura, Portugal: Springer-Verlag.

KEY TERMS AND DEFINITIONS

Application Programmable Interface (API): Application programmable interface is a specification intended to be followed by software components. It describes ways that a particular task can perform. Exposing APIs allows third party to build software system in a specifically intended way.

Augmented Reality (AR): Augmented reality is a term that describes the seamless integration of virtual objects into the real-world. For example Google Glass project where users see the real world through special glasses that provide users with virtual information.

BioCondition: BioCondition is a vegetation assessment framework developed by Queensland Department of Resource Management. It uses surrogate to measure the health or condition of an environment.

Citizen Science: Citizen science is a technique that employs public participation in research projects. The main purpose is to gather large amount of data that targets a specific scientific issue.

Crowd Sourcing: Similar to citizen science, crowd sourcing is a technique that outsources tasks, traditionally performed by experts, to a large group of people. The difference between citizen science and crowd sourcing is the latter is not necessary a scientific project.

Geotagging: Geotagging is the process of adding geographical information (e.g. latitude and longitude) to various digital media, such as photograph.

Participatory Sensing: Participatory sensing is a form of citizen science, it employs the community who is interested in the issue to participate in the data collection process, in which typically uses mobile phone as the instrument for collecting data.

Regional Ecosystem (RE): Regional ecosystems are vegetation communities that are consistently associated with a particular combination of geology, landform, and soil.

Compilation of References

000. *Webhost.* (1997). Retrieved October 2012 from http://www.000Webhost.com

A&D Medical. (n.d.). *A&D medical.* Retrieved March 5, 2011, from http://www.andonline.com

Ableiter, D. (2008). *Smart caching for efficient information sharing in distributed information systems.* (M.S. thesis). Naval Postgraduate School, Monterey, CA. Retrieved from http://www.dtic.mil

Abowd, G. D., Anind, D. K., Brown, P. J., Davies, N., Smith, M., & Steggles, P. (1999). Towards a better understanding of context and context-awareness. In *Proceedings of the 1st International Symposium on Handheld and Ubiquitous Computing* (pp. 304-307). London: ACM.

Abowd, G. D., Atkeson, C. D., Hong, J., Long, S., Kooper, R., & Pinkerton, M. (1997). Cyberguide: A mobile context-aware tour guide. *Wireless Networks, 3*(5), 421–433. doi:10.1023/A:1019194325861.

Abt, C. C. (1987). *Serious games.* University Press of America.

Abuan, L. A. (2009). *Information sharing for medical triage tasking during mass casualty/humanitarian operations.* (M.S. thesis). Naval Postgraduate School, Monterey, CA. Retrieved from http://www.dtic.mil

ActiHealth. (n.d.). Retrieved May 1, 2011, from http://www.actihealth.com

Activity Starter DatePicker. (2011). *Jefferson software, version 0.2.1.* Retrieved October 2012 from https://play.google.com/store/apps/details?id=com.jsoft.android.util

Adam, S., Henry, A. K., & Jeffrey, P. B. (2012). Finding your friends and following them to where you are. In *Proceedings of the Fifth ACM International Conference on Web Search and Data Mining* (pp. 723-732). ACM.

Adams, E. (2010). *Fundamentals of game design* (2nd ed.). Berkeley, CA: New Riders.

Adomavicius, G., Sankaranarayanan, R., Sen, S., & Tuzhilin, A. (2005). Incorporating contextual information in recommender systems using a multidimensional approach. *ACM Transactions on Information Systems, 23*(1), 103–145. doi:10.1145/1055709.1055714.

Adomavicius, G., & Tuzhilin, A. (2005). Toward the next generation of recommender systems: A survey of the state-of-the-art and possible extensions. *IEEE Transactions on Knowledge and Data Engineering, 17*(6), 734–749. doi:10.1109/TKDE.2005.99.

Agboma, F., & Liotta, A. (2006). *User centric assessment of mobile contents delivery.* Paper presented at 4th International Conferences on Advances in Mobile Computing and Multimedia. Yogyakarta, Indonesia.

Agboma, F., & Liotta, A. (2007). Addressing user expectations in mobile content delivery. *Mobile Information Systems, 3*(3-4), 153–164.

Agboma, F., & Liotta, A. (2012). Quality of experience management in mobile content delivery systems. *Telecommunication Systems, 49*(1), 85–98. doi:10.1007/s11235-010-9355-6.

AgileDelta. (2012). *AgileDelta technologies: Software and services for XML, web services and the mobile internet.* Retrieved June 19, 2012, from http://www.agiledelta.com

AgilePlatform. (2012). Retrieved October 2012 from http://www.outsystems.com/agile-platform/

Alferez, M., Santos, J., Moreira, A., Garcia, A., Kulesza, U., Araújo, J., & Amaral, V. (2009). Multi-view composition language for software product line requirements. In *Proceedings of the 2nd International Conference on Software Language Engineering (SLE'2009).* Denver, CO.

Alves, A. O., Rodrigues, F., & Pereira, F. C. (2011). Tagging space from information extraction and popularity of points of interest. In *Proceedings of the 2nd International Conference on Ambient Intelligence* (pp. 115-125). IEEE.

Alves, V., et al. (2005). Extracting and evolving mobile games product lines. In *Proceedings of SPLC'05* (LNCS), (vol. 3714, pp. 70-81). Berlin: Springer.

Alves, M., Pires, P., Delicato, F., & Campos, M. (2008). CrossMDA: A model-driven approach for aspect management. *Journal of Universal Computer Science*, *14*(8), 1314–1343.

American College of Radiology. (2002). *ACR standard for teleradiology*. Retrieved from http://imaging.stryker.com/images/ACR_Standards-Teleradiology.pdf

Anastasios, N., Salvatore, S., Cecilia, M., & Massimiliano, P. (2011). An empirical study of geographic user activity patterns in foursquare. In *Proceedings of the 5th International AAAI Conference on Weblogs and Social Media* (pp. 570-573). AAAI.

Angelides, M. C. (2003). Multimedia content modeling and personalization. *IEEE MultiMedia*, *10*(4), 12–15. doi:10.1109/MMUL.2003.1237546.

Anquetil, N., Kulesza, U., Mitschke, R., Moreira, A., Royer, J., Rummler, A., & Sousa, A. (2009). A model-driven traceability framework for software product lines. *Software & Systems Modeling*.

Anthonysamy, P., & Somé, S. S. (2008). Aspect-oriented use case modeling for software product lines. In *Proceedings of the AOSD Workshop on Early Aspects*, (pp. 1-8). New York, NY: ACM.

Antin, J., & Churchill, E. (2011). Badges in social media: A social psychological perspective. In *Proceedings of the 29th ACM SIGCHI International Conference on Human-Computer Interaction*. Vancouver, Canada: ACM.

App. Inventor. (2010). Retrieved October 2012 from http://www.appinventor.mit.edu/

Ariadne GPS: Mobility and Map Exploration for All. (n.d.), Retrieved from http://www.ariadnegps.eu

Aron, A., Aron, E., & Coups, E. J. (Eds.). (2009). *Statistics for psychology* (6th ed.). Harlow, UK: Pearson Education.

Aspect, J. Team. (2009). *The AspectJ programming guide*. Retrieved from http://eclipse.org/aspectj/

Atwell, J. (2005). *Mobile technologies and learning: A technology update and m-learning project summary* (Tech. Rep. No. 1.). London, UK: Technology Enhanced Learning Research Center, Learning and Skills Development Agency.

Backs, R. W., & Walrath, L. C. (1992). Eye movement and pupillary response indices of mental workload during visual search of symbolic displays. *Applied Ergonomics*, *23*(4), 243–254. doi:http://dx.doi.org/10.1016/0003-6870(92)90152-L

Baldauf, M., Dustdar, S., & Rosenberg, F. (2007). A survey on context-aware systems. *International Journal of Ad Hoc and Ubiquitous Computing*, *2*(4), 263–277. doi:10.1504/IJAHUC.2007.014070.

Baltrunas, L., Ludwig, B., Peer, S., & Ricci, F. (2012). Context relevance assessment and exploitation in mobile recommender systems. *Personal and Ubiquitous Computing*, *16*(5), 507–526. doi:10.1007/s00779-011-0417-x.

Bång, M., Svahn, M., & Gustafsson, A. (2009). Persuasive design of a mobile energy conservation game with direct feedback and social cues. In *Proceedings of the 3rd International Conference of the Digital Games Research Association*. West London, UK: ACM Press.

Barragáns-Martinez, A. B., Rey-López, M., Costa-Montenegro, E., Mikic-Fonte, F. A., Burguillo, J. C., & Peleteiro, A. (2010). Exploiting social tagging in a web 2.0 recommender system. *IEEE Internet Computing*, *14*(6), 23–30. doi:10.1109/MIC.2010.104.

Bashashati, A., Fatourechi, M., Ward, R. K., & Birch, G. E. (2007). A survey of signal processing algorithms in brain–computer interfaces based on electrical brain signals. *Journal of Neural Engineering*, *4*(2), R32–R57. doi:10.1088/1741-2560/4/2/R03 PMID:17409474.

Bazeley, P. (2004). Issues in mixing qualitative and quantitative approaches to research. In Buber, R., Gadner, J., & Richards, L. (Eds.), *Applying qualitative methods to marketing management research* (pp. 141–156). Basingstoke, UK: Palgrave Macmillan.

Bellotti, F., Berta, R., DeGloria, A., & Margarone, M. (2006). Guiding visually impaired people in the exhibition. In *Proceedings of the Virtuality, Mobile Guide Workshop*. Retrieved February 26, 2011, from http://hcilab.uniud.it/sigchi/doc/Virtuality06/Bellotti&al.pdf

Berenson, R. A., Grossman, J. M., & November, E. A. (2009). Does telemonitoring of patients--the eICU--improve intensive care? *Health Affairs*, *28*(5), 937–947. doi:10.1377/hlthaff.28.5.w937.

Big Buck Bunny. (2008). Retrieved from http://www.bigbuckbunny.org/

Biotronik. (n.d.). *Biotronik home monitoring*. Retrieved March 7, 2011, from http://www.biotronik.com/biohm/home

Bluetooth SIG. (n.d.), Retrieved from http://www.bluetooth.com/Pages/Low-Energy.aspx

Bluetooth, S. I. G. (n.d.). *The official bluetooth® technology web site*. Retrieved April 18, 2011, from http://www.bluetooth.com

Blythe, M. A., Overbeeke, K., Monk, A. F., & Wright, P. C. (2004). *Funology: From usability to enjoyment*. London: Springer.

Breslow, M. J. (2005). *Building a better delivery system: The eICU solution: A technology-enabled care paradigm for ICU performance*. Washington, DC: National Academies Press.

Brower, V. (2005). When mind meets machine. *EMBO reports, 6*(2), 108–110. doi:http://dx.doi.org/10.1038/sj.embor.7400344

Brown, K., & Ellis, M. (2004, January 30). *Best practices for web services versioning*. Retrieved August 6, 2011, from http://www.ibm.com/developerworks/Webservices/library/ws-version/

BrowseAloud Software. (n.d.), Retrieved from http://www.browsealoud.com

Buchanan, R. (1992). Wicked problems in design thinking. *Design Issues*, *8*(2), 5–21. doi:10.2307/1511637.

Buchinger, S., Kriglstein, S., & Hlavacs, H. (2009). *A comprehensive view on user studies: Survey and open issues for mobile TV*. Paper presented at 7th European Conference on Interactive Television EuroITV'09. Leuven, Belgium.

Buchinger, S., Kriglstein, S., Brandt, S., & Hlavacs, H. (2011). A survey on user studies and technical aspects of mobile multimedia applications. *Entertainment Computing*, *2*(3), 175–190. doi:10.1016/j.entcom.2011.02.001.

Burger, A. (2012). *Report: Mobile video growing, fueling demand for multiscreen options*. Retrieved July 22, 2012, from http://www.telecompetitor.com/report-mobile-video-growing-fueling-demand-for-multiscreen-options/

Burke, J., Estrin, D., Hansen, M., Parker, A., Ramanathan, N., Reddy, S., et al. (2006). Participatory sensing. In *Proceedings of the Workshop on World-Sensor-Web: Mobile Device Centric Sensor Networks and Applications* (pp. 117-134). Boulder, CO: ACM.

Butek, R. (2005). *Which style of WSDL should I use?* Retrieved May 1, 2011, from https://www.ibm.com/developerworks/Webservices/library/ws-whichwsdl/

Buur, J., & Soendergaard, A. (2000). Video card game: an augmented environment for user centred design discussions. In *Proceedings of DARE 2000 on Designing Augmented Reality* (pp. 63-69). New York: ACM.

Buxton, B., & Greenberg, S. (2009). Usability evaluation considered harmful (some of the time). In *Proceedings of the Twenty-Sixth Annual SIGCHI Conference on Human Factors in Computing Systems –CHI 2009* (pp. 111-120). New York: ACM.

Campbell, A. T., Choudhury, T., Hu, S., Lu, H., Mukerjee, M. K., Rabbi, M., et al. (2010). NeuroPhone: Brain-mobile phone interface using a wireless EEG headset. In *Proceedings of the Second ACM SIGCOMM Workshop on Networking, Systems, and Applications on Mobile Handhelds* (pp. 3–8). New York, NY: ACM. doi:10.1145/1851322.1851326

Canadian Association of Radiologists. (2008). *CAR standards for irreversible compression in digital diagnostic imaging within radiology*. Retrieved from http://www.car.ca/uploads/standards%20guidelines/Standard_Lossy_Compression_EN.pdf

Carlsson, C., & Walden, P. (2007). *Mobile TV-to live or die by content*. Paper presented at 40th Annual Hawaii International Conference on System Sciences. Hawaii, HI.

Carlsson, C., Carlsson, J., Puhakainen, J., & Walden, P. (2006). *Nice mobile services do not fly: Observations of mobile services and the Finnish consumers*. Paper presented at 19th Bled eCommerce Conference. Bled, Slovenia.

CDMA Development Group. (n.d.). *CDG*. Retrieved June 19, 2012, from http://www.cdg.org

Cha, H. J., Kim, Y. S., Park, S. H., & Yoon, T. B. (2006). Learning styles diagnosis based on user interface behaviours for the customisation of learning interfaces in an intelligent tutoring system. *Lecture Notes in Computer Science*, *4053*, 513–524. doi:10.1007/11774303_51.

Chandon, P., Hutchenson, J. W., Ardlow, E. T., & Young, S. H. (2007). *Measuring value point of purchase marketing with commercial eye tracking data* (Research Paper No.2007/22/MKT/ACGRD). Paris, France: INSEAD Business School Global Research and Development Alliance.

Chipchase, J., Yanqing, C., & Jung, Y. (2006). *Personal television: A qualitative study of mobile TV users in South Korea*. Paper presented at Mobile Human Computer Interaction 2006. Espoo, Finland.

Chiu, M.-C., Chang, S.-P., Chang, Y.-C., Chu, H.-H., Chen, C. C.-H., Hsiao, F.-H., & Ko, J.-C. (2009). Playful bottle: A mobile social persuasion system to motivate healthy water intake. In *Proceedings of the 11th ACM International Conference on Ubiquitous Computing* (pp. 185-194). Orlando, FL: ACM.

Chowdhury, A. R., Falchuk, B., & Misra, A. (2010). MediAlly: A provenance-aware remote health monitoring middleware. In *Proceedings of the IEEE International Conference on Pervasive Computing and Communications* (pp. 125-134). IEEE.

Clements, P., & Northrop, L. (2001). *Software product lines: Practices and patterns*. Reading, MA: Addison-Wesley Professional.

Clotfelter, C. T., & Towl, J. E. (2007). *Twiddlenet: Metadata tagging and data dissemination in mobile device networks*. (M.S. thesis). Naval Postgraduate School, Monterey, CA. Retrieved from http://www.dtic.mil

Coan, J. A., & Allen, J. J. B. (2004). Frontal EEG asymmetry as a moderator and mediator of emotion. *Biological Psychology*, *67*(1-2), 7–49. doi:10.1016/j.biopsycho.2004.03.002 PMID:15130524.

Cockburn, A., & Mckenzie, B. (2001). What do web users do? An empirical analysis of web use. *International Journal of Human-Computer Studies*, *54*(6), 903–922. Retrieved from http://www.sciencedirect.com/science/article/pii/S1071581901904598 doi:10.1006/ijhc.2001.0459.

Cohen, J. (1960). A coefficient of agreement for nominal scales. *Educational and Psychological Measurement*, *20*(1), 37–46. doi:10.1177/001316446002000104.

Continua Health Alliance. (2010). *Homepage*. Retrieved June 27, 2012, from http://www.continuaalliance.org

Costabile, M. F., De Angeli, A., Lanzilotti, R., Ardito, C., Buono, P., & Pederson, T. (2008). *Explore! Possibilities and challenges of mobile learning*. Paper presented at the Twenty-Sixth Annual SIGCHI Conference on Human Factors in Computing Systems. Florence, Italy.

Coughlan, J., & Manduchi, R. (2007). *Functional assessment of a camera phone-based wayfinding system operated by blind and visually-impaired users*. Paper presented at the IEEE-BAIS Symposium on Research on Assistive Technology. Dayton, OH.

Coyle, M., Freyne, J., Brusilovsky, P., & Smyth, B. (2008). *Social information access for the rest of us: An exploration of social*. Springer. doi:10.1007/978-3-540-70987-9_12.

Creswell, J. W. (2003). *Research design: Qualitative, quantitative, and mixed method approaches*. Thousand Oaks, CA: Sage Publications.

Cristiana, B., Carlo, C., Elisa, Q., Fabio, A. S., & Letizia, T. (2007). A data-oriented survey of context models. *SIGMOD Record*, *36*(4), 19–26. doi:10.1145/1361348.1361353.

Crowley, K., Sliney, A., Pitt, I., & Murphy, D. (2010). Evaluating a brain-computer interface to categorise human emotional response. In Proceedings of Advanced Learning Technologies (ICALT), (pp. 276–278). IEEE. doi: doi:10.1109/ICALT.2010.81.

Crowley, K., Sliney, A., Pitt, I., & Murphy, D. (2011). Capturing and using emotion-based BCI signals in experiments: How subject's effort can influence results. In *British Computer Society Human Computer Interaction* (pp. 1–6). Swinton, UK: British Computer Society. doi: http://dl.acm.org/citation.cfm?id=2305316.2305341

Cummings, J., Krsek, C., Vermoch, K., & Matuszewski, K., & University HealthSystem Consortium ICU Telemedicine Task Force. (2007). Intensive care unit telemedicine: Review and consensus recommendations. *American Journal of Medical Quality*, *22*(4), 239–250. doi:10.1177/1062860607302777 PMID:17656728.

Czarnecki, K., & Eisenecker, W. (2000). *Generative programming: Methods, tools, and applications*. Reading, MA: Addison-Wesley Publishing.

D'Atri, E., Medaglia, C. M., Serbanati, A., & Ceipidor, U. (2007). *A system to aid blind people in mobility: A usability test and results*. Paper presented at the Second International Conference on Systems (ICONS). Sainte-Luce, Martinique.

Dağtas, S., Pekhteryev, G., Sahinoğlu, Z., Çam, H., & Challa, N. (2008). Real-time and secure wireless health monitoring. *International Journal of Telemedicine and Applications*, (1): 1–10. doi:10.1155/2008/135808 PMID:18497866.

De Moor, K., Ketyko, I., Joseph, W., Deryckere, T., De Marez, L., Martens, L., & Verleye, G. (2010). Proposed framework for evaluating quality of experience in a mobile, testbed-oriented living lab setting. *Mobile Networks and Applications*, *15*(3), 378–391. doi:10.1007/s11036-010-0223-0.

de Oliveira, R., Cherubini, M., & Oliver, N. (2010). MoviPill: Improving medication compliance for elders using a mobile persuasive social game. In *Proceedings of the 12th ACM International Conference on Ubiquitous Computing*, (pp. 251–260). Copenhagen, Denmark: ACM.

Dellman, D. (2000). *Mail and Internet survey: The tailored design method*. New York: John Wiley & Sons Ltd..

Denso-Wave. (1994). *Creator of QRcode*. Retrieved October 2012 from http://www.densowave.com/qrcode/aboutqr-e.html

Denzin, N. K., & Lincoln, Y. S. (Eds.). (2000). *Handbook of qualitative research*. Thousand Oaks, CA: Sage Publications.

DeOliveira, A., & Amazonas, A. (2008). *Proposal of an adaptive mlearning architecture on the characteristics of mobile devices, to learning styles, to the performance and knowledge of students*. Paper presented at the International Technology, Education, and Development Conference (INTED). Valencia, Spain.

Department of Health and Human Services. (2002). *Standards for privacy of individually identifiable health information, final rule*. Retrieved from http://www.hhs.gov/ocr/privacy/hipaa/administrative/privacyrule/privrulepd.pdf

Department of Health and Human Services. (2003). *Health insurance reform: Security standards, final rule*. Retrieved from http://www.hhs.gov/ocr/privacy/hipaa/administrative/securityrule/securityrulepdf.pdf

Deterding, S., Dixon, D., Khaled, R., & Nacke, L. (2011). From game design elements to gamefulness: Defining gamification. In *Proceedings of the 15th International Academic MindTrek Conference: Envisioning Future Media Environments*, (pp. 9–15). Tampere, Finland: ACM.

Dey, I. (Ed.). (1993). *Qualitative data analysis - A user-friendly guide for social scientists*. New York, NY: Routledge. doi:10.4324/9780203412497.

Dick, T. J. (1979). Mixing qualitative and quantitative methods: Triangulation in action. *Qualitative Methodology*, 602-611.

DigiSoft. (n.d.). *DigiSoft: Innovative software solutions*. Retrieved August 8, 2011, from http://www.digisoftdirect.com

Dixon, C. S., & Morrisson, R. (2008). A pseudolite-based maritime navigation system: Concept through to demonstration. *Journal of Global Positioning Systems*, *7*(1), 9–17. doi:10.5081/jgps.7.1.9.

Dokovsky, N., van Halteren, A., & Widya, I. (2004). *BANip: Enabling remote healthcare monitoring with body area networks. Scientific Engineering of Distributed Java Applications*. Berlin, Germany: Springer.

Doukas, C., Maglogiannis, I., Anagnostopoulos, I., & Peraki, K. (2007). A context-aware telemedicine platform for monitoring patients in remote areas. *The Journal on Information Technology in Healthcare*, *5*(4), 255–262.

Draper, S. (1999). Analysing fun as a candidate software requirement. *Personal Technologies*, *3*(3), 117–122. doi:10.1007/BF01305336.

Dumas, B., Lalanne, D., & Oviatt, S. (2009). Multimodal interfaces: a survey of principles, models and frameworks. In Lalanne, D., & Kohlas, J. (Eds.), *Human machine interaction (LNCS)* (*Vol. 5440*, pp. 3–26). Berlin: Springer. doi:10.1007/978-3-642-00437-7_1.

EAN13. (1973). *George J. Laurer, ISO/IEC 15420:2000 - Information technology – Automatic identification and data capture techniques - Bar code symbology*. EAN/UPC.

Edmunds, H. (Ed.). (1999). *The focus group research handbook*. Chicago, IL: NTC/Contemporary Publishing Group, Inc..

Eggen, R., & Sunku, S. (2003). Efficiency of SOAP versus JMS. In *Proceedings of the International Conference on Internet Computing* (pp. 99-105). IEEE.

Elgazzar, K., Aboelfotoh, M., Martin, P., & Hassanein, H. S. (2012). Ubiquitous health monitoring using mobile web services. In *Proceedings of The 3rd International Conference on Ambient Systems, Networks and Technologies (ANT)*. Niagara Falls, Canada: ANT.

EN 13606 Association. (n.d.). *The CEN/ISO 13606 association site*. Retrieved April 21, 2011, from http://www.en13606.org

Endrei, M., Gaon, M., Graham, J., Hogg, K., & Mulholland, N. (2006, May 1). *Moving forward with web services backward compatibility*. Retrieved August 6, 2011, from http://www.ibm.com/developerworks/java/library/ws-soa-backcomp/

Eronen, L. (2001). *Combining quantitative and qualitative data in user research on digital television*. Paper presented at 1st Panhallenic Conference on Human Computer Intteraction. Patras, Greece.

Eronen, L. (2006). Five qualitative research methods to make iTV applications universally accessible. *Universal Access in the Information Society*, *5*(2), 219–238. doi:10.1007/s10209-006-0031-2.

ETSI. (2011). *ETSI medical*. Retrieved May 1, 2011, from http://www.etsi.org/WebSite/technologies/Medical.aspx

Eunjoon, C., Seth, A. M., & Jure, L. (2011). Friendship and mobility: user movement in location-based social networks. In *Proceedings of the 17th ACM SIGKDD International Conference on Knowledge Discovery and Data mining* (pp. 1082-1090). ACM.

Eyre, T. J., Kelly, A. L., Neldner, V. J., Wilson, B. A., Ferguson, D. J., Laidlaw, M. J., & Franks, A. J. (Eds.). (2011). *BioCondition: A condition assessment framework for terrestrial biodiversity in Queensland: Assessment manual*. Brisbane, Australia: Biodiveristy and Ecosystem Sciences.

FA0. (2011). *Global food losses and food waste*. Düsseldorf, Germany: Interpack2011. Retrieved October 2012 from http://www.fao.org/docrep/014/mb060e/mb060e00.pdf

Fabrizio, S. (2002). Machine learning in automated text categorization. *ACM Computing Surveys*, *34*(1), 1–47. doi:10.1145/505282.505283.

Felder, R. M., & Solomon, B. A. (n.d.). *Index of learner styles*. Retrieved from http://www.engr.ncsu.edu/learningstyles/ilsWeb.html

Felder, R. M., & Silverman, L. K. (1988). Learning and teaching styles in engineering education. *English Education*, *78*(7), 674–681.

Felder, R. M., & Spurlin, J. (2005). Applications, reliability, and validity of the index of learning styles. *International Journal of Engineering*, *21*(1), 103–112.

Fielding, R. T. (2000). *Architectural styles and the design of network-based software architectures*. (Doctoral Dissertation). Retrieved from http://www.ics.uci.edu/~fielding/pubs/dissertation/top.htm

Figueiredo, E., et al. (2008). Evolving software product lines with aspects: An empirical study on design stability. In *Proceedings of the 30th ICSE'08*, (pp. 261-270). ICSE.

Flatla, D. R., Gutwin, C., Nacke, L. E., Bateman, S., & Mandryk, R. L. (2011). Calibration games: Making calibration tasks enjoyable by adding motivating game elements. In *Proceedings of the 24th Annual ACM Symposium on User Interface Software and Technology*, (pp. 403–412). Santa Barbara, CA: ACM.

Floerkemeier, C., & Lampe, M. (2004). Issues with RFID usage in ubiquitous computing applications. *Lecture Notes in Computer Science*, *3001*, 188–193. doi:10.1007/978-3-540-24646-6_13.

Frank, N. Magid Associates Inc. (2009). *The OMVC mobile TV study: Live, local programming will drive demand for mobile TV*. Retrieved July 22, 2010, from http://mobiletvworld.com/documents/OMVC%20Mobile%20TV%20Study%20December%202009.pdf

Froehlich, J., Chen, M., Consolvo, S., Harrison, B., & Landay, J. (2007). *My experience: A system for in situ tracing and capturing of user feedback on mobile phones.* Paper presented at the International Conference on Mobile Systems, Applications, and Services 2007. San Juan, Puerto Rico.

Fujiki, Y., Kazakos, K., Puri, C., Pavlidis, I., Starren, J., & Levine, J. (2007). NEAT-o-games: Ubiquitous activity-based gaming. In *Proceedings of the 25th ACM SIGCHI International Conference on Human-Computer Interaction*, (pp. 2369–2374). San Jose, CA: ACM.

GamePlay Talk. (n.d.). Retrieved from http://www.gesturetekmobile.com/pressreleases/press_mar312008.php

Gediminas, A., & Alexander, T. (2005). Toward the next generation of recommender systems: A survey of the state-of-the-art and possible extensions. *IEEE Transactions on Knowledge and Data Engineering*, *17*(6), 734–749. doi:10.1109/TKDE.2005.99.

Gerst, M., Bunduchi, R., & Graham, I. (2005). Current Issues with RFID standardization. *Interop*. Retrieved January 2012, from http://www.york.ac.uk/res/e-society/projects/24/interop2005.pdf

Goecks, J., & Shavlik, J. (2000). Learning users' interests by unobtrusively observing their normal behavior. In *Proceedings of the 5th International Conference on Intelligent User Interfaces (IUI '00)*. ACM. DOI=10.1145/325737.325806http://doi.acm.org/10.1145/325737.325806

Gomaa, H. (2004). *Designing software product lines with UML: From use cases to pattern-based software architectures.* Redwood City, CA: Addison Wesley Longman Publishing.

Goni, A., Burgos, A., Dranca, L., Rodriguez, J., Illarramendi, A., & Bermudez, J. (2009). Architecture, cost-model and customization of real-time monitoring systems based on mobile biological sensor data-streams. *Computer Methods and Programs in Biomedicine*, *96*(2), 141–157. doi:10.1016/j.cmpb.2009.04.010 PMID:19481289.

Gonzalez, M. C., Hidalgo, C. A., & Barabasi, A. (2008). Understanding individual human mobility patterns. *Nature*, *453*(7196), 779–782. doi:10.1038/nature06958 PMID:18528393.

Gordon, V. S., & Bieman, J. M. (1995). Rapid prototyping: lessons learned. *IEEE Software*, *12*(1), 85–95. doi:10.1109/52.363162.

Graf, S. (2007). *Adaptivity in learning management systems focusing on learning styles.* (Unpublished Doctoral Dissertation). Vienna University of Technology, Vienna, Austria.

Graf, S., & Kinshuk. (2006). *An approach for detecting learning styles in learning management systems.* Paper presented at the Sixth International Conference on Advanced Learning Technologies. Kerkrade, The Netherlands.

Graf, S., & Liu, T. C., Kinshuk, Chen, N.S., & Yang, S.J.H. (2009). Learning styles and cognitive traits – Their relationship and its benefits in web-based educational systems. *Computers in Human Behavior*, *25*(1), 1280–1289. doi:10.1016/j.chb.2009.06.005.

Greenfield, J., Short, K., Cook, S., & Kent, S. (2004). *Software factories: Assembling applications with patterns, models, frameworks, and tools.* New York: Wiley.

Griswold, W. et al. (2006). Modular software design with crosscutting interfaces. *IEEE Software*, *23*(1), 51–60. doi:10.1109/MS.2006.24.

Grudin, J. (1994b). Groupware and social dynamics: Eight challenges for developers. *Communications of the ACM*, *37*(1), 92–105. doi:10.1145/175222.175230.

Guan, Z., Wang, C., Bu, J., Chen, C., Yang, K., Cai, D., & He, X. (2010). Document recommendation in social tagging services. In *Proceedings of the Nineteenth International Conference on World Wide Web* (pp. 391-400). IEEE.

Guindon, R. (1990). Designing the design process: exploiting opportunistic thoughts. *Human-Computer Interaction*, *5*(2), 305–344. doi:10.1207/s15327051hci0502&3_6.

Guo, J. F., Xu, G., Cheng, X. Q., & Li, H. (2009). Named entity recognition in query. In *Proceedings of the 32th Annual International ACM SIGIR Conference* (pp. 267-274). ACM.

Gustafsson, A., Katzeff, C., & Bang, M. (2009). Evaluation of a pervasive game for domestic energy engagement among teenagers. *Computers in Entertainment*, *7*(4), 1–19. doi:10.1145/1658866.1658873.

Hannemann, J., & Kiczales, G. (2002). Design pattern implementation in Java and aspectJ. In *Proceedings of the 17th ACM SIGPLAN Conference on Object-Oriented Programming, Systems, Languages, and Applications*, (pp. 161-173). New York, NY: ACM.

Hassenzahl, M., & Tractinsky, N. (2006). User experience - A research agenda. *Behaviour & Information Technology, 25*(2), 91–97. doi:10.1080/01449290500330331.

Hautefeuille, M., O'Mahony, C., O'Flynn, B., Khalil, K., & Peters, F. (2008). A MEMS-based wireless multisensor module for environmental monitoring. *Microelectronics and Reliability, 48*(1), 906–910. doi:10.1016/j.microrel.2008.03.007.

Hays, C., Avesani, P., & Veeramachaneni, S. (2007). An analysis of the use of tags in a blog recommender system. In *Proceedings of International Joint Conference on Artificial Intelligence* (pp. 2772-2777). IEEE.

Health Level Seven International. (2012). *Homepage.* Retrieved June 27, 2012, from http://www.hl7.org/

Helal, A., Moore, S., & Ramachandran, B. (2001). *Drishti: An integrated navigation system for visually impaired and disabled.* Paper presented at the Fifth International Symposium on Wearable Computers. Zurich, Switzerland.

Henze, N., Schinke, T., & Boll, S. (2009). What is that? Object recognition from natural features on a mobile phone. In *Proceedings of the Workshop on Mobile Interaction with the Real World.* Retrieved October 30, 2011 from http://mirw09.offis.de/paper/

Henze, N., & Boll, S. (2010). Push the study to the app. store: Evaluating on-screen visualizations for maps in the android market. In *Proceedings of MobileHCI '10: Human-Computer Interaction with Mobile Devices and Services* (pp. 373–374). Lisbon, Portugal: ACM. doi:10.1145/1851600.1851671.

Heyer, C., & Brereton, M. (2008). Reflective agile iterative design. In *Proceedings of the Social Interaction with Mundane Technologies Conference.* Brisbane, Australia: QUT.

Heyer, C., & Brereton, M. (2010). Design from the everyday: continuously evolving, embedded exploratory prototypes. In *Proceedings of 8th ACM Conference on Designing Interactive Systems* (pp. 1-10). Aarhus, Germany: ACM.

Hix, D., & Hartson, R. H. (1993). *Developing user interfaces: Ensuring usability through product & process.* New York: John Wiley & Sons.

Holst, O. R. (Ed.). (1969). *Content analysis for the social sciences and humanities.* Reading, MA: Addison-Wesley.

Horlings, R., Datcu, D., & Rothkrantz, L. J. M. (2008). Emotion recognition using brain activity. In *Proceedings of the 9th International Conference on Computer Systems and Technologies and Workshop for PhD Students in Computing* (pp. 6:II.1–6:1). New York, NY: ACM. doi:10.1145/1500879.1500888

Huber, J., Steimle, J., & Mühlhäuser, M. (2010). *Toward more efficient user interfaces for mobile video browsing: An in-depth exploration of the design space.* Paper presented at 16th international conference on Multimedia. Firenze, Italy.

Humanware. (n.d.). Retrieved from http://www.humanware.com/en-usa/products/blindness/talking_gps/trekker/_details/id_88/trekker_talking_gps.html

Hung, K., & Zhang, Y. (2003). Implementation of a WAP-based telemedicine system for patient monitoring. *Transactions on Information Technology in Biomedicine, 7*(2), 101–107. doi:10.1109/TITB.2003.811870 PMID:12834165.

Hunicke, R., LeBlanc, M., & Zubek, R. (2004). MDA: A formal approach to game design and game research. In *Proceedings of the AAAI Workshop on Challenges in Game AI* (pp. 04–04). AAAI.

Hutchinson, H., Mackay, W., Westerlund, B., Bederson, B. B., Druin, A., & Plaisant, C. Sundblad, Y. (2003). Technology probes: Inspiring design for and with families. In *Proceedings of the ACM CHI 2003 Human Factors in Computing Systems Conference* (pp. 17-24). New York: ACM.

IEEE. (2007). *1451.0-2007 - IEEE standard for a smart transducer interface for sensors and actuators - Common functions, communication protocols, and transducer electronic data sheet (TEDS) formats.* Retrieved May 31, 2011, from http://standards.ieee.org/findstds/standard/1451.0-2007.html

IEEE-SA. (2011). *The IEEE standards association.* Retrieved June 27, 2012, from http://standards.ieee.org/

Indrajit, I., & Verma, B. (2007). DICOM, HL7 and IHE: A basic primer on healthcare standards for radiologists. *Indian Journal of Radiology and Imaging*, *17*(2), 66–68. doi:10.4103/0971-3026.33610.

Indulska, J., & Sutton, P. (2003). Location management in pervasive systems. In *Proceedings of the Australasian Information Security Workshop Conference on ACSW Frontiers 2003* (vol. 21, pp. 143–151). Darlinghurst, Australia: Australian Computer Society, Inc.

International Telecommunication Union. (2011). *Key global telecom indicators for the world telecommunication service sector*. Retrieved July 6, 2012, from www.itu.int/ITU-D/ict/statistics/at_glance/KeyTelecom.html

International, I. H. E. (2011). *IHE.net home*. Retrieved June 27, 2012, from http://www.ihe.net/

Internet Stroke Center. (2011). *Radiology image library | internet stroke center*. Retrieved April 17, 2011, from http://www.strokecenter.org/radiology/

ITU. (n.d.). *What really is a third generation (3G) mobile technology*. Retrieved April 28, 2011, from http://www.itu.int/ITU-D/tech/FORMER_PAGE_IMT2000/DocumentsIMT2000/What_really_3G.pdf

ITU-R. (2004). *Methodology for the subjective assessment of quality for television pictures*. Recommendation BT. 500-11. Retrieved from http://www.itu.int/rec/R-REC-BT.500/en

ITU-T. (1999). *Subjective video quality assessment methods for multimedia applications*. Retrieved from http://www.itu.int/md/T01-SG09-040510-D-0108/en

ITU-T. (2001). *Tutorial - Objective perceptual assessment of video quality*. Retrieved from http://www.itu.int/ITU-T/studygroups/com09/docs/tutorial_opavc.pdf

Ivetic, D., & Dragan, D. (2009). Medical image on the go! *Journal of Medical Systems*, *35*(4), 499–516. doi:10.1007/s10916-009-9386-2 PMID:20703540.

Iwasa, K. (2004). *Web services reliable messaging (WS-reliability 1.1)*. Retrieved from http://docs.oasis-open.org/wsrm/ws-reliability/v1.1/wsrm-ws_reliability-1.1-spec-os.pdf

Jacob, R. J. K., & Karn, K. S. (2003). Eye tracking in human – computer interaction and usability research : Ready to deliver the promises. In Hyona, J., Radach, R., & Deubel, H. (Eds.), *The Mind's Eye: Cognitive and Applied Aspects of Eye Movement Research* (pp. 573–603). Academic Press.

Jacobson, I., & Ng, P.-W. (2004). *Aspect-oriented software development with use cases*. Reading, MA: Addison-Wesley Professional.

James, D., Benjamin, L., Liu, J., Palash, N., Taylor, V. V., & Ullas, G. Dasarathi, S. (2010). The YouTube video recommendation system. In *Proceedings of the 4th ACM Conference on Recommender Systems* (pp. 293-296). ACM.

Janssen, G., & De Mey, H. (2010). Celeration of executive functioning while solving the tower of Hanoi: Two single case studies using protocol analysis. *International Journal of Psychology & Psychological Therapy*, *10*(1), 19–40.

Jassal, S., Brissenden, J., Raisbeck, A., & Roscoe, J. (1998). Comparative cost-analysis of two different chronic care facilities for end-stage renal disease patients. *Geriatric Nephrology and Urology*, *8*(2), 69–76. doi:10.1023/A:1008378422292 PMID:9893214.

JBOSS AOP. (2009). Retrieved from http://www.jboss.org/community/docs/DOC-10201

Jose, B., & Mark, L. (2010). A comparison of scoring metrics for predicting the next navigation step with Markov model-based systems. *International Journal of Information Technology and Decision Making*, *9*(4), 547–573. doi:10.1142/S0219622010003956.

JPEG. (n.d.). *JPEG2000*. Retrieved April 18, 2011, from http://www.jpeg.org/jpeg2000

Jumisko-Pyykkö, S., & Häkkinen, J. (2005). *Evaluation of subjective video quality on mobile devices*. Paper presented at 13th ACM International Conference on Multimedia. Singapore.

Jumisko-Pyykkö, S., & Häkkinen, J. (2006). *I would like see the face and at least hear the voice: Effects of screen size and audio-video bitrate ratio on perception of quality in mobile television*. Paper presented at 4th European Interactive TV Conference. London, UK.

Jumisko-Pyykkö, S., & Hannuksela, M. M. (2008). *Does context matter in quality evaluation of mobile television?* Paper presented at 10th International Conference on Human Computer Interaction with Mobile Devices and Services. Amsterdam, The Netherlands.

Jumisko-Pyykkö, S., Häkkinen, J., & Nyman, G. (2007). *Experienced quality factors – Qualitative evaluation approach to audiovisual quality.* Paper presented at SPIE Multimedia on Mobile Device. New York, NY.

Jumisko-Pyykkö, S., Strohmeier, D., Utriainen, T., & Kunze, K. (2010). *Descriptive quality of experience for mobile 3D video.* Paper presented at Nordic Conference on Human Computer Interaction 2010. Reykyavik, Iceland.

Jumisko-Pyykkö, S., Weitzel, M., & Strohmeier, D. (2008). *Designing for user experience: What to expect from mobile 3D TV and video?* Paper presented at 1st International Conference on Designing Interactive User Experiences for TV and Video. Silicon Valley, CA.

Jumisko-Pyykkö, S., Vadakital, V. K. M., & Hannuksela, M. M. (2008). Acceptance threshold: A bidimensional research method for user-oriented quality evaluation studies. *International Journal of Digital Multimedia Broadcasting*, 20.

Just, M. A., & Carpenter, P. A. (1980). A theory of reading: From eye fixations to comprehension. *Psychological Review*, *87*(1), 329–355. doi:10.1037/0033-295X.87.4.329 PMID:7413885.

Kalliola, K. (2011). *High accuracy indoor positioning system on BLE.* Retrieved October 14th, 2011, from http://hermia-fi-bin.directo.fi/@Bin/b85924038a216224b98 7d81d837f0959/1327586874/application/pdf/865170/HighAccuracyIndoorPositioningBasedOnBLE_Kalliola_270411.pdf

Kangasharju, J. (2008). *XML messaging for mobile devices.* (Doctoral dissertation). Department of Computer Science, University of Helsinki, Helsinki, Finland. Retrieved from http://www.doria.fi/

Kanjo, M., Benford, S., Paxton, M., Camberlain, A., Fraser, D., & Woodgate, D. et al. (2003). MobGeoSen: Facilitating personal geosensor data collection and visualization using mobile phones. *Personal and Ubiquitous Computing*, *12*(8), 599–697. doi:10.1007/s00779-007-0180-1.

Kaywa. (2012). Retrieved October 2012 from http://qrcode.kaywa.com/

Keir, M., Hann, C., Chase, G., & Chen, X. (2007). A new approach to accelerometer-based head tracking for augmented reality & other applications. In M. Keir (Ed.), *IEEE International Conference on Automation Science and Engineering* (pp. 603-608). Scottsdale, AZ: IEEE Computer Press.

Ketabdar, H., Roshandel, M., & Yuksel, K. A. (2010). Towards using embedded magnetic field sensor for around mobile device 3D interaction. Paper presented Mobile HCI 2010. New York, NY.

Ketomo, H. (2002). mLearning for kindergarten's mathematics teaching. In *Proceedings of the IEEE International Workshop on Wireless and Mobile Technologies in Education*, (vol. 1, pp. 167-168). IEEE.

Ketyko, I. N., De Moor, K., Joseph, W., Martens, L., & De Marez, L. (2010). *Performing QoE-measurements in an actual 3G network.* Paper presented at 2010 IEEE International Symposium on Broadband Multimedia Systems and Broadcasting (BMSB). Shanghai, China.

Kiczales, G., Irwin, J., Lamping, J., Loingtier, J., Lopes, C. V., Maeda, C., & Mendhekar, A. (1997). Aspect-oriented programming.[). Berlin: Springer-Verlag.]. *Proceedings of the ECOOP*, *1241*, 220–242.

Kim, E., Pyo, S., Park, E., & Kim, M. (2011). An automatic recommendation scheme of TV program contents for (IP) TV personalization. *IEEE Transactions on Broadcasting*, *57*(3), 674–684. doi:10.1109/TBC.2011.2161409.

Kim, H. K., Kim, J. K., & Ryu, Y. U. (2009). Personalized recommendation over a customer network for ubiquitous shopping. *IEEE Transactions on Service Computing*, *2*(2), 140–151. doi:10.1109/TSC.2009.7.

Klopfer, E., & Squire, K. (2008). Environmental detectives—The development of an augmented reality platform for environmental simulations. *Educational Technology Research and Development*, *56*(2), 203–228. doi:10.1007/s11423-007-9037-6.

Knoche, H., & McCarthy, J. D. (2004). *Mobile users' needs and expectations of future multimedia services.* Paper presented at Wireless World Research Forum (WWRF)12. Toronto, Canada.

Knoche, H., & McCarthy, J. D. (2005). *Design requirements for mobile TV*. Paper presented at Mobile Human Computer Interaction 2005. Salzburg, Austria.

Knoche, H., & Sasse, M. A. (2006). *Breaking the news on mobile TV: User requirements of a popular mobile content*. Paper presented at Multimedia on Mobile Devices II. New York, NY.

Knoche, H., & Sasse, M. A. (2008b). *The sweet spot: How people trade off size and definition on mobile devices*. Paper presented at 16th ACM International Conference on Multimedia. Vancouver, Canada.

Knoche, H., McCarthy, J. D., & Sasse, M. A. (2005). *Can small be beautiful? Assessing image resolution requirements for mobile TV*. Paper presented at 13th Annual ACM International Conference on Multimedia. Singapore.

Knoche, H., McCarthy, J. D., & Sasse, M. A. (2008). How low can you go? The effect of low resolutions on shot types in mobile TV. *Multimedia Tools and Applications*, *36*(1), 145–166. doi:10.1007/s11042-006-0076-5.

Knoche, H., & Sasse, M. A. (2008a). Getting the big picture on small screens: Quality of experience in mobile TV. In Ahmad, A. M. A., & Ibrahim, I. K. (Eds.), *Multimedia Transcoding in Mobile and Wireless Networks* (pp. 31–46). Hershey, PA: Information Science Reference. doi:10.4018/978-1-59904-984-7.ch003.

Knoche, H., & Sasse, M. A. (2009). The big picture on small screens: Delivering acceptable video quality in Mobile TV. *ACM Transactions on Multimedia Computing. Communications and Applications*, *5*(3), 27.

Kolb, D. A. (1999). *The Kolb learning style inventory: Version 3*. Boston: Hay Group.

Konstantas, D., Halteren, A. V., Bults, R., Wac, K., Jones, V., Widya, I., & Herzog, R. (2004). MobiHealth: Ambulant patient monitoring over next generation public wireless networks. In Demiris, G. (Ed.), *E-Health: Current Status and Future Trends* (pp. 107–122). Boca Raton, FL: IOS.

Krasner, G. E., & Pope, S. T. (1988). A description of the model-view-controller user interface paradigm in the smalltalk-80 system. *Journal of Object Oriented Programming*, *1*(3), 26–49.

Kratz, S. G., & Rohs, M. (2009). Hoverow: Expanding the design space of around-device interaction. Paper presented Mobile HCI 2009. New York, NY.

Krippendorff, K. H. (Ed.). (2004). *Content analysis: An introduction to its methodology*. Thousand Oaks, CA: Sage Publications.

Krishnendu, R. (2012). App. inventor for android: report from a summer camp.[Raleigh, NC: ACM.]. *Proceedings of SIGCSE*, *12*, 283–288.

Kristensen, M. D., & Bouvin, N. O. (2008). Developing cyber foraging applications for portable devices. In *Proceedings of the IEEE International Conference on Portable Information Devices*. Garmisch-Partenkirchen, Germany: IEEE.

Krueger, C. (2003). Easing the transition to software mass customization. In *Proceedings of the 4th International Workshop on Software Product-Family Engineering* (LNCS), (vol. 2290, pp. 282-293). Berlin: Springer-Verlag.

Kulesza, U., Alves, V., Garcia, A., Lucena, C., & Borba, P. (2006a). Improving extensibility of object-oriented frameworks with aspect-oriented programming. In *Proceedings of the 9th International Conference on Software Reuse (ICSR)*, (pp. 231-245). Torino, Italy: ICSR.

Kulesza, U., Alves, V., Garcia, A., Neto, A., Cirilo, E., Lucena, C., & Borba, P. (2007). Mapping features to aspects: A model-based generative approach. In A. Moreira & J. Grundy (Eds.), *Early Aspects: Current Challenges and Future Directions, 10th International Workshop* (LNCS), (vol. 4765, pp. 155-174). Berlin: Springer-Verlag.

Kulesza, U., Coelho, R., Alves, V., Neto, A., Garcia, A., Lucena, C., & Borba, P. (2006b). Implementing framework crosscutting extensions with EJPs and AspectJ. In *Proceedings of Brazilian Symposium on Software Engineering*, (pp. 117-192). IEEE.

Kurkovsky, S., & Harihar, K. (2006). Using ubiquitous computing in interactive mobile marketing. *Personal and Ubiquitous Computing*, *10*(4), 227–240. doi:10.1007/s00779-005-0044-5.

Kuznetsov, S., & Eric, P. (2010). *Participatory sensing in public spaces: Activating urban surfaces with sensor probes.* Paper presented at the 8th ACM Conference on Designing Interactive Systems. Aarhus, Denmark.

Kwon, H. J., & Hong, K. S. (2011). Personalized smart TV program recommender based on collaborative filtering and a novel similarity method. *IEEE Transactions on Consumer Electronics, 57*(3), 1416–1423. doi:10.1109/TCE.2011.6018902.

Kyriacou, E. C., Pattichis, C. S., & Pattichis, M. S. (2009). An overview of recent health care support systems for eEmergency and mHealth applications. In *Proceedings of the Annual International Conference of the IEEE Engineering in Medicine and Biology Society* (pp. 1246-1249). IEEE.

Landay, J. (2009). *I give up on CHI/UIST.* Retrieved 06 2012, from http://dubfuture.blogspot.com.au/2009/11/i-give-up-on-chiuist.html

Lane, N. D., Miluzzo, E., Hong, L., Peebles, D., Choudhury, T., & Campbell, A. T. (2010). A survey of mobile phone sensing. *IEEE Communications Magazine, 48*(9), 140–150. doi:10.1109/MCOM.2010.5560598.

Lave, J. (1991). *Situated learning: Legitimate peripheral participation.* New York: Cambridge U.P. doi:10.1017/CBO9780511815355.

Lawson, S. (2011). *Video dominates mobile traffic, survey shows.* Retrieved July 22, 2012, from http://www.computerworld.com/s/article/9218660/Video_dominates_mobile_traffic_survey_shows

Lazar, J., Feng, J. H., & Hochheiser, H. (2010). *Research methods in human-computer interation.* Chichester, UK: John Wiley and Sons Ltd..

Lee, I., Kim, J., & Kim, J. (2005). Use contexts for the mobile Internet: a longitudinal study monitoring actual use of mobile Internet services. *International Journal of Human-Computer Interaction, 18*(3), 269–292. doi:10.1207/s15327590ijhc1803_2.

Lee, M., & Gatton, T. M. (2010). Wireless health data exchange for home healthcare monitoring systems. *Sensors (Basel, Switzerland), 10*(4), 3243–3260. doi:10.3390/s100403243 PMID:22319296.

Lee, W. P., & Wang, J. H. (2004). A user-centered remote control system for personalized multimedia channel selection. *IEEE Transactions on Consumer Electronics, 50*(4), 1009–1015. doi:10.1109/TCE.2004.1362492.

Lehner, F., Nösekabel, H., & Lehmann, H. (2003). Wireless e–learning and communication environment: WELCOME at the University of Regensburg. *e-Service Journal, 2*(3), 23-41.

Levy, M., & Sandler, M. (2009). Music information retrieval using social tags and audio. *IEEE Transactions on Multimedia, 11*(3), 383–395. doi:10.1109/TMM.2009.2012913.

Lian, D. F., & Xie, X. (2011). Collaborative activity recognition via check-in history. In *Proceedings of 3rd ACM SIGSPATIAL International Workshop on Location-Based Social Networks* (pp. 45-48). ACM.

Lieberman, H. (2003). *The tyranny of evaluation.* Retrieved 05 2012, from http://Web.media.mit.edu/~lieber/Misc/Tyranny-Evaluation.html

Lindlof, T. R., & Taylor, B. C. (2002). *Qualitative communication research methods* (2nd ed.). Thousand Oaks, CA: SAGE.

Liu, T., Wang, H., Liang, J., Chan, T., Ko, H., & Yang, J. (2003). Wireless and mobile technologies to enhance teaching and learning. *Journal of Computer Assisted Learning, 19*, 1–14. doi:10.1046/j.0266-4909.2003.00038.x.

Lopez-Herrejon, R. E., & Batory, D. (2006). Modeling features in aspect-based product lines with use case slices: An exploratory case study.[MoDELS.]. *Proceedings of MoDELS Workshops, 2006*, 6–16.

Lublinsky, B. (n.d.). *Versioning in SOA.* Retrieved August 6, 2011, from http://msdn.microsoft.com/en-us/library/bb491124.aspx

Macgregor, G., & McCulloch, E. (2006). Collaborative tagging as a knowledge organisation and resource discovery tool. *Library Review, 55*(5), 291–300. doi:10.1108/00242530610667558.

Maglio, P., Matlock, T., Campbell, C., Zhai, S., & Smith, B. (2000). *Gaze and speech in attentive user interfaces.* Springer. doi:10.1007/3-540-40063-X_1.

Maglogiannis, I., Doukas, C., Kormentzas, G., & Pliakas, T. (2009). Wavelet-based compression with ROI coding support for mobile access to DICOM images over heterogeneous radio networks. *IEEE Transactions on Information Technology in Biomedicine*, *13*(4), 458–466. doi:10.1109/TITB.2008.903527 PMID:19586812.

Mahmoud, Q. (2004, November). *Java messaging service*. Retrieved March 5, 2011, from http://java.sun.com/developer/technicalArticles/Ecommerce/jms/index.html

Mäki, J. (2005). *Finnish mobile tv pilot*.

Malone, T. W. (1981). Toward a theory of intrinsically motivating instruction. *Cognitive Science*, *5*(4), 333–369. doi:10.1207/s15516709cog0504_2.

Mandryk, R. L., Inkpen, K. M., & Calvert, T. W. (2006). Using psychophysiological techniques to measure user experience with entertainment technologies. *Behaviour & Information Technology*, *25*(2), 141–158. doi:10.1080/01449290500331156.

Manning, M., & Munro, D. (Eds.). (2007). *The survey researcher's SPSS cookbook* (2nd ed.). Sydney, Australia: Pearson Education Australia.

Martin, D., & Sommerville, I. (2004). Ethnomethodology, patterns of cooperative interaction and design. *ACM Transactions on Computer-Human Interaction*, *11*(1), 59–89. doi:10.1145/972648.972651.

Mau, S., Melchior, N. A., Makatchev, M. M., & Steinfield, A. (n.d.). *BlindAid: An electronic travel aid for the blind* (Technical Report Ref CMU-RI-TR-07-39). Pittsburgh, PA: The Robotics Institute, Carnegie Mellon University.

McCarthy, J. (2002, June 1). *Reap the benefits of document style web services*. Retrieved March 5, 2011, from http://www.ibm.com/developerworks/Webservices/library/ws-docstyle.html

McCarthy, J. D., Sasse, M. A., & Miras, D. (2004). *Sharp or smooth? Comparing the effects of quantization vs. frame rate for streamed video*. Paper presented at SIGCHI Conference on Human Factors in Computing Systems. Vienna, Austria.

McGovern, J., Tyagi, S., Stevens, M., & Mathew, S. (2003). Service-oriented architecture. In *Java Web Services Architecture* (pp. 35–63). New York: Elsevier Science. doi:10.1016/B978-155860900-6/50005-1.

Meawed, F. E., & Stubbs, G. (2005). *An initial framework for implementing and evaluating probabilistic adaptivity in mobile learning*. Paper presented at the Fifth IEEE International Conference on Advanced Learning Technologies (ICALT'05). Kaohsiung, Taiwan.

MedApps. (2011). *MedApps mobile health monitoring*. Retrieved March 7, 2011, from http://www.medapps.net/HealthPAL.html

Mehigan, T., & Pitt, I. (2011). *Learning on the move: Harnessing accelerometer devices to detect learner styles for mobile learning*. Paper presented at I-HCI 11, 5th conference of the Irish Human Computer Interaction Community. Cork, Ireland.

Mehigan, T., & Pitt, I. (2012). *Harnessing wireless technologies for campus navigation by blind students and visitors*. Paper presented at the 13th International Conference on Computers Helping People with Special Needs (ICCHP). Linz, Austria.

Mehigan, T., Barry, M., & Pitt, I. (2009). *Individual learner styles inference using eye tracking technology*. Paper presented at I-HCI 09, 3rd Conference of the Irish HCI Community. Dublin, Ireland.

Mehigan, T., Barry, M., Kehoe, A., & Pitt, I. (2011). *Using eye tracking technology to identify visual and verbal learners*. Paper presented at IEEE International Conference on Multimedia & Expo. Barcelona, Spain.

Mehigan, T. J., & Pitt, I. (2010). Individual learner style inference for the development of adaptive mobile learner systems. In Guy, R. (Ed.), *Mobile Learning - Pilot Projects and Initiatives* (pp. 167–183). Santa Rosa, CA: Informing Science Press.

Mehigan, T. J., & Pitt, I. (2012). Detecting learning style through biometric technology for mobile GBL. *International Journal of Game-Based Learning*, *2*(2), 55–74. doi:10.4018/ijgbl.2012040104.

Menkovski, V., Exarchakos, G., & Liotta, A. (2011). The value of relative quality in video delivery. *Journal of Mobile Multimedia*, *7*(3), 151–162.

Menkovski, V., & Liotta, A. (2012). Adaptive psychometric scaling for video quality assessment. *Signal Processing Image Communication*, *27*(8), 788–799. doi:10.1016/j.image.2012.01.004.

Michael, D. R., & Chen, S. (2006). *Serious games: Games that educate, train and inform*. Boston, MA: Thomson Course Technology.

Mieure, M. (2012). *Gamification: A guideline for integrating and aligning digital game elements into a curriculum*. (Masters Thesis). Bowling Green State University, Bowling Green, OH.

Mikic, F., Anido, L., Valero, E., & Picos, J. (2007). *Accessibility and mobile learning standardisation, introducing some ideas about device profile (DP)*. Paper presented at the Second International Conference on Systems, ICOS'07. Sainte-Luce, Martinique.

Miller, D. J., & Robertson, D. P. (2010). Using a games console in the primary classroom: Effects of 'brain training' programme on computation and self-esteem. *British Journal of Educational Technology*, *41*(2), 242–255. doi:10.1111/j.1467-8535.2008.00918.x.

Miluzzo, E., Lane, N. D., Lu, H., & Campbell, A. T. (2010). *Research in the app. store era: Experiences from the cenceme app. deployment on the iPhone*. Paper presented at the First Workshop on Research in the Large at UbiComp 2010. Copenhagen, Denmark.

Miyauchi, K., Sugahara, T., & Oda, H. (2008). *Relax or study? A qualitative user study on the usage of mobile TV and video*. Paper presented at 6th European Interactive TV Conference. Salzburg, Austria.

Mobile Pocket Speak. (n.d.). Retrieved from http://www.codefactory.es/en/products.asp?id=336

Molina, G. G., Nijholt, A., & Twente, U. (2009). Emotional brain-computer interfaces. In *Proceedings of Affective Computing and Intelligent Interaction and Workshops* (pp. 1–9). IEEE.

Montola, M., Nummenmaa, T., Lucero, A., Boberg, M., & Korhonen, H. (2009). Applying game achievement systems to enhance user experience in a photo sharing service. In *Proceedings of the 13th International MindTrek Conference: Everyday Life in the Ubiquitous Era*, (pp. 94–97). Tampere, Finland: ACM.

Moore, M. (1999). The evolution of telemedicine. *Future Generation Computer Systems*, *15*(2), 245–254. doi:10.1016/S0167-739X(98)00067-3.

Morgan, D. L. (Ed.). (1998). *The focus group guidebook*. Thousand Oaks, CA: Sage Publications, Inc..

Mostow, J., Chang, K., & Nelson, J. (2011). Toward exploiting EEG input in a reading tutor. Paper presented AIED 2011. New York, NY.

MotoDev. (2012). Retrieved October 2012 from http://developer.motorola.com/

Muir, K., Buchan, A., von Kummer, R., Rother, J., & Baron, J. (2006). Imaging of acute stroke. *The Lancet Neurology*, *5*(9), 755–768. doi:10.1016/S1474-4422(06)70545-2 PMID:16914404.

Mulizzo, A., Campell, T., & Wang, E. (2010). *EyePhone: Activating mobile phones with your eyes*. Paper presented at MobiHeld 2010. New Delhi, India.

Müller, A., Helms, T. M., Wildau, H., Schwab, J. O., & Zugck, C. (2011). Remote monitoring in patients with pacemakers and implantable cardioverter-defibrillators: New perspectives for complex therapeutic management. In M. Kumar Das (Ed.), *Modern Pacemakers - Present and Future*. Retrieved from http://www.intechopen.com/articles/show/title/remote-monitoring-in-patients-with-pacemakers-and-implantable-cardioverter-defibrillators-new-perspe

Murugappan, M., Rizon, M., Nagarajan, R., Yaacob, S., Hazry, D., & Zunaidi, I. (2008). *Time-frequency analysis of EEG signals for human emotion detection*. Springer. doi:10.1007/978-3-540-69139-6_68.

Nabi, M., Kiah, M., Zaidan, B., Zaidan, A., & Alam, G. (2010). Suitability of using SOAP protocol to secure electronic medical record databases transmission. *International Journal of Pharmacology*, *6*, 959–964. doi:10.3923/ijp.2010.959.964.

Nakata, K., Maeda, K., Umedu, T., Hiromori, A., Yamaguchi, H., & Higashino, T. (2009). Modeling and evaluation of rescue operations using mobile communication devices. In *Proceedings of the ACM/IEEE/SCS 23rd Workshop on Principles of Advanced and Distributed Simulation* (pp. 64-71). ACM/IEEE/SCS.

NEMA. (n.d.). *Digital imaging and communications in medicine*. Retrieved June 27, 2012, from http://medical.nema.org/

Neubert, S., Arndt, D., Thurow, K., & Stoll, R. (2010). Mobile real-time data acquisition system for application in preventive medicine. *Telemedicine and e-Health Journal, 16*(4), 504-509.

NeuroSky. (2009). *Whitepaper: NeuroSky's eSense meters and detection of mental state September 2009*. Retrieved from http://company.neurosky.com/files/neurosky_esense_whitepaper.pdf

Niagara Health System. (n.d.). *Niagara health system – Services – Complex care (chronic care)*. Retrieved July 6, 2011, from http://www.niagarahealth.on.ca/services/chronic_care.html

Nicholas, D. L., Dimitrios, L., Zhao, F., & Andrew, T. C. (2010). Hapori: Context-based local search for mobile phones using community behavioral modeling and similarity. In *Proceedings of the 12th International Conference on Ubiquitous Computing* (pp. 109-118). IEEE.

Nikolidakis, S. A., Georgakakis, E., Giotsas, V., Vergados, D. D., & Douligeris, C. (2010). A secure ubiquitous healthcare system based on IMS and the HL7 standards. In *Proceedings of the 3rd International Conference on Pervasive Technologies Related to Assistive Environments*. Samos, Greece: IEEE.

Ninos, K., Spiros, K., Glotsos, D., Georgiadis, P., Sidiropoulos, K., & Dimitropoulos, N. et al. (2010). Development and evaluation of a PDA-based teleradiology terminal in thyroid nodule diagnosis. *Journal of Telemedicine and Telecare, 16*(5), 232–236. doi:10.1258/jtt.2010.090512 PMID:20423934.

Nunes, C., et al. (2009). Comparing stability of implementation techniques for multi-agent system product lines. In *Proceedings of the 3th European Conference on Software Maintenance and Reengineering (CSMR'09)*. Kaiserslautern, Germany: CSMR.

Nurminen, A., Ernest, K., & Eduardo, V. (2011). *HYDROSIS: A mixed reality platform for on-site visualization of environmental data*. Paper presented at the 10th International Conference on Web and Wireless Geographical Information System. Kyoto, Japan.

O'Reilly. (2009). *Google wave: What might email look like if it were invented today*. Retrieved 08 2012, from http://radar.oreilly.com/2009/05/google-wave-what-might-email-l.html

Obrist, M., Meschtscherjakov, A., & Tscheligi, M. (2010). User experience evaluation in the mobile context. In Mobile, T. V. (Ed.), *Customizing Content and Experience* (pp. 195–204). London: Springer-Verlag. doi:10.1007/978-1-84882-701-1_15.

O'Hara, K., Black, A., & Lipson, M. (2006). *Everyday practices with mobile video telephony*. Paper presented at SIGCHI Conference on Human Factors in Computing Systems. Montréal, Canada.

O'Hara, K., Mitchell, A. S., & Vorbau, A. (2007). *Consuming video on mobile devices*. Paper presented at SIGCHI on Human Factors in Computing Systems. San Jose, CA.

Oikonomidis, I., Kyriazis, N., & Argyros, A. (2011). Efficient model-based 3D tracking of hand articulations using Kinect. In *Proceedings of the British Machine Vision Conference* (pp. 101.1-101.11). BMVA Press.

Olofsson, J. K., Nordin, S., Sequeira, H., & Polich, J. (2008). Affective picture processing: An integrative review of ERP findings. *Biological Psychology, 77*(3), 247–265. doi:10.1016/j.biopsycho.2007.11.006 PMID:18164800.

Olsen, R. D., Jr. (2007). Evaluating user interface systems research. In *Proceedings of the 20th Annual ACM Symposium on User Interface Software and Technology* (pp. 251-258). Providence, RI: ACM.

Open, O. A. S. I. S. (2006). *Web services security: SOAP message security 1.1 (WS-security 2004)*. Retrieved from http://www.oasis-open.org/committees/download.php/16790/wss-v1.1-spec-os-SOAPMessageSecurity.pdf

Oracle. (n.d.). *Remote method invocation home*. Retrieved May 31, 2011, from http://www.oracle.com/technetwork/java/javase/tech/index-jsp-136424.html

Oren, E., Michael, J. C., Doug, D., Ana-Maria, P., Tal, S., & Stephen, S. et al. (2005). Unsupervised named-entity extraction from the Web: An experimental study. *Artificial Intelligence, 165*(1), 91–134. doi:10.1016/j.artint.2005.03.001.

Orgad, S. (2006). *This box was made for walking*. Retrieved Sep 20, 2009, from http://www.nokia.com/NOKIA_COM_1/Press/Press_Events/mobile_tv_report,_november_10,_2006/Mobil_TV_Report.pdf

Orwat, C., Graefe, A., & Faulwasser, T. (2008). Towards pervasive computing in health care – A literature review. *BMC Medical Informatics and Decision Making*, *8*(26). PMID:18565221.

Park, E., & Nam, H. S. (2009). A service-oriented medical framework for fast and adaptive information delivery in mobile environment. *IEEE Transactions on Information Technology in Biomedicine*, *13*(6), 1049–1056. doi:10.1109/TITB.2009.2031495 PMID:19775976.

Petrovic, O., Fallenböck, M., Kittl, C., & Langl, A. (2006). Mobile TV in Austria. *Schriftenreihe der Rundfunk und Telekom Regulierungs-GmbH, 2*.

PhoneGap. (2012). Retrieved October 2012 from http://www.phonegap.com/

Picard, R. (1995). *Affective computing*. Cambridge, MA: MIT Press.

Pielot, M., Henze, N., Nickel, C., Menke, C., Samadi, S., & Boll, S. (2009). Evaluation of camera phone based interaction to access information related to posters. In *Mobile Interaction with the Real World* (pp. 61–72). Citeseer.

Pittenger, D. J. (1993). The utility of the Myers- Briggs type indicator. *Review of Educational Research*, *63*(4), 467–488. doi:10.3102/00346543063004467.

Popescu, E. (2008). An artificial intelligence course used to investigate students' learning style. *Lecture Notes in Computer Science*, *5145*, 122–131. doi:10.1007/978-3-540-85033-5_13.

Prensky, M. (2001). *Digital game-based learning*. New York: McGraw-Hill.

Prixing. (2011). Retrieved October 2012 from http://www.prixing.fr

Pudota, N., Dattolo, A., Baruzzo, A., Ferrara, F., & Tasso, C. (2010). Automatic keyphrase extraction and ontology mining for content-based tag recommendation. *International Journal of Intelligent Systems*, *25*(12), 1158–1186. doi:10.1002/int.20448.

Puming, L. (2006). *Service agreements and facilities for m-health vital sign monitoring*. (M.S. thesis). Department of Computer Science, University of Twente, Enschede, Netherlands. Retrieved from http://essay.utwente.nl/

Qiang, L., Stankovic, J. A., Hanson, M. A., Barth, A. T., Lach, J., & Zhou, G. (2009). *Accurate, fast fall detection using gyroscopes and accelerometer-derived posture information*. Paper presented at the 6th International Workshop on Wearable and Implantable Body Sensor Networks. Berkeley, CA.

Quinn, C. (2000). *mLearning: Mobile, wireless, in-your-pocket learning*. Retrieved from www.Linezine.com

Raento, M., Oulasvirta, A., Petit, R., & Toivonen, H. (2005). ContextPhone: A prototyping platform for context-aware mobile applications. *IEEE Pervasive Computing Special Issue on Smartphone*, *4*(2), 51–59. doi:10.1109/MPRV.2005.29.

Rashvand, H. F., Salcedo, V. T., Sanchez, E. M., & Iliescu, D. (2008). Ubiquitous wireless telemedicine. *IET Communications*, *2*(2), 237–254. doi:10.1049/iet-com:20070361.

Rebolledo-Mendez, G., Dunwell, I., Martínez-Mirón, E. A., Vargas-Cerdán, M., de Freitas, S., Liarokapis, F., & García-Gaona, A. (2009). Assessing NeuroSky's usability to detect attention levels in an assessment exercise. *Assessment*, *5610*, 1–10. doi: doi:10.1007/978-3-642-02574-7_17.

Redhead, F., & Brereton, M. (2008). Nnub: A display for local communications. In *Proceedings of the Workshop on Public and Situated Displays to Support Communities at the 22nd Conference of the Computer-Human Interaction Special Interest Group of Australia on Computer-Human Interaction 2008*. Cairns, Australia: ACM.

Redhead, F., Dekker, A., & Brereton, M. (2010). NNUB: The neighbourhood nub digital noticeboard system. In *Proceedings of the 22nd Conference of the Computer-Human Interaction Special Interest Group of Australia on Computer-Human Interaction 2010* (pp. 418-419). Brisbane, Australia: ACM.

RemotEye. (n.d.). *DICOM viewer – RemotEye*. Retrieved August 8, 2011, from http://www.neologica.it/eng/RemotEye.php

Reynolds, C., & Picard, R. (2004). Affective sensors, privacy, and ethical contracts. In *Proceedings of the 2004 Conference on Human Factors and Computing Systems - CHI '04*. ACM. doi:10.1145/985921.985999

RFID Issues. (n.d.). Retrieved from http://www.slais.ubc.ca/courses/libr500/04-05-wt2/www/T_Gnissios/problems.htm

RFID Journal. (n.d.). Retrieved from http://www.rfid-journal.com/article/view/7224

Ricci, F., Rokach, L., & Shapira, B. (2011). Introduction to recommender systems handbook. In Ricci, F., Rokach, L., Shapira, B., & Kantor, P. (Eds.), *Recommender systems handbook* (pp. 1–35). Berlin: Springer. doi:10.1007/978-0-387-85820-3_1.

Ries, M., Nemethova, O., & Rupp, M. (2008). *On the willingness to pay in relation to delivered quality of mobile video streaming*. Paper presented at International Conference on Consumer Electronics. New York, NY.

Robitza, W., Buchinger, S., Hummelbrunner, P., & Hlavacs, H. (2010). *Acceptance of mobile TV channel switching delays*. Paper presented at Workshop on Quality of Multimedia Experience. Trondheim, Norway.

Rochelle, J. (2003). Unlocking the learning potential of wireless mobile devices. *Journal of Computer Assisted Learning*, *19*, 260–273.

Rodden, T., Chervest, K., Davies, N., & Dix, A. (1998). *Exploiting context in HCI design for mobile systems*. Paper presented at the Workshop on Human Computer Interaction with Mobile Devices. Glasgow, UK.

Rogers, Y., Sharp, H., & Preece, J. (2011). *Interaction design: Beyond human - computer interaction* (3rd ed.). New York, NY: John Wiley & Sons.

Rosenberg, J., Schulzrinne, H., Camarillo, G., et al. (2002). *SIP: Session initiation protocol*. Retrieved November 18, 2011, from http://www.ietf.org/rfc/rfc3261.txt

Rothkrantz, L. J. M., Wiggers, P., van Wees, J. W. A., van Vark, R., & Vark, R. J. V. (2004). *Voice stress analysis*. Springer.

Roto, V. (2006). *Web browsing on mobile phones – Characteristics of user experience*. (Doctoral Thesis). Helsinki University of Technology, Helsinki, Finland.

Rouillard, J. (2012). *The pervasive fridge: A smart computer system against uneaten food loss*. Paper presented at the Seventh International Conference on Systems, ICONS 2012. Saint Gilles, Reunion Island.

Rouillard, J. (2009). Multimodal and multichannel issues in pervasive and ubiquitous computing. In Kurkovsky, S. (Ed.), *Multimodality in Mobile Computing and Mobile Devices: Methods for Adaptable Usability* (pp. 1–23). Hershey, PA: Idea Group. Inc. doi:10.4018/978-1-60566-978-6.ch001.

Russell, J. A. (1980). A circumplex model of affect. *Journal of Personality and Social Psychology*, *39*(6), 1161–1178. doi:10.1037/h0077714.

Ryan, R. M., Rigby, C. S., & Przybylski, A. (2006). The motivational pull of video games: A self-determination theory approach. *Motivation and Emotion*, *30*(4), 344–360. doi:10.1007/s11031-006-9051-8.

Salvucci, D. D., & Goldburg, J. H. (2000). Identifying fixations and saccades in eye tracking protocols. In *Proceedings of the Eye Tracking Research and Applications Symposium* (pp. 71–78). New York: ACM Press.

Sánchez, P., Fuentes, L., Stein, D., Hanenberg, S., & Unland, R. (2008). Aspect-oriented model weaving beyond model composition and model transformation. *Proceedings of MoDELS*, *2008*, 766–781.

Santoro, C., Paterno, F., Ricci, G., & Leporini, B. (2007). *A multimodal mobile museum guide for all*. Paper presented at Mobile Interaction with the Real World (MIRW). Singapore.

Sasse, M. A., & Knoche, H. (2006). *Quality in context-An ecological approach to assessing QoS for mobile TV*. Paper presented at 2nd ISCA/DEGA Tutorial & Research Workshop on Perceptual Quality of System. Berlin, Germany.

Satyanarayanan, M. (2001). Pervasive computing: Vision and challenges. *IEEE Personal Communications*, *8*(4), 10–17. doi:10.1109/98.943998.

Savran, A., Ciftci, K., Chanel, G., Mota, J. C., & Viet, L. H. Sankut, B., Rombaut, M. (2006). Emotion detection in the loop from brain signals and facial images. In *Proceedings of eNTERFACE 2006*. Retrieved from http://www.enterface.net/results/

Sawyer, S., & Tapia, A. (2005). The sociotechnical nature of mobile computing work: Evidence from a study of policing in the United States. *International Journal of Technology and Human Interaction*, *1*(3), 1–14. doi:10.4018/jthi.2005070101.

Schadler, T., McCarthy, J. C., Brown, M., Martyn, H., & Brown, R. (2012). *Mobile is the new face of engagement – An information workplace report*. Retrieved October 2012, from http://www.forrester.com/Mobile+Is+The+New+Face+Of+Engagement/fulltext/-/E-RES60544?objectid=RES60544

Schatz, R., & Egger, S. (2008). *Social interaction features for mobile TV services*. Paper presented at 2008 IEEE International Symposium on Broadband Multimedia Systems and Broadcasting. Las Vegas, NV.

Schatz, R., Egger, S., & Platzer, A. (2011). *Poor, good enough or even better? Bridging the gap between acceptability and QoE of mobile broadband data services*. Paper presented at IEEE International Conference on Communications ICC 2011. Kyoto, Japan.

Schilling, J. (2006). On the pragmatics of qualitative assessment: Designing the process for content analysis. *European Journal of Psychological Assessment*, *22*(1), 28–37. doi:10.1027/1015-5759.22.1.28.

Schloeffel, P., Beale, T., Hayworth, G., Heard, S., & Lesli, H. (n.d.). *The relationship between CEN 13606, HL7, and openEHR*. Retrieved April 6, 2011, from http://www.oceaninformatics.com/Media/docs/Relationship-between-CEN-13606-HL7-CDA--openEHR-2ba3675f-2136-4069-ac5c-152139c70bd0.pdf

Schmidt, A., Beigl, M., & Gellerson, H. W. (1999). There is more to context than location. *Computers & Graphics*, *23*, 893–901. doi:10.1016/S0097-8493(99)00120-X.

Schneiderman, R. (2010). DSPs evolving in consumer electronics applications. *IEEE Signal Processing Magazine*, *27*(3), 6–10. doi:10.1109/MSP.2010.936031.

Shnayder, V., Chen, B., Lorincz, K., Jones, T. R. F. F., & Welsh, M. (2005). Sensor networks for medical care. In *Proceedings of the 3rd International Conference on Embedded Networked Sensor Systems*. San Diego, CA: IEEE.

Shotton, J., Fitzgibbon, A., Cook, M., Sharp, T., Finocchio, M., & Moore, R. Blake, A. (2011). Real-time human pose recognition in parts from single depth images. In *Proceedings of IEEE International Conference on Computer Vision and Pattern Recognition*. IEEE.

Sillitti, A., Vernazza, T., & Succi, G. (2002). Service oriented programming: A new paradigm of software reuse. *Lecture Notes in Computer Science*, *2319*, 268–280. doi:10.1007/3-540-46020-9_19.

Sliney, A., & Murphy, D. (2011). Using serious games for assessment. In M. Ma, A. Oikonomou, & L. C. Jain (Eds.), Serious Games and Edutainment Applications (pp. 225–243). Springer. Retrieved fromhttp://dx.doi.org/doi:10.1007/978-1-4471-2161-9_12.

Sliney, A., Murphy, D., & O'Mullane, J. (2009). Secondary assessment data within serious games. In O. Petrovic & A. Brand (Eds.), Serious Games on the Move (pp. 225–233). Springer. Retrieved fromhttp://dx.doi.org/doi:10.1007/978-3-211-09418-1_15.

Social Security Advisory Board. (2009). *The unsustainable cost of health care*. Retrieved from http://www.ssab.gov/documents/TheUnsustainableCostofHealth-Care_graphics.pdf

Södergård, C. (2003). *Mobile television-technology and user experiences. Report on the Mobile-TV project*. Finland: VTT.

Song, W., & Tjondronegoro, D. (2010). *A survey on usage of mobile video in Australia*. Paper presented at Australian Human-Computer Interaction Conference 2010. Brisbane, Australia.

Song, W., Tjondronegoro, D., & Docherty, M. (2011a). *Quality delivery of mobile video: In-depth understanding of user requirements*. Paper presented at Australian Human-Computer Interaction Conference 2011. Canberra, Australia.

Song, W., Tjondronegoro, D., & Docherty, M. (2011b). *Saving bitrate vs. pleasing users: Where is the break-even point in mobile video quality?* Paper presented at ACM Multimedia 2011. Scottscale, AZ.

Song, W., Tjondronegoro, D., & Docherty, M. (2010). Exploration and optimisation of user experience in viewing videos on a mobile phone. *International Journal of Software Engineering and Knowledge Engineering*, *8*(20), 1045–1075. doi:10.1142/S0218194010005067.

Spada, S., Sánchez-Montañés, M., Paredes, P., & Carro, R. M. (2008). Towards inferring sequential-global dimension of learning styles from mouse movement patterns. *Lecture Notes in Computer Science*, *5149*, 337–340. doi:10.1007/978-3-540-70987-9_48.

Spring AOP Aspect Library. (2009). Retrieved from http://www.springframework.org

Strauss, A., & Corbin, J. (Eds.). (1990). *Basis of qualitative research: Grounded theory procedures and techniques*. Thousand Oaks, CA: Sage Publications.

Stroop, J. R. (1935). Studies of interference in serial verbal reactions. *Journal of Experimental Psychology*, *28*(1), 643–662. doi:10.1037/h0054651.

Subramanya, A. (2001). Image compression technique. *IEEE Potentials*, *20*(1), 19–23. doi:10.1109/45.913206.

Sullivan, K. et al. (2005). Information hiding interfaces for aspect-oriented design. In *Proceedings of ESEC/FSE'2005* (pp. 166–175). Lisbon, Portugal: ESEC.

Sveriges Television AB (SVT). (n.d.). Retrieved Feb 2010, from ftp://vqeg.its.bldrdoc.gov/HDTV/SVT_MultiFormat/

Tacchi, J., Foth, M., & Hearn, G. (2007). *Ethnographic action research*. Brisbane, Australia: QUT.

Tahat, A. A. (2009). Mobile messaging services-based personal electrocardiogram monitoring system. *International Journal of Telemedicine and Applications*, 85929. PMID:19707531.

Tasaka, S., Yoshimi, H., & Hirashima, A. (2008). *The effectiveness of a QoE-based video output scheme for audio-video IP transmission*. Paper presented at ACM Multimedia Information System. Vancouver, Canada.

Technical University of Munich. (n.d.). Retrieved Feb 2010 ftp://ftp.ldv.e-technik.tu-muenchen.de/pub/test_sequences/

Tee, Z., Ang, L. M., Seng, K. P., Kong, J. H., Lo, R., & Khor, M. Y. (2009). *SmartGuide system to assist visually impaired people in a university environment*. Paper presented at the Third International Convention on Rehabilitation Engineering & Assistive Technology (ICREAT). Singapore.

Temple Go Mobile. (n.d.). Retrieved from http://www.here2there.com/pdf/H2T%20University%20PDF.pdf

Tewissen, F., Lingnau, A., Hoppe, U., Mannhaupt, G., & Nischk, D. (2001). *Collaborative writing in a computer-intergrated classroom for early learning*. Hoboken, NJ: Lawrence Erlbaum Associates.

The Apache Software Foundation. (2009). *Jini_Architecture_Specification – River wiki*. Retrieved June 19, 2012, from http://wiki.apache.org/river/Jini_Architecture_Specification

The Institute of Electrical and Electronics Engineers. (1990). *IEEE standard glossary of software engineering terminology*. Retrieved from http://ieeexplore.ieee.org/stampPDF/getPDF.jsp?tp=&arnumber=159342

The Nielsen Company. (2011a). *The cross-platform report, Quarter 1 2011*. Retrieved Feb 11, 2012, from http://www.tvb.org/media/file/nielsen_cross-platform_report_Q1-2011.pdf

The Nielsen Company. (2011b). *Telstra smartphone index*. Retrieved Jul 22, 2011, from http://sensisdigitalmedia.com.au/Files/Mobile/Nielsen_Telstra_Smartphone_Index_June2011_Presentation.pdf

The Nielsen Company. (2012). *Australian multi-screen report (Quarter 1, 2012) - Trends in video viewership beyond conventional television sets*. Retrieved Feb 11, 2012, from http://www.nielsen.com/content/dam/corporate/au/en/reports/2012/MultiScreenReportQ12012_FINAL.pdf

Thormählen, T., & Seidel, H. (2008). 3D-modeling by ortho-image generation from image sequences. *ACM Transactions on Graphics*, *27*(3), 1–5. doi:10.1145/1360612.1360685.

Thüring, M., & Mahlke, S. (2007). Usability, aesthetics and emotions in human–technology interaction. *International Journal of Psychology*, *42*(4), 253–264. doi:10.1080/00207590701396674.

TIGSource. (2009). *Minecraft (alpha)*. Retrieved 06 27, 2012, from http://forums.tigsource.com/index.php?PHPSESSID=8a7200c9319b16be007dabdc9d31706b&topic=6273.0

Ting, S. L. J., Kwok, S. K., Lee, W. B., Tsang, A. H. C., & Lee, Y. H. (March). *Dynamic mobile RFID-based knowledge hunting system in ubiquitous learning environment*. Paper presented at INTED'08. Valencia, Spain.

Tobii. (n.d.). Retrieved from www.tobii.com

Tokyo. (2006, March). Oldies get a gaming goodies. *The Sunday Morning Herald*.

Tominaga, T., Hayashi, T., Okamoto, J., & Takahashi, A. (2010). *Performance comparisions of subjective quality assessment methods for mobile video*. Paper presented at Second International Workshop on Quality of Multimedia Experience 2010. Trondheim, Norway.

Traxler, J. (2005). MLearning its here but what is it? *Interactions: University of Warwick Centre for Academic Practice*, *25*(9), 1–6.

Trefzger, J. (2005). *Mobile TV launch in Germany: Challenges and implications*. Cologne, Germany: Institute for Broadcasting Economics Cologne University.

Trompler, C., Muhlhauser, M., & Wegner, W. (2002). *Open client lecture interaction: An approach to wireless learners-in-the-loop*. Paper presented at the International Conference on New Educational Environment. Lucerne, Switzerland.

U.S National Library of Medicine. (2011). *Home-monitoring in implantable cardioverter defibrillator (ICD) patients (monitor-ICD)*. Retrieved April 28, 2011, from http://clinicaltrials.gov/ct2/show/NCT00787683

UCC President's Report 2010. (n.d.). Retrieved from http://www.ucc.ie/en/news/newsarchieve/2010PressRelease/fullstory-91963-en.html

Vangenck, M., Jacobs, A., Lievens, B., Vanhengel, E., & Pierson, J. (2008). *Does mobile television challenge the dimension of viewing television? An explorative research on time, place and social context of the use of mobile television content*. Paper presented at European Interaction TV Conference 2008. Salzburg, Austria.

Varshney, U. (2009). Wireless health monitoring: State of the art. In *Pervasive Healthcare Computing* (pp. 119–146). New York: Springer US. doi:10.1007/978-1-4419-0215-3_6.

Vassileva, J. (2012). Motivating participation in social computing applications: A user modeling perspective. *User Modeling and User-Adapted Interaction*, *22*(1), 177–201. doi:10.1007/s11257-011-9109-5.

Vergados, D. D. (2010). Service personalization for assistive living in a mobile ambient healthcare-networked environment. *Personal and Ubiquitous Computing Journal*, *14*(6), 575–590. doi:10.1007/s00779-009-0278-8.

Villaverde, J. E., Godoy, D., & Amandi, A. (2006). Learning styles' recognition in e-learning environments with feed-forward neural networks. *Journal of Computer Assisted Learning*, (22): 197–206. doi:10.1111/j.1365-2729.2006.00169.x.

von Ahn, L., & Dabbish, L. (2004). Labeling images with a computer game. In *Proceedings of the 22nd ACM SIGCHI International Conference on Human-Computer Interaction*. Vienna, Austria: ACM.

W3C. (n.d.a). *SOAP version 1.2 part 1: Messaging framework (2nd Ed.)*. Retrieved May 31, 2011, from http://www.w3.org/TR/soap12-part1/

W3C. (n.d.b). *Web services architecture*. Retrieved May 31, 2011, from http://www.w3.org/TR/ws-arch/

Wachs, J. P., Kolsch, M., Stern, H., & Edan, Y. (2011). Vision-based hand-gesture applications. *Communications of the ACM*, *54*(2), 60–71. doi:10.1145/1897816.1897838 PMID:21984822.

Wafa, T. (2010). How the lack of prescriptive technical granularity in HIPAA has compromised patient privacy. *North Illinois University Law Review*, *30*(3).

Wagner, D., Reitmayr, G., Mulloni, A., Drummond, T., & Schmalstieg, D. (2008). *Pose tracking from natural features on mobile phones*. Paper presented at the 7th IEEE/ACM International Symposium on Mixed and Augmented Reality. Cambridge, UK.

Walsh, M., Gaffney, M., Barton, J., O'Flynn, B., O' Mathuna, C., Hickey, A., & Kellett, J. A. (2011). *Medical study on wireless inertial measurement technology as a tool for identifying patients at risk of death or imminent clinical deterioration*. Paper presented at Pervasive Health. Dublin, Ireland.

Wang, H., Peng, D., Wang, W., Sharif, H., Chen, H., & Khoynezhad, A. (2010). Resource-aware secure ECG healthcare monitoring through body sensor networks. *IEEE Wireless Communications*, *17*(1), 12–19. doi:10.1109/MWC.2010.5416345.

Wang, S. C., Chung, T. C., & Yan, K. Q. (2011). A new territory of multi-user variable remote control for interactive TV. *Multimedia Tools and Applications*, *51*(3), 1013–1034. doi:10.1007/s11042-009-0435-0.

Weibel, N. (2002). *Web services technologies SOAP vs. JINI.* Retrieved from http://www.rudibelotti.com/doc/projects/Webservices/Webservices.pdf

Weiser, M. (1991). The computer for the 21st century. *Scientific American, 265*(3), 66–75. doi:10.1038/scientificamerican0991-94 PMID:1754874.

Wi-Fi. (n.d.). *Wi-Fi alliance.* Retrieved April 18, 2011, from http://www.wi-fi.org

Wilkinson, T., Haines, S., & Williams, C. (2001). Jini in military system applications. In *Proceedings of the RTO Information Systems Technology Panel (IST) Symposium on Information Management Challenges in Achieving Coalition Interoperability,* (pp. 154-172). Quebec, Canada: IST.

Willett, P., Barnard, J. M., & Downs, G. M. (1998). Chemical similarity searching. *Journal of Chemical Information and Modeling, 38*(6), 983–996. doi:10.1021/ci9800211.

WiMAX. (n.d.). *What is WiMAX? | general | WiMAX FAQ.* Retrieved April 18, 2011, from http://www.wimax.com/general/what-is-wimax

Winder, J. (2001). *Net content: From free to fee.* Retrieved Aug 19, 2009, from http://hbr.harvardbusiness.org/2001/07/net-content-from-free-to-fee/ar/1

Wither, J., DiVerdi, S., & Höllerer, T. (2009). Annotation in outdoor augmented reality. *Computers & Graphics, 33*(6), 679–689. doi:10.1016/j.cag.2009.06.001.

Wolber, D. (2011). App. inventor and real-world motivation. In *Proceedings of the 42nd ACM Technical Symposium on Computer Science Education,* (pp. 601–606) Dallas, TX: ACM.

Wolber, D. (2012). *App. inventor capabilities and limitations.* Retrieved June 2012 from http://www.appinventor.org/capabilities-limitations

Wolber, D., Abelson, H., Spertus, E., & Looney, L. (2011). *App. inventor: Create your own android apps.* O'Reilly Media.

Wolpaw, J. R., Birbaumer, N., McFarland, D. J., Pfurtscheller, G., & Vaughan, T. M. (2002). Brain-computer interfaces for communication and control. *Clinical Neurophysiology, 113*(6), 767–791. doi:http://dx.doi.org/10.1016/S1388-2457(02)00057-3

Wu, X., Zhang, L., & Yu, Y. (2006). Exploring social annotations for the semantic web. In *Proceedings of the Fifteenth International Conference on World Wide Web* (pp. 417-426). IEEE.

Young, T. (2005). *Using AspectJ to build a software product line for mobile devices.* (MSc Dissertation). University of British Columbia, Vancouver, Canada.

Zhang, Y., & Wildemuth, B. M. (2009). Qualitative analysis of content. In Wildemuth, B. (Ed.), *Applications of social research methods to questions in information and library science* (pp. 308–319). Westport, CT: Libraries Unlimited.

Zhuang, J. F., Mei, T., Steven, C. H., & Li, S. P. (2011). When recommendation meets mobile: Contextual and personalized recommendation on the go. In *Proceedings of the 13th International Conference on Ubiquitous Computing* (pp. 153-162). IEEE.

Ziegler, R., Wojciech, M., Hanspeter, P., & Leonard, M. (2003). *3D reconstruction using labeled image regions.* Paper presented at Eurographics Symposium on Geometry Processing. Aachen, Germany.

ZigBee. (n.d.). *ZigBee alliance.* Retrieved April 18, 2011, from http://www.zigbee.org

ZXing. (2009). *Zebra crossing.* Retrieved October 2012 from http://code.google.com/p/zxing

About the Contributors

Dian Tjondronegoro is currently an Associate Professor in Service Sciences Discipline, Faculty of Science and Engineering, Queensland University of Technology. His research interests are: (1) Mobile Multimedia: delivering rich-media contents on smart phones, such as iPhone, with a particular focus on streaming video; (2) Video Analysis and Summarisation: extracting semantic contents from video using audiovisual features including motion, face, objects, and events; (3) Cross-Media Content Annotation and Clustering: analysing text, image, and video contents to support clustering of semantically related contents for enhancing user experience. Dr. Tjondronegoro has obtained Bachelor of IT with first class honours in 2002 (QUT), PhD in 2005 (Deakin). He was a tutor, associate lecturer, lecturer, and senior lecturer in Faculty of IT, QUT in 2004-2011.

* * *

Muhammad H. Aboelfotoh is a PhD candidate in the School of Computing at Queen's University of Kingston. His research focuses on means to effectively utilize mobile devices to deliver patient-centric health care services in a more cost effective manner. He obtained his MSc from the same school. His MSc work involved the completion of an automated networked software security testing framework. He holds a BEng in computer engineering from Kuwait University. His computer engineering graduation project involved working with a team to come up with a document that serves as a usability standard for e-government Websites of member countries of the Gulf Cooperation Council (GCC). His research interests are in ubiquitous health care services, personal electronic health records, distributed systems, cloud computing, information privacy, and network and software security.

Thais Batista received her PhD from the Pontifical Catholic University of Rio de Janeiro (PUC-Rio) in 2000. She is an Associate Professor of the Federal University of Rio Grande do Norte (UFRN) since 1996. From 2004 to 2005, she was a visiting researcher at the Lancaster University, UK. She participates in several research projects with funding from International and Brazilian government agencies (CNPq, CAPES, ANP, RNP). Her primary research interest is on software architecture, aspect-oriented development, middleware, distributed systems and software engineering techniques applied to the development of ubiquitous systems. She is a Researcher Fellow of the National Council for Scientific and Technological Development.

Roberta Coelho is an Associate Professor at Federal University of Rio Grande do Norte, Brazil. She holds a Ph.D. from the Informatics Department of the Pontifical Catholic University of Rio (PUC-Rio) (2004-2008) and worked as a researcher at Lancaster University, where she conducted empirical studies in the context of reliability of AO applications. Her research interests include static analysis, exception handling, dependability, and empirical software engineering.

Katie Crowley is a researcher with the Interaction Design, E-Learning, and Speech (IDEAS) Research Group at the Department of Computer Science, University College Cork, Ireland. Her research interests include voice analysis; bio-signals and biometrics; physiological measurement and digital signal processing. Her primary research focus is the analysis of emotional speech, particularly speech under stress, using psychophysiological HCI methods such as multimodal biometric affective feedback.

Andrew Dekker is a researcher and practitioner within the fields of ubiquitous computing and interaction design. Andrew is currently completing a PhD in Computer Science on investigating tools and methods to support designer/client communication within Web design. He also holds degrees in Multimedia Design, Information Environments, and Information Technology, with a research focus on understanding how novel interfaces and interaction techniques can deliver an improved user experience to traditional computing devices. Andrew also works as a mobile and Web designer and developer within industry, working primarily with educational and commercial enterprises to design new techniques to engage with users.

Flávia C. Delicato received her PhD from Federal University of Rio de Janeiro in 2005. From 2006 to 2011, she was an Associate Professor of the Federal University of Rio Grande do Norte. She is currently an Associate Professor of the Federal University of Rio de Janeiro, where she teaches for undergraduate and post-graduate courses and works as researcher. During 2009 she was a Visitor Researcher at the Málaga University, Spain. In 2010, she was a visiting academic at the University of Sydney, Australia. She participates in several research projects with funding from International and Brazilian government agencies (CNPq, CAPES, ANP, RNP, Fundación Carolina). Her primary research interest is on middleware, wireless sensor networks and Software Engineering techniques applied to the development of ubiquitous systems. She is a Researcher Fellow of the National Council for Scientific and Technological Development. She integrates the Centre for Distributed and High Performance Computing at University of Sydney.

Michael Docherty, after working as an Architect in Adelaide and Canberra, a developing interest in computing led to various consultancies culminating in three years as the software manager of a computer company in Adelaide. Michael joined the University of Queensland, Department of Architecture, in 1984. In 1999, joined the School of Information Technology, and subsequently appointed Director (Head) of the Information Environments Program, UQ. In 2005 moved to QUT as Head of Discipline, Communication Design. Michael now teaches and researchers in the areas 3D Virtual Environments, mobile and Web based communication and knowledge management.

Zachary Fitz-Walter is a PhD student at the Faculty of Science and Technology, Queensland University of Technology with a research focus on user experience, mobile applications and video games. His primary interests lie in the field of Human Computer Interaction, focusing on interactions and experiences with mobile devices. His PhD thesis aims to explore how game design elements, embedded into non-game contexts using mobile technology, can be used as tools of motivation and engagement.

Hossam S. Hassanein is leading research in the areas of wireless and mobile networks architecture, protocols and services. His record spans more than 500 publications in journals, conferences, and book chapters, in addition to numerous keynotes and plenary talks in flagship venues. Dr. Hassanein has received several recognition and best papers awards at top international conferences. He is also the founder and director of the Telecommunications Research Lab at Queen's University School of Computing, with extensive international academic and industrial collaborations. Dr. Hossam Hassanein is a senior member of the IEEE, and is a former chair of the IEEE Communication Society Technical Committee on Ad hoc and Sensor Networks (TC AHSN). Dr. Hassanein is an IEEE Communications Society Distinguished Speaker (Distinguished Lecturer 2008-2010).

Che Kao-Li received his B.A. degree from the Department of Information Management, National Chyi University, and his M.A. degree from the Department of Information Management, National Sun Yat-sen University, Taiwan. He is a software engineer in the Chimei Innolux Corporation. His research interests include data mining, multimedia broadcasting, and social networking.

Uirá Kulesza is an Associate Professor at the Department of Informatics and Applied Mathematics (DIMAp), Federal University of Rio Grande do Norte (UFRN), Brazil. He obtained his PhD in Computer Science at PUC-Rio – Brazil (2007), in cooperation with University of Waterloo and Lancaster University. His main research interests include: aspect-oriented development, software product lines, and design/implementation of model-driven generative tools. He has co-authored over 100 referred papers in journals, conferences, and book chapthers. He worked as a post-doc researcher member of the AMPLE project (2007-2009) – Aspect-Oriented Model-Driven Product Line Engineering (www.ample-project.net) at the New University of Lisbon, Portugal. He is currently a Researcher Fellow of the National Council for Scientific and Technological Development (CNPq).

Chin Loong Law is a Masters by Research student in Queensland University of Technology. His research interests include innovative mobile applications in environmental monitoring, computer programming, and applications of augmented reality for computer-human interaction.

Wei-Po Lee received his Ph.D. in artificial intelligence from University of Edinburgh, United Kingdom. His is currently an associate professor at the Department of Information Management, National Sun Yat-sen University, Kaohsiung, Taiwan. He is interested in intelligent autonomous systems, machine learning, mobile multimedia, social networking, and entertainment computing.

Shipeng Li joined Microsoft Research Asia (MSRA) in May 1999. He is now a Principal Researcher and Research Manager of the Media Computing group. He also serves as the Research Area Manager coordinating the multimedia research activities at MSRA. His research interests include multimedia processing, analysis, coding, streaming, networking and communications; digital right management; advertisement; user intent mining; eHealth; etc. From Oct. 1996 to May 1999, Dr. Li was with Multimedia Technology Laboratory at Sarnoff Corporation as a Member of Technical Staff. Dr. Li has been actively involved in research and development in broad multimedia areas. He has authored and co-authored 6 books/book chapters and 250+ referred journal and conference papers. He holds 120+ granted US patents. Dr. Li received his B.S. and M.S. in Electrical Engineering from the University of Science and Technology of China, Hefei, China in 1988 and 1991, respectively. He received his Ph.D. in Electrical Engineering from Lehigh University, Bethlehem, PA, USA in 1996. He was a faculty member in Department of Electronic Engineering and Information Science at University of Science and Technology of China in 1991-1992.

Carlos Lucena is a Full-Professor of Computer Science at the Pontifical Catholic University of Rio (PUC-Rio) (since 1982); Adjunct Professor of Computer Science and Senior Research Associate, Computer Systems Group, University of Waterloo, Ontario, Canada (since 1993); full member of the Brazilian Academy of Sciences. He presented (and published papers) and acted as Program Committee Member in about 30 international conferences (example of sponsors: IEEE, ACM, AFIPS, IFIP, IFAC). Moreover, he acted twice as area chairman for IFIP World Congresses (1980, 1992) and four times as Program Committee member for the International Conference on Software Engineering (ICSE).

Justin Marrington is a researcher in interaction design and software design at the University of Queensland, focusing on mobile application design, ubiquitous computing, and digital media. Whilst completing his studies, Justin collaborated on a wide variety of research-focused technology projects, including a health-focused social network, smartphone-powered transit surveys, and experimental platforms for digitally marking university assignments. His undergraduate thesis project was a locative publishing platform. Justin currently teaches programming and user-centred design at UQ, and remains an active participant in multiple research projects. He also runs introductory prototyping workshops for high-school students, on the Arduino microcontroller and MIT App Inventor.

Patrick Martin holds a BSc and a PhD from the University of Toronto and a MSc from Queen's University. He joined Queen's University in 1984 and is currently a Professor in the School of Computing. He served as Associate Director of the School from 2002 to 2007 and Acting Director in 2004. He has been involved in numerous collaborative projects with industry and is a Visiting Scientist with IBM's Centre for Advanced Studies. He has supervised over 80 graduate students at Queen's and is the author of over 100 peer-reviewed journal and conference papers. His research interests include database system performance, cloud computing, Web services and services management, and autonomic computing systems.

Tracey J. Mehigan is currently a researcher with the Interaction Design, E-learning and Speech (IDEAS) Research Group at the Department of Computer Science, University College Cork, Ireland. Her principle area of research focuses on the use of mobile computing for eLearning purposes. She has a particular interest in the area of adaptive learning systems and the development of systems to facilitate the inclusion of those with disabilities and special needs into mainstream and ubiquitous collaborative learning environments. She has authored a number of articles within the area of mobile learning and HCI.

Tao Mei received the B.E. and the Ph.D. degrees from the University of Science and Technology of China, Hefei, in 2001 and 2006, respectively. He is now a Researcher with Microsoft Research Asia. His current research interests include multimedia information retrieval, computer vision, and multimedia applications such as advertising, social networking, and mobile computing. He has published over 120 referred papers in these areas. He serves as an Associate Editor for *Neurocomputing* and *Journal of Multimedia*. He received the best paper awards in ACM Multimedia 2007 and 2009, the best paper award from ACM ICIMCS 2012, the best poster paper award from IEEE MMSP 2008, and the best demo award from ACM Multimedia 2007. He is a senior member of IEEE and ACM.

Camila Nunes received her bachelor degree in Computer Science from the Federal University of Alagoas (UFAL) in 2007. She obtained her Master and PhD degree in Computer Science from the Pontifical Catholic University of Rio de Janeiro (PUC-Rio), Rio de Janeiro, Brazil. Currently, she has been working as software engineering researcher at the General Electric (GE) Global Research Center in Brazil. Her areas of interest are experimental software engineering, software architecture, software design and recovery, software product lines.

Paulo F. Pires is an Associate Professor at the Department of Computer Science at Federal University of Rio de Janeiro, Brazil. He received his M.Sc. and Ph.D. in Computer Science and Systems Engineering in Computer Science, in 1997 and 2002, respectively, both from the Federal University of Rio de Janeiro, Brazil. In 2000, he was a visiting researcher at the CLIP lab in University of Maryland (USA) and in 2009 he at Málaga University, Spain. He spent the year of 2010 at the University of Sydney, Australia, as visiting scholar on a sabbatical leave. He holds a technological innovation productivity scholarship from the Brazilian Research Council (CNPq) since 2010 and is a member of the Brazilian Computer Society (SBC). His main research interests include model driven development, software product lines, infrastructures for Web service composition, and application of Software Engineering techniques in emerging domains, as embedded, ubiquitous, and pervasive systems.

Ian Pitt lectures in Usability Engineering and Interactive Media at University College Cork, Ireland. He took his D.Phil at the University of York, UK, then worked in the Institute for Simulation and Graphics at Otto-von-Guericke University, Magdeburg, Germany, before moving to Cork in 1997. He is the leader of the Interaction Design, E-Learning, and Speech (IDEAS) Research Group at UCC, which is currently working on a variety of projects relating to multi-modal human computer interaction across various application domains. His own research interests centre around the use of speech and non-speech sound in computer interfaces and the design of computer systems for use by blind and visually-impaired people.

Paul Roe received his Master of Engineering from the University of York in 1987 and his PhD from the University of Glasgow in 1991. He is currently a full professor in the Science and Engineering Faculty at Queensland University of Technology. He leads the Microsoft-QUT eResearch Centre. The eResearch centre is collaboration between the Queensland State Government, QUT and Microsoft Research which is investigating smart tools for eResearch. He has published more than 70 papers and received over $4M in competitive research funding. His research is focused on smart tools, which enable new forms and scales of research, particularly in the areas of distributed computing such as wireless sensors networks and end user programming languages.

José Rouillard is an Associate Professor in Computer Science at the University of Lille 1 in the LIFL laboratory. The LIFL (Laboratoire d'Informatique Fondamentale de Lille) is a Research Laboratory in the Computer Science field of the University of Sciences and Technologies of Lille (USTL) linked to the CNRS, and in partnership with the INRIA Lille - Nord Europe. José Rouillard obtained his PhD in 2000 from the University of Grenoble (France) in the field of Human-Computer Interfaces. He is interested in Human-Computer Interaction, plasticity of user interfaces, multi-modality, multi-channel interfaces, and human-machine dialogue. He has written one book on VoiceXML in 2004, another one on Software Oriented Architecture (SOA) in 2007 and more than 80 scientific articles. He is now engaged in research on mobility, pervasive/ubiquitous computing, and natural human-machine dialogue.

Jitao Sang is Assistant Professor in National Lab of Pattern Recognition, Institute of Automation, Chinese Academy of Sciences. He received the B.E. degree from the SouthEast University and PhD degree from Institute of Automation, Chinese Academy of Sciences, in 2007 and 2012, respectively. In 2010 and 2011, he was an intern student in the China-Singapore Institute of Digital Media (CSIDM) and Microsoft Research Asia (MSRA), respectively. His current research interests include multimedia content analysis, social media mining, computer vision, and pattern recognition.

Wei Song is currently working as a research fellow in Queensland University of Technology (QUT), involved in multimedia, quality of experience, and video delivery research funded by Smart Services CRC. She obtained her PhD degree with the research on user experience of mobile video in QUT. She received her B.E degree on computer engineering in 2003 and M.S. degree on Signal and Information Processing in 2008, in China. Wei Song was working as a technician, an associate engineer of telecommunication in China UniCom Co. Ltd. in 1996-2005.

Stephen Viller is a researcher and educator in people centred design methods, particularly applied to the design of social, domestic, and mobile computing technologies and understanding people in their everyday settings. He has over 20 years' experience in the fields of Computer Supported Cooperative Work (CSCW), Interaction Design, and Human-Computer Interaction (HCI) where he has focused on bridging between disciplines and perspectives. He has concentrated on qualitative methods, particularly observational fieldwork, contextual interviews, diary studies and field trips, but also increasingly on more 'designerly' approaches such as cultural probes, low fidelity prototypes, rapid prototyping, and sketching. Stephen is a senior lecturer in the Interaction Design group in the School of Information Technology and Electrical Engineering at the University of Queensland, where he is also the Program Director for Multimedia and Interaction Design. His publications span various interdisciplinary journals and conferences in HCI/CSCW. He has a BSc Computation (UMIST), MSc Cognitive Science (Manchester) and PhD Computing (Lancaster).

Peta Wyeth is a Senior Lecturer at the Games Research and Interaction Design Lab, Science and Engineering Faculty, Queensland University of Technology. Peta's major research focus is in the domain of human-computer interaction with an emphasis on creating effective technology for educational purposes. She is particularly interested in building intelligent, ubiquitous technology that children can use in meaningful, purposeful, and appropriate ways. From a methodological perspective, the development of such systems involves the use of novel interaction design techniques, which are used to inform the design of technology to meet the unique needs of children.

Changsheng Xu (M'97-SM'99) is Professor in National Lab of Pattern Recognition, Institute of Automation, Chinese Academy of Sciences and Executive Director of China-Singapore Institute of Digital Media. His research interests include multimedia content analysis/indexing/retrieval, pattern recognition and computer vision. He has hold 30 granted/pending patents and published over 200 refereed research papers in these areas. Dr. Xu is an Associate Editor of *ACM Transactions on Multimedia Computing, Communications and Applications,* and *ACM/Springer Multimedia Systems Journal.* He served as Program Chair of ACM Multimedia 2009. He has served as associate editor, guest editor, general chair, program chair, area/track chair, special session organizer, session chair, and TPC member for over 20 IEEE and ACM prestigious multimedia journals, conferences, and workshops.

Jinglan Zhang is a senior lecturer in Queensland University of Technology. She received her PhD in Information Technology in 2003 from Queensland University of Technology. Dr. Zhang has worked as an Engineer in Computer Aided Design and Computer Aided Engineering for 8 years and a researcher in Information Technology for more than 10 years. She has published more than 30 refereed papers. Her broad research area falls in Artificial Intelligence and Computer Software. In particular, her research interests include Mobile Computing, Data Mining, Visual and Acoustic Information (Graphics, Images, and Sound), Processing and Retrieval, and Decision Support Systems.

Index

N

P

Q

R

S

T

U

V

W

X

Y